INTELLIGENTSIA SCIENCE

The Russian Century, 1860-1960

EDITED BY
Michael D. Gordin, Karl Hall,
and Alexei Kojevnikov

OSIRIS | 23

A Research Journal Devoted to the
History of Science and Its Cultural Influences

Osiris

Series editor, 2002–2012
KATHRYN OLESKO, *Georgetown University*

Volumes 17 to 27 in this series are designed to dissolve boundaries between history and the history of science. They cast science in the framework of larger issues prominent in the historical discipline but infrequently treated in the history of science, such as the development of civil society, urbanization, and the evolution of international affairs. They aim to open up new categories of analysis, to stimulate fresh areas of investigation, and to explore novel ways of synthesizing major historical problems that demand consideration of the role science has played in them. They are written not only for historians of science, but also for historians and other scholars who wish to integrate issues concerning science into courses on broader themes, as well as for readers interested in viewing science from a general historical perspective. Special attention is paid to the international dimensions of each volume's topic.

17   LYNN K. NYHART & THOMAS H. BROMAN, EDS., *Science and Civil Society*
18   SVEN DIERIG, JENS LACHMUND, & J. ANDREW MENDELSOHN, EDS., *Science and the City*
19   GREGG MITMAN, MICHELLE MURPHY, & CHRISTOPHER SELLERS, EDS., *Landscapes of Exposure: Knowledge and Illness in Modern Environments*
20   CAROLA SACHSE & MARK WALKER, EDS., *Politics and Science in Wartime: Comparative International Perspectives on the Kaiser Wilhelm Institute*
21   JOHN KRIGE & KAI-HENRIK BARTH, EDS., *Global Power Knowledge: Science and Technology in International Affairs*
22   GREG EGHIGIAN, ANDREAS KILLEN, & CHRISTINE LEUENBERGER, EDS., *The Self as Project: Politics and the Human Sciences*
23   MICHAEL D. GORDIN, KARL HALL, & ALEXEI KOJEVNIKOV, EDS., *Intelligentsia Science: The Russian Century, 1860–1960*

Cover Illustration:

The "Tsiolkovskii corner" at the Moscow exhibition on interplanetary travel in 1927. Konstantin Tsiolkovskii was the patriarch of Soviet space exploration. Copyright Ron Miller. Reproduced with the permission of Ron Miller.

INTELLIGENTSIA SCIENCE: THE RUSSIAN CENTURY, 1860–1960

MICHAEL D. GORDIN AND KARL HALL: *Introduction: Intelligentsia Science Inside and Outside Russia* — 1

INTELLIGENTSIA AS SOCIAL ORGANIZATION

MICHAEL D. GORDIN: *The Heidelberg Circle: German Inflections on the Professionalization of Russian Chemistry in the 1860s* — 23

ANDY BYFORD: *Turning Pedagogy into a Science: Teachers and Psychologists in Late Imperial Russia (1897–1917)* — 50

SONJA D. SCHMID: *Organizational Culture and Professional Identities in the Soviet Nuclear Industry* — 82

INTELLIGENTSIA AS POLITICAL AGENT

ALEXEI KOJEVNIKOV: *The Phenomenon of Soviet Science* — 115

OLGA VALKOVA: *The Conquest of Science: Women and Science in Russia, 1860–1940* — 136

NILS ROLL-HANSEN: *Wishful Science: The Persistence of T. D. Lysenko's Agrobiology in the Politics of Science* — 166

SLAVA GEROVITCH: *Stalin's Rocket Designers' Leap into Space: The Technical Intelligentsia Faces the Thaw* — 189

INTELLIGENTSIA AS UTOPIA

KIRILL ROSSIIANOV: *Taming the Primitive: Elie Metchnikov and His Discovery of Immune Cells* — 213

KARL HALL: *The Schooling of Lev Landau: The European Context of Postrevolutionary Soviet Theoretical Physics* — 230

ASIF SIDDIQI: *Imagining the Cosmos: Utopians, Mystics, and the Popular Culture of Spaceflight in Revolutionary Russia* — 260

NOTES ON CONTRIBUTORS — 289

INDEX — 291

Introduction:
Intelligentsia Science Inside and Outside Russia

*Michael D. Gordin and Karl Hall**

ABSTRACT

The concept of "intelligentsia" has had a long life both as an actors' category and as a tool of historical analysis to understand the role of the educated elite in Russia and the Soviet Union. For a variety of reasons, this term has not achieved much prominence in exploring the assimilation of science and technology into Russian culture. This essay examines the term's convoluted historiography and makes a case for the utility of its revival—understanding "intelligentsia" not as a single concept or group but as a heterogeneous and evolving social institution. The entire volume is then introduced with these features in mind.

INTRODUCTION

Andrei Sakharov was no sport of nature. The moral authority that he had acquired by the time of his death in December 1989 was the product of more than his extraordinary scientific aptitude and a rare willingness to speak truth to power. Since the demise of Communism, he has rightly been claimed as a paragon of the Russian intelligentsia tradition stretching back to the nineteenth century. His scientific biographer has demonstrated Sakharov's clear filiations with the laboratory milieu cultivated by the Moscow University physicist P. N. Lebedev (1866–1912), whose career had likewise fallen victim to overweening government authority.[1] The widespread post-Soviet eagerness to recover a usable historical continuity with a more distant Russian past is in no small part a symptom of the often tragic fortunes of the "intelligentsia"—a term whose meaning has long been contested—under Soviet power.

Yet Sakharov's colleagues' invocations of a uniquely heroic figure for his era elide the social institutions and traditions that are indispensable to any historical appeal to the intelligentsia legacy. An invaluable collection of reminiscences titled *He Lived among Us* . . . handily illustrates the multilayered nature of the problem in

---

* Michael D. Gordin, Department of History, 129 Dickinson Hall, Princeton University, Princeton, NJ, 08544; mgordin@princeton.edu. Karl Hall, History Department, Central European University, Nádor u. 9, H-1051, Budapest, Hungary; hallk@ceu.hu.

We would like to thank Alexei Kojevnikov for his initiative in proposing this volume, his organizational acumen in coordinating the initial conference from which these papers were drawn in fall 2004 in Athens, Georgia, and his contributions to and suggestions on the first version of this introduction. We also greatly appreciate the support of the University of Georgia and Princeton University in making this volume possible. Comments from an anonymous referee and especially the judicious interventions of Kathryn Olesko have made this introduction—and this volume—substantially better.

[1] Gennady E. Gorelik, with Antonina W. Bouis, *The World of Andrei Sakharov: A Russian Physicist's Path to Freedom* (New York, 2005).

© 2008 by The History of Science Society. All rights reserved. 0369-7827/08/2008-0001$10.00

capturing this seemingly singular *intelligent* (i.e., a member of the intelligentsia; pl. *intelligenty*) against a larger historical backdrop.² Despite the vaguely messianic connotations, the title actually refers to a poem by Aleksandr S. Pushkin, a touchstone reference for every incarnation of the Russian intelligentsia and a particular favorite of Sakharov's. In 1937, at the centenary of Pushkin's death, a young Andrei Sakharov had witnessed the lavish state-sponsored celebrations of the national poet and even then had been quite conscious of the irony, amid the Great Purges, of officially praising the poet's disdain for an oppressive tsarist regime. And that is the historical parallel his colleagues sought to emphasize—artist and scientist each resisting arbitrary state power. However, the poem has more than that to say about the scientist in Russian culture.

Pushkin's "He Lived among Us" does not celebrate political nonconformity or moral uprightness but rather dresses down the Polish national poet Adam Mickiewicz for his manifest ingratitude in supporting the Polish rebellion against the Russian empire of 1830–1831. The Polish exile had enjoyed extended hospitality in both Moscow and St. Petersburg only a few years previously, and Pushkin himself had translated several of Mickiewicz's poems into Russian. But after Mickiewicz's passage to sedition, wrote Pushkin, "our peaceable guest has become an enemy to us—and poison." The tone is not angry—simply wounded. Among those colleagues invoking Sakharov's Pushkin, however, only one pointed to the contents of the poem itself. He recalled that Sakharov had explicitly drawn an analogy between Pushkin's altered relationship with Mickiewicz and his own with a fellow dissident who had strayed into the thickets of anti-Semitic Russian nationalism.³ Identifying with Pushkin, Sakharov somewhat ruefully recognized the righteous impulse to banish the wayward scholars from their midst, yet he seemed insensible to a reading of the poem in which the rebellious Pole is the protodissident and Pushkin the unself-conscious proponent of an imperial Great Russian state. That is to say, Sakharov's status as a scientist did not exempt him from participating in a lengthy tradition of competing universalisms in Russia. Scientists were part and parcel of the agonistic strain in the intelligentsia tradition. One of the tasks of this volume is not only to highlight the role the intelligentsia played in the formation of professional identities and even research agendas in the Russian empire and the Soviet Union but also to recover some of the international context of this distinctive Russian phenomenon: the intelligentsia scientist.

Yet Sakharov was just as much a weapons scientist, a father of the Soviet hydrogen bomb, as he was a dissident *intelligent* dipping into the well of intellectual as oppositionist. Making sense of science in the Russian context requires considering both aspects as part of a single milieu. Western eyes might detect a contradiction here, and one way of addressing it, attractive to both physicists and historians, is to treat Sakharov's career as a quid pro quo: the delivery of technical expertise in return for the toleration of limited intelligentsia values inside a physics microcosm. This should be the starting point for historical explanation, not its conclusion.⁴ A

---

² B. L Al'tshuler et al., eds., *On mezhdu nami zhil . . . : Vospominaniia o Sakharove* (Moscow, 1996).
³ The colleague was the mathematician I. R. Shafarevich. M. L. Levin, "Progulki s Pushkinym," in ibid., 330–75, on 346; cf. Andrei Sakharov, *Memoirs*, trans. Richard Lourie (London, 1990), 327.
⁴ To give a prominent example, émigré cosmologist Andrei Linde claimed that the atomic bomb was the reason "the scientific culture associated with the Soviet school of physics became a culture of free political thought." See his letter to the editor, *Physics Today* 45 (June 1992): 13. This view is affirmed by David Holloway, who argues that the bomb saved "a small island of intellectual autonomy in a so-

functional trade-off between knowledge and power will not suffice to explain "intelligentsia science." Drawing back from the iconic Sakharov to focus on a wider field of scientific endeavor, we see a long line of intelligentsia "descent with modification" that suggests more contextual hybridity than singular mutation. We argue for the study of Russian *culture* as a fruitful incubator for metaphorical models in the history of science. The structure of Russian cultural institutions, and more specifically the hallmark institution of the intelligentsia, both before and after the revolution of 1917, offers a deep study of how elites organize themselves around the themes of knowledge and political action. By repeatedly using this institution as the structural level of their investigations, the authors here push for a new scale of understanding the microcomposition of what have traditionally been painted in broader strokes. We wish to do more than just revive the now-unfashionable category of "intelligentsia" to demonstrate its centrality for specific questions in Russian history. We hope to show precisely how the analysis of these kinds of institutions can offer a model for not merely bringing the history of science closer to mainstream political, social, and economic history but placing science and technology right at the heart of national and international developments.

THE RUSSIAN INTELLIGENTSIA AS CATEGORY AND CRITIQUE

Andrew Dickson White is best known to historians of science as the author of *History of the Warfare of Science with Theology in Christendom*, which he completed in 1895 while posted to St. Petersburg as the American ambassador to the Romanov court. He had first served in Russia in the 1850s as a young man, before going on to a distinguished career that included a stint as the first president of Cornell University. It is only in recent decades that his influential account of "the great, sacred struggle for the liberty of science" against the (autocratic) forces of religious dogmatism has been displaced by more nuanced views of the multivalent relations between scientific thought and Christian culture.[5] For White, this vaunted scientific autonomy was achieved through the exceptional agency of great men striving against deeply pernicious character traits instilled by theological training. Although his writings on science and on Russia scarcely intersected, White objected to the "barbarism" of the Russian tsars and similarly ascribed deep-rooted character traits (some positive, most negative) to Russians as products of this autocratic setting. This dynamic extended to intellectuals as well, with White asserting that "the evolution of Tolstoi's ideas has evidently been mainly determined by his environment." Only two scientists warranted mention in White's account of Russia. In the case of D. I. Mendeleev, with his "epoch-making discovery" of the periodic system of elements, White regarded this achievement as made possible because the scientist "[gave] himself wholly to some well-defined purpose" and was not distracted by that same Russian milieu. By contrast, in the case of geographer and biologist Prince Peter Kropotkin, that environment could not provide the necessary intellectual discipline, and he gave himself over

---

ciety where the state claimed control of all intellectual life." See his *Stalin and the Bomb* (New Haven, Conn., 1994), 213; Holloway, "Physics, the State, and Civil Society in the Soviet Union," *Historical Studies in the Physical and Biological Sciences* 30 (1999): 173–93.

[5] See, in particular, John Hedley Brooke, Margaret J. Osler, and Jitse M. van der Meer, eds., *Science in Theistic Contexts*, *Osiris* 16 (2001); Andrew D. White, *History of the Warfare of Science with Theology in Christendom* (New York, 1896).

to "speculations of an abstract sort, with no chance of discussing his theories until they are full-grown and have taken fast hold upon him."[6] Kropotkin's long career abroad went unmentioned, in the presumption that these weaknesses had been indelibly inscribed while he was still in Russia.

White's protostructuralist and dualist mode of interpretation has proven surprisingly durable (if unacknowledged among historians of Russian science), as evidenced by Loren Graham's plea in 1998 for historians of science to treat Russia in symmetrical terms and renounce the "deviationist" model once and for all.[7] The fact that Graham even needed to raise this point after more than a century indicates that something is indeed amiss in the framework through which most historians have tended to view Russian scientific activity. We believe that closer study of the evolving identities of the (scientific and technical) intelligentsia—including Russian scientists abroad—can do a good deal more than transcend tired realist/constructivist or internalist/externalist dichotomies in the field.

By the time anglophone historians of science first adopted the term "intelligentsia" in the mid-1930s, it had already been suitably deracinated and was used to describe the sort of men who congregated at the London coffeehouse frequented by Benjamin Franklin and Joseph Priestley in the late eighteenth century.[8] Occasionally, even the political overtones were stripped away to render the term nearly synonymous with what social scientists would call a "cohort."[9] Yet the original usage a century earlier connoted anything but quietism. Its very coinage is emblematic of the boundary problems we seek to address in this volume, for while it entered English directly from Russian in the 1920s, it was originally appropriated into Russian much earlier from Polish, which was itself a mediator for analogous terms in France, Germany, and Italy in the 1830s and 1840s. This cluster of parallel terms was initially used rather broadly in Europe to designate educated and "progressive" individuals, but the various Western European analogues gradually fell out of use. The Polish and Russian usage of "intelligentsia," by contrast, continued to accrete complex associations with word and deed in the tsarist context.[10]

The men discussing the term were largely from the Western-oriented minor service nobility. The small size and privileged location of these early discussions was in part

---

[6] Andrew D. White, *Autobiography of Andrew Dickson White*, vol. 2 (New York, 1905), 494–5, 498–9. On White as one of the first American experts on Russia, see David C. Engerman, *Modernization from the Other Shore: American Intellectuals and the Romance of Russian Development* (Cambridge, Mass., 2003), 43–5.

[7] Loren R. Graham, *What Have We Learned about Science and Technology from the Russian Experience?* (Stanford, Calif., 1998).

[8] W. Cameron Walker, "The Beginnings of the Scientific Career of Joseph Priestley," *Isis* 21 (1934): 81–97, 84.

[9] The Invisible College "was merely a club of young men, some of the intelligentsia of that day, brought together by common college backgrounds and many common interests for rest and refreshment from the political, theological and military turmoil of the times." Dorothy Stimson, "Comenius and the Invisible College," *Isis* 23 (1935): 373–88, on 374.

[10] For the *Begriffsgeschichte*, see Otto Wilh. Müller, *Intelligencija: Untersuchungen eines politischen Schlagwortes* (Frankfurt am Main, 1971); S. O. Shmidt, "K istorii slova 'intelligentsii,'" in *Rossiia, zapad, vostok: Vstrechnye techeniia; K 100-letiiu so dnia rozhdeniia akademika M. P. Alekseeva*, ed. V. E. Bagno et al. (St. Petersburg, 1996), 409–17. The most ambitious recent treatment is the collection of essays T. B. Kniazevskaia, ed., *Russkaia intelligentsiia: Istoriia i sud'ba* (Moscow, 2000); see, especially, the contribution of Iu. S. Stepanov. The late eighteenth-century sources of the "service to the people" mentality are best chronicled by Marc Raeff in *Origins of the Russian Intelligentsia: The Eighteenth-Century Nobility* (New York, 1966); for the nineteenth century and beyond, see Richard Pipes, ed., *The Russian Intelligentsia* (New York, 1961).

a function of the repressive political circumstances—larger gatherings and men of lesser social station would have been much more vulnerable to the political police. It was in these small circles (*kruzhki*) that discontent with the regime of Tsar Nicholas I (r. 1825–1855) mixed with discussions of class identity and *Bildung*, incorporating the normative element of the participants' moral behavior vis-à-vis enserfed peasant Russia. What began as salon politics quickly underwent an astonishing intellectual ferment that for practical (read: censorship) as much as aesthetic reasons was manifested mostly in literary forms, whether as Aesopian criticism or as *belles lettres*. In this first generation of nascent intelligenty, only Prince Vladimir F. Odoevskii demonstrated any active interest in natural philosophy (though many members of the intelligentsia embraced some variant of Comtean positivism), and his scientific pursuits found few outlets outside Russia.[11]

The generation of the 1860s transformed "intelligentsia" into a central notion of Russian popular discourse, cementing its association with revolutionary politics—and with natural science.[12] Science became the cornerstone of the intelligentsia's ideological and political projects, either as an alternative to socialism or, more often, as its nominal raison d'être. During the Great Reforms of the 1860s, with their modest abatement of censorship, figures such as Petr Lavrov strove to articulate a role for "critically thinking individuals" in Russian society, sparking myriad debates about who counted as an intelligent. Although the roles of bureaucrat or writer or officer remained compatible with a primary identity as a nobleman, the normative intelligentsia role soon became militantly déclassé in the very period when more finely differentiated social roles, including that of the scientist, were becoming possible.[13] The intelligentsia nonetheless maintained its embattled self-conception and oppositional ethic through the end of the century, even as it fractured into liberal, radical, Marxist, and even reactionary groupings and subgroupings. Liberal geochemist Vladimir Vernadskii, for example, took a decidedly evolutionary view of scientific discovery; yet in a course offered at Moscow University, he also comfortably associated receptivity to new scientific truths with a heretical mindset.[14] The comparative "respectability" of scientists in social terms has tended to obscure the complexity of their roles in this period's confrontation between a suspicious state and those intellectual "subversives" claiming to represent the greater public good. The oppositional boundaries

---

[11] Dimitri Bayuk, "Literature, Music, and Science in Nineteenth Century Russian Culture: Prince Odoyevskiy's Quest for a Natural Enharmonic Scale," *Science in Context* 15 (2002): 183–207. We are not making a claim about the place of *Naturphilosophie* in Russia more generally, and its importance for nurturing a culture of science in Russia under Nicholas I should be acknowledged. But its most accomplished adepts had little direct influence on subsequent arguments about "Russian science" and should not be regarded as intelligentsia scientists *avant la lettre*.

[12] For the classic set of Slavophiles, Westernizers, and other representatives of the politically oriented intelligentsia in this period, see the essays gathered in Isaiah Berlin, *Russian Thinkers*, ed. Henry Hardy and Aileen Kelly (London, 1978). Berlin's rather sanguine interpretation of the Russian intelligentsia, especially the central figure of Aleksandr Herzen, has recently received rather wide exposure through the popular success of Tom Stoppard's dramatic trilogy, *The Coast of Utopia* (London, 2002).

[13] On the transformations in the estate social structure of this period, see Elise Kimerling Wirtschafter, *Social Identity in Imperial Russia* (DeKalb, Ill., 1997); Edith W. Clowes, Samuel D. Kassow, and James L. West, eds., *Between Tsar and People: Educated Society and the Quest for Public Identity in Late Imperial Russia* (Princeton, N.J., 1991); and Harley D. Balzer, ed., *Russia's Missing Middle Class: The Professions in Russian History* (Armonk, N.Y., 1996).

[14] V. I. Vernadskii, "O nauchnom mirovozzrenii," in *Sbornik po filosofii estestvoznaniia* (Moscow, 1906), 104–57, 145.

proved porous over time, and it was plausible for participants in prominent state and industrial initiatives in oil exploration, university education, and industrialization to engage simultaneously in the philosophical and political debates about science then raging in intelligentsia circles.

The Russian intelligentsia at the fin de siècle was relentlessly self-critical, and the marginal figures of the scientific intelligentsia found themselves shunted aside both at that time and in later histories of the period. Indeed, the critics of the intelligentsia have defined it as much as its members, most famously evidenced by the *Signposts* volume of 1909.[15] While idealist philosophers N. A. Berdiaev, S. N. Bulgakov, and S. L. Frank dominated this polemic, P. B. Struve's critique was most effective in ensuring that "intelligentsia" would be bound to "revolution" in subsequent Western historiography. For Struve, the intelligentsia was not simply the educated class (with ready analogues in other European countries), but "the offspring of the interaction of Western socialism with the particular conditions of our cultural, economic, and political development." Moreover, it was *defined* by its hostile relation to the state, an oppositional stance that Struve labeled "renegadism" (*otshchepenstvo*).[16] That oppositional stance did not seek to replace the "mystery" of the state with another Hegelian foundation but rather to deny the state entirely in favor of a rational and empirical foundationalism with no spiritual component whatsoever. The main thrust of this critique helped cement the conviction among subsequent observers that the Russian intelligentsia necessarily shared a particular political and philosophical stance toward the state. That view produced one of two possible consequences for the historical treatment of scientists, depending on which strategy one followed.

The first strand of argument drew upon broader debates about modernization theory in the social sciences and treated the apparent fading of the (oppositional) intelligentsia in the Soviet era as a symptom of universal processes of industrialization. If there was no more "renegadism" under the Soviets, it was because intellectuals globally were increasingly subject to professionalization and standardization of social roles—not least among them, Soviet scientists.[17] This then licensed a dual terminology that mirrored actual Soviet usage, following Stalin's redefinition of the term "intelligentsia" to refer to what in the West would later be termed "white-collar workers": clerks, bureaucrats, and the like. These interpretations split the intelligentsia into two groups: the fading prerevolutionary remnants that had slid into irrelevance to the extent that they did not identify completely with Soviet power (a "quasi-intelligentsia"), and a modernizing technical elite without explicit political aspirations (i.e., "technocrats"—Stalin's newly redefined "intelligentsia"). In more optimistic readings, professionalization would even slowly overtake the party appa-

---

[15] N. A. Berdiaev et al., *Vekhi* (Moscow, 1909; repr., Sverdlovsk, 1991). On literary critics of the intelligentsia, cf. Gary Saul Morson, "What Is the Intelligentsia? Once More, an Old Russian Question," *Academic Questions* 6 (Summer 1993): 20–38.

[16] P. B. Struve, "Intelligentsiia i revoliutsiia," in Berdiaev et al., *Vekhi* (cit. n. 15), 148–65, on 164, 151. On the significance of Struve as a marker for countercurrents in Russian culture, see Richard Pipes, *Struve: Liberal on the Left, 1870–1905* (Cambridge, Mass., 1970); and Pipes, *Struve: Liberal on the Right, 1905–1944* (Cambridge, Mass., 1980).

[17] "Alienation has been dethroned . . . from its position as the dominant trait of the old intelligentsia of a century ago. The singularly modernizing USSR has gone far in incorporating the new intelligentsia," which for Fischer unproblematically includes scientists and engineers. George Fischer, "The Intelligentsia and Russia," in *The Transformation of Russian Society: Aspects of Social Change since 1861*, ed. Cyril E. Black (Cambridge, Mass., 1960), 253–74, on 273–4.

ratus.[18] Within such frameworks, scientists and engineers, being the standard-bearers of modernization, became the least likely objects for study with respect to the Soviet intelligentsia, understood in the pre-Stalinist sense. For it was science and technology that would eventually return Russia from its Communist detour back on to the historical trajectory of Western modernity.[19]

The second and more influential strand of argument within Russian studies emphasized the longer-term historical specificity of the Russian intelligentsia but was no less universalist in explaining its ultimate fate. Following Struve, these scholars emphasized the essential Europeanness of the Russian intelligentsia, and then portrayed the revolution as either apotheosis or apocalypse (usually the latter), but effectively as an end point for the conditions that had made the intelligentsia possible in the first place: the intelligentsia made the revolution, which then swallowed its parents. The earliest Western students of the Bolshevik regime thought the revolution had destroyed the intelligentsia a decade later—yet did not seem to notice the simultaneous explosive growth in the ranks of scientists alongside the supposed death of the intelligenty.[20] If the intelligentsia had been destroyed by the revolution, the rise of science and technology after the Bolshevik victory was a priori irrelevant to any intelligentsia tradition. Martin Malia was especially eloquent in extending the *Signposts* position that the intelligentsia was by definition alienated and, after 1917, marginalized and virtually nonexistent.[21] Malia's magisterial synthesis of Western views of Russia remains exemplary in demonstrating that the history of the Russian intelligentsia must be told in pan-European (if not global) terms, and several of the essays in this volume consciously heed that injunction. We differ from Malia in two important respects, however. First, the grand temporal arc of intellectual legacies that he traces needs to be complemented by accounts of medium-scale conjunctural effects (e.g., scientists at war; scientists as bearers of national/patriotic aspirations) and of small-scale interactions at the personal and institutional level. These levels of analysis comprise the specifically Russian and Soviet contexts in which intelligentsia social identities were formed. Second, when these effects are taken into account, we believe that 1917 will

---

[18] Robert V. Daniels, "Intellectuals and the Russian Revolution," *American Slavic and East European Review* 20 (1961): 270–8, on 277. More pessimistic readings, following Leon Trotsky (Lev Trotskii), saw this development as a new Thermidor and a prelude to Soviet Bonapartism: Leon Trotsky, *The Revolution Betrayed: What Is the Soviet Union and Where Is It Going?* trans. Max Eastman (Garden City, N.Y., 1937).

[19] Albert Parry, *The New Class Divided: Science and Technology Versus Communism* (New York, 1966). Although political scientists showed no interest in the internal structure of the scientific community, they were quick to criticize the purported "cleavage" between technically trained managers and party apparatchiki. Cf. Stephen White, "Contradiction and Change in State Socialism," *Soviet Studies* 26 (1974): 41–55.

[20] René Fülöp-Miller, *Geist und Gesicht des Bolschewismus: Darstellung und Kritik des kulturellen Lebens in Sowjet-Russland* (Zurich, 1926), 19, 327–9; Friedrich Braun, "Über die russische 'Intelligenz,'" in *Kultur- und Universalgeschichte: Walter Goetz zu seinem 60. Geburtstage*, ed. colleagues et al. (Leipzig, 1927), 362–75; David Joravsky, "Soviet Scientists and the Great Break," in Pipes, *Russian Intelligentsia* (cit. n. 10), 122–40; Joravsky, *Soviet Marxism and Natural Science, 1917–1932* (New York, 1961). We owe the Braun reference to Jutta Scherrer.

[21] Martin Malia, "What Is the Intelligentsia?" in Pipes, *Russian Intelligentsia* (cit. n. 10), 1–18; Malia, *Russia under Western Eyes: From the Bronze Horseman to the Lenin Mausoleum* (Cambridge, Mass., 1999). Cf. James H. Billington, "The Renaissance of the Russian Intelligentsia," *Foreign Affairs* 35 (1957): 525–30, in which Khrushchev's "liberalizing" measures are seen as a (still premature) opportunity for the revival of a (by definition) oppositional intelligentsia. See also Billington, *The Icon and the Axe: An Interpretive History of Russian Culture* (New York, 1966).

cease to loom so large and that subtler enduring continuities between imperial and Soviet phenomena will become apparent.[22]

Others have demanded thoroughgoing social histories of the Russian intelligentsia that encompass more than socialist revolutionaries, but their calls seem to exclude natural scientists and engineers.[23] Jutta Scherrer has rightly called for more explicit historical connections between proclaimed self-identity and actual socialization, for studying lifestyles in order to get at questions of corporate identity among the intelligentsia. But we can go beyond the officers, civil servants, and upper levels of the clergy she identifies as illuminating boundary test cases for such research.[24] Scientists also offer abundant possibilities for testing hypotheses about Russia, and this most international of social identities should surely play a role in further comparative studies of European intellectuals in the spirit of ambitious synthetic works such as those of Christophe Charle or Denis Sdvizhkov.[25]

So why have the scientists been left out of this history? For a variety of reasons, both scientists themselves and Soviet historians of intelligentsia found it useful to employ the term in ways that obscured or effaced its historical specificity vis-à-vis science in Russia. Lenin himself underwrote the misleading dichotomy between "re-educating old specialists" and "training new cadres" to make a new socialist intelligentsia, and Soviet historians duly followed suit, making the new "technical intelligentsia" into something socially and politically distinct from its prerevolutionary avatars. In the rush to bleed "intelligentsia" of its specificity, Soviet historians advanced the notion of a complementary "nonprofessional intelligentsia" composed of a broad variety of "proletarian" pursuits not necessarily captured by the labels "engineer" and "technician."[26] This expansive category surely helped swell the ranks of the Soviet-educated elites for bookkeeping purposes, but was this really the most appropriate term for operators of high-speed excavators?[27] Even when the focus was on natural scientists, there was every motivation to restrict oneself to cautious demographic analyses demonstrating the social breadth, rather than the historical depth, of the scientific intelligentsia.[28] The intelligentsia, it seemed, had lost its taste for

---

[22] For continuities in the exceptional cases of Bogdanov, Lunacharskii, Bukharin, and Gastev, see Jutta Scherrer, "Die sozialistische Intelligenzija: Ihre Rolle in der russischen Kulturrevolution vor und nach 1917," *Comparativ* 6 (1995): 94–119.

[23] Daniel R. Brower, "The Problem of the Russian Intelligentsia," *Slavic Review* 26 (1967): 638–47; Michael Confino, "On Intellectuals and Intellectual Traditions in Eighteenth- and Nineteenth-Century Russia," *Daedalus* 101, no. 2 (1972): 117–49; Alexander Gella, ed., *The Intelligentsia and the Intellectuals: Theory, Method, and Case Study* (Beverly Hills, Calif., 1976).

[24] Jutta Scherrer, "Russkaia dorevoliutsionnaia intelligentsiia v zapadnoi istoriografii," in *Intelligentsiia v istorii: Obrazovannyi chelovek v predstavleniiakh i sotsial' noi deistvitel' nosti*, ed. D. A. Sdvizhkov (Moscow, 2001), 9–30.

[25] Christophe Charle, *Les intellectuels en Europe au XIXe siècle: Essai d' histoire comparée* (Paris, 1996); Denis Sdvizhkov, *Das Zeitalter der Intelligenz: Zur vergleichenden Geschichte der Gebildeten in Europa bis zum Ersten Weltkrieg* (Göttingen, 2006).

[26] Western journalists could help reinforce this usage of intelligentsia; see, e.g., "Soviet Now Seeks Aid of the Intelligentsia: Need of Engineers and Technical Experts in Industry Brings Offer of Reconciliation," *New York Times*, 13 Dec. 1924.

[27] M. P. Kim, ed., *Sovetskaia intelligentsiia (Istoriia formirovaniia i rosta 1917–1965 gg.)* (Moscow, 1965), 418; S. A. Fediukin, *Velikii oktiabr' i intelligentsiia* (Moscow, 1972); P. V. Alekseev, *Revoliutsiia i nauchnaia intelligentsiia* (Moscow, 1987).

[28] D. M. Gvishiani et al., *The Scientific Intelligentsia in the USSR: Structure and Dynamics of Personnel*, trans. Jane Sayers (Moscow, 1976).

ideology.²⁹ Old Regime scientists sympathetic to the Soviets were also inclined toward inclusive definitions, but they were not necessarily being disingenuous when they celebrated the vast qualitative increase since 1917, or the increasing diversity of the scientific community.³⁰ Indeed, the point is not to claim that all scientists and engineers under state socialism were "genuine" intelligentsia—whatever that would mean—but rather to show concretely how such social and ideological distinctions were scrambled over time among scientists themselves. The term "intelligentsia" may have lost its specificity, or its corporate identity, but that does not mean that we as historians cannot locate traditions, institutions, and ideologies that bear more than family resemblances to the classic conception of the term.

This introduction began with Sakharov and the intelligentsia, so it is only fair that the other great anti-Soviet dissident—and fierce critic of the intelligentsia—should have his say, in passing. Aleksandr Solzhenitsyn has been instrumental in hardening the 1917 divide regarding the composition of the intelligentsia, although his identifying criteria were neither intellectual nor political but profoundly moral. The ability to function successfully under Soviet power was, in Solzhenitsyn's scheme of things, evidence of collaboration and compromise and thus barred one from membership in the true "intelligentsia," the exemplars of which were drawn in his case from the Slavophiles of the 1840s and their ilk.³¹ Soviet scientists were especially vulnerable to this charge, given how closely they were tied to the state as civil servants and technical experts—and despite his sharp differences with Solzhenitsyn, Andrei Sakharov shared some of the same sense of disappointment in his peers. In Solzhenitsyn's eyes, the Soviet successors of the Russian intelligentsia had sold out their ideas for material comforts. This polemic carried a great deal of truth, but Sakharov's own experience already suggests that it was overstated.³²

Scientists outside Russia have themselves contributed to the marginalization of the term, tending toward usages that imply the least politically problematic relation between science and power. The earliest usages of the term among Western scientists sometimes made it even more inclusive than "intellectuals," without attributing anything particularly Russian to the meaning.³³ Émigré Russian scientists were likewise naturally uninterested in fostering any historical distinctions in social roles between

---

²⁹ An American Sovietologist in the age of *The Organization Man* described an "ersatz intelligentsia" that wants "simply to describe life as they see it without constant reference to ideology." L., "The Soviet Intelligentsia," *Foreign Affairs* 36 (1957): 122–30.

³⁰ B. Keller, "Proletarskaia revoliutsiia i sovetskaia intelligentsiia," *Front nauki i tekhniki*, 1937, no. 11:49–58.

³¹ A. I. Solzhenitsyn, "Obrazovanshchina" (1974), reprinted in Kniazevskaia, *Russkaia intelligentsiia* (cit. n. 10), 125–49; cf. M. L. Gasparov, "Intellektualy, intelligenty, intelligentnost'," in ibid., 5–14, on 7. On the Slavophiles, see Andrzej Walicki, *The Slavophile Controversy: History of a Conservative Utopia in Nineteenth-Century Russian Thought*, trans. Hilda Andrews-Rusiecka (Oxford, 1975).

³² Sakharov's post-Soviet canonization was thus not simply evidence of bad conscience among his scientific peers (by Solzhenitsyn's lights) but also a sign of some more durable social processes peculiar to the Soviet era.

³³ For U.S. National Research Council representative Vernon Kellogg, the "intelligentsia" (his scare quotes, marking the contemporary novelty of the term) was "the university faculties and the Russian professional and scientific men in general." Kellogg, "The Present Status of University Men in Russia," *Science* 54 (25 Nov. 1921): 510–1. This tendency to conflate intelligentsia with intellectuals may also be found in the important early study by Karl Nötzel, *Die Grundlagen des geistigen Russlands: Versuch einer Psychologie des russischen Geisteslebens* (Leizpig, 1923; repr., New York, 1970), e.g., 260.

them and their Western counterparts. Yet while condemning the Soviet system, they invariably insisted that "Russia has always generously contributed her share to the science of the world."[34] Eager to catalog Russian achievements, they were loath to evaluate the scientist in Soviet context. The left-leaning P. M. S. Blackett adopted a Soviet-style "service intellectuals and technicians" definition that made (scientific) intelligentsia into a source of (political) progress, again without any historical differentiation of social roles—every scientist became an intelligent willy-nilly in the modern era.[35] The Marxist Léon Rosenfeld, by contrast, took intelligentsia to have a political bent everywhere, not just in Russia, and he did not shy away from anachronistic usage.[36]

Perhaps more representative in the postwar era was the experience of the biologist Conway Zirkle, an early visitor to the Soviet Union who became sharply critical of it at the height of the cold war during the depredations of Lysenkoists. Reviewing James Bryant Conant's *Science and Common Sense*, Zirkle happily agreed with its premise that education in science builds good liberal-democratic citizens but worried that it was not enough to persuade laypeople that science is "organized knowledge," when they still might fail to appreciate its "conceptual schemes" properly. Witness the Marxist dupes of the 1930s in Great Britain, inveighed Zirkle. "In the intellectual climate of the nineteen-thirties and early forties, becoming a sucker was almost an occupational hazard of the intelligentsia."[37] Not surprisingly, accusations of Western co-optation did not lend themselves to further analysis of the original Russian problem. Other, less politically engaged, scientists with extensive knowledge of the Soviet setting preferred a definition of "producers of high culture," thereby distancing scientists from the discussion.[38] A dissident scientist such as Zhores Medvedev might hedge the matter by placing scientists among the intelligentsia but still distinguishing them from the "culturally creative intelligentsia."[39] More conservative and elitist appropriations of the term by anglophone scientists have been ignored by historians of science, notwithstanding the awkward parallels with Leninist practice.[40] In short, the anglophone meanings of intelligentsia among scientists have been so promiscuous as to discourage reflection about its relevance in historical perspective.

---

[34] V. N. Ipatieff, "Modern Science in Russia," *Russian Review* 2, no. 2 (1943): 68–80, on 68. Cf. George Vernadsky, "Rise of Science in Russia, 1700–1917," *Russian Review* 28, no. 1 (1969): 37–52. Although George was a historian by training, his father, Vladimir, was an eminent geochemist and liberal who remained in the Soviet Union and wrote regularly to his son. This article was already quite dated at the moment of publication but very much in keeping with the style of Ipatieff's earlier essay.

[35] P. M. S. Blackett, "Science, Technology, and World Advancement," *Nature* 193 (3 Feb. 1962): 416–20.

[36] Léon Rosenfeld, "Voltaire on Newton," *Nature* 218 (11 May 1968): 607.

[37] Conway Zirkle, "Science and Common Sense," *Isis* 42 (1951): 269–71, on 270.

[38] E.g., Cambridge physicist David Shoenberg wrote that Kapitza had "a wide circle of friends among the 'intelligentsia': artists, sculptors, writers, musicians, actors, film directors and so on." Shoenberg, "Piotr Leonidovich Kapitza," *Biographical Memoirs of the Fellows of the Royal Society* 31 (1985): 326–74, on 364.

[39] Zhores Medvedev, *The Medvedev Papers* (London, 1971), 170–1.

[40] A favorite example comes, appropriately enough, from an Anglophile Canadian witness to the demise of the British Empire: "Always a firm believer in an oligarchy of the intelligentsia, Rowan looked upon the events following the Second World War as conclusive evidence of the depravity of democracy. He was convinced that mankind is heading toward total self-destruction and that it can be saved only by placing itself under the guidance of an intellectual few." W. Ray Salt, "Prof. William Rowan," *Nature* 180 (7 Sept. 1957): 463.

"SCIENCE IN RUSSIA" VERSUS "RUSSIAN SCIENCE"

Yet "intelligentsia," precisely because it is an international category closely identified with educated elites that simultaneously carries a strong Russian contextual specificity, offers the potential to recast a typically thorny way of carving up the history of the sciences. Of all the ways to divide up the history of modern science, nation-states offer perhaps the most common, yet the least defended, variant. One is accustomed to perusing histories of American biology, Brazilian physics, French chemistry, and so on—but often without any explicit attempt to argue why this is a fruitful way to analyze each of those different fields.[41] After all, modern science bears the distinctive character of international communication (whether in the traditional forms of publication or travel or the relatively more recent innovation of postdoctoral exchanges and conferences). Even the linguistic bounds usually used to parcel off various sciences in academic studies seem beside the point: Russians often published in French, Britons in German, and almost everyone now publishes in English. Why, then, does one continually find the history of science carved into segments whose traditional boundary lines have been set by the vagaries of nineteenth- and twentieth-century politics?

This volume offers both a reasoned instantiation of such nation-state studies—paired with an implicit defense of their utility—and a critique of their unavoidable limitations. The series of essays presented here derive from recent research on the history of science in the Russian empire and its successor, the Soviet Union, from 1860 to 1960. The essays suggest ways in which it is still valuable to examine modern science using this particular unit of analysis, even (or especially) when the science in question crosses international and cultural borders.

The crux of the transition that took place in this period was the transformation of what one might call "science in Russia" (that is, science that happened in the geopolitical space defined by Russia, which had been taking place at the very least since the early eighteenth century) into "Russian science"—the assimilation and adaptation of scientific traditions and institutions into Russian culture so that they became an integral part of Russian culture. The peculiar features of both science in Russia and Russian science have provoked persistent study at a moderate level since at least the end of the Second World War. The boom of interest at that time was, of course, no accident and proved intimately linked to three specific developments of the late 1940s. One was the onset of the cold war between the United States and the Soviet Union, which led to the academic institutionalization of Russian/Soviet studies within universities in the West. Another was the proliferation of nuclear weapons (the first Soviet atomic test took place at the end of August 1949) and the consequent recognition of the growing importance of science and technology as modes of competition in the cold war, highlighted most dramatically by the October 4, 1957, launch of *Sputnik*. Third was the abolition of genetics in the Soviet Union in 1948 under the auspices of both Joseph Stalin and Trofim D. Lysenko, which added an ideological aspect to the superpower conflict specifically in the area of science.

As a consequence of these three undeniably important developments, early studies

---

[41] Even raising the issue may strike the contemporary reader as a bit old-fashioned. Defenses of national categories in the science do, in fact, tend to be rather dated. For one of the more persuasive efforts, see Maurice Crosland, "History of Science in a National Context," *British Journal for the History of Science* 10 (1977): 95–113.

of science in Russia acquired certain features that still dominate the views of its past among historians and political scientists who do not specialize in the area. The first, almost so obvious as to be unstated, is that the history of science in Russia was completely absorbed within a history of the Soviet Union. It was the Soviet Union that interested historians and political scientists, and the history of Soviet science was a subset of that interest. The necessary consequence of this was that the trends, methodologies, and vicissitudes of Sovietology proved dominant in this field—insulating it until very recently from some of the leading methodologies in history of science as a profession.[42] This attention to the Soviet experience to the exclusion of almost all else meant that—with a few exceptions—most of the research done exhibited a staggering indifference to the "background" of prerevolutionary Russia or interpreted it mainly as "prehistory" to the relevant Soviet developments.[43] If, until the last decade or so, the history of imperial Russia was centered (as it was) around finding precursors for the success of Vladimir Lenin's Bolshevik Party in 1917, the history of science in the Soviet Union implicitly accepted some of the regime's propaganda at face value and understood that most of the developments of Soviet science emerged without significant continuities to the almost desolate prerevolutionary intellectual landscape.[44]

A second feature of the cold war historiography of science in the Soviet Union was an overemphasis on *ideological interference* or distortion of the "normal" course of science—"normal" being defined as the development of modern science in Western Europe and North America. According to the internal logic of this argument, Lysenko represented a feature of the Soviet system repressing the ordinary course of normal science, so our attention should be focused on understanding such features of the Soviet system and the repressive aspects of ideology. Borrowing an age-old template

---

[42] Alexei Kojevnikov, "A New History of Russian Science," *Science in Context* 15 (2002): 177–82; Engerman, *Modernization from the Other Shore* (cit. n. 6); and Jonathan Coopersmith, "The Dog That Did Not Bark in the Night: The 'Normalcy' of Russian, Soviet, and Post-Soviet Science and Technology Studies," *Technology and Culture* 47 (2006): 623–37.

[43] The pattern was already set before the cold war, when one of the earliest students (and celebrants) of Soviet science, J. D. Bernal, claimed in 1939, "the great Russian scientists, such as Lomonosov, Mendeleyev, Kovalevsky or Pavlov, managed to do their work rather in spite of than because of the official organization [the state-sponsored Academy of Sciences]." The Soviet Union thus had to "build up a new and greater science" after the revolution. Bernal's examples are taken over wholesale from the Soviet physicist A. F. Joffe, who maintained that "Russian physics was, until the Revolution, one of the most backward and weakest branches of world science." J. D. Bernal, *The Social Function of Science* (London, 1939), 222–3, 236; cf. A. F. Joffe, "O nedostatkakh i nekotorykh problemakh razvitiia sovetskoi fiziki," *Front nauki i tekhniki,* 1934, no. 4:32–6. Another physicist, P. P. Lazarev, made similar assertions in "O mezhdunarodnykh nauchnykh snosheniiakh," *Nauchnyi rabotnik,* 1926, no. 3:3–10, 4.

[44] Important exceptions in English include Daniel Todes, *Darwin without Malthus: The Struggle for Existence in Russian Evolutionary Thought* (New York, 1989); Daniel Todes, *Pavlov's Physiology Factory: Experiment, Interpretation, Laboratory Enterprise* (Baltimore, 2002); Alexander Vucinich, *Science in Russian Culture,* 2 vols. (Stanford, Calif., 1963–70); and Michael D. Gordin, *A Well-Ordered Thing: Dmitrii Mendeleev and the Shadow of the Periodic Table* (New York, 2004). To be sure, Loren R. Graham, who is almost single-handedly responsible for the maintenance of a community of Western historians concerned with Russian science, offers considerable attention to the eighteenth- and nineteenth-century peaks of science in Russia in his survey history *Science in Russia and the Soviet Union: A Short History* (New York, 1993), but even his level of analysis acquires greater depth when it comes to the Soviet period. For a more typical example of Soviet-centric studies, see Alexander Vucinich's history of the Soviet Academy of Sciences, *Empire of Knowledge: The Academy of Sciences of the USSR (1917–1970)* (Berkeley, Calif., 1984), which in a few pages dispenses with the 200 years of institutional momentum that preceded the sovietization of the academy in the early 1930s.

from scientists' understanding of their own history, the conventional understanding of the Galileo Affair (righteousness under the heel of dogmatic orthodoxy) was literally resurrected as the Lysenko Affair.[45] The history of genetics and biology served as a model for other fields (e.g., psychology, chemistry), and where it failed to destroy a field—as in theoretical physics—it was necessary to explain the case in terms of a *deviation* from the supposed Soviet norm.[46]

These specific features are far from adequate for a comprehensive understanding of the nuanced development of science in Russia and the Soviet Union, and have been either substantially modified or abandoned by specialists on this region. It is the manifest goal of this volume not only to fracture some of the old preconceptions about this field among the uninitiated but also to offer new frameworks for how studying this specific region of the world can illuminate a broad swath of central historical and historiographical questions.

It stands to reason that historians of science in Russia have been arguing for alternative justifications for the study of this area and conclusions that differ from the Sovietological model outlined above. All of the essays in this volume draw from this tradition of alternative explications but seek to recast them in the light of new empirical evidence from Russian archives and intellectual developments within historical studies generally. The eventual result of an active engagement with these emerging frameworks may be a reinvigoration of how we should understand the advantages and drawbacks of reliance on concepts such as the nation-state, ideology, and militarization in our understanding of the place of science in modernity. We believe one concept that can help effect this reorientation is the "intelligentsia."

INTELLIGENTSIA SCIENCE IN THE RUSSIAN CENTURY

Several themes run through the various essays in this volume, which we can parse in one way in terms of time and space. First, the issue of chronology. The start and end years of this volume are somewhat unusual in the historiography of Russian science (and in Russian history more generally). Typically, Russian history is presented as beginning or ending with the Bolshevik Revolution of October 1917, or the end of World War II in 1945, or the death of Stalin in 1953. We have eschewed all of these periodizations. Instead, the essays in this volume begin just before the emancipation of the serfs in 1861, the conventional beginning date for the last great effort of modernization of the Russian empire, although those efforts famously failed to preserve tsarism. Beginning here, and ending a century later, we focus—with one exception—

---

[45] The classic instances of this presentation of Lysenkoism are: David Joravsky, *The Lysenko Affair* (Cambridge, Mass., 1970); Valery N. Soyfer, *Lysenko and the Tragedy of Soviet Science*, trans. Leo Gruliow and Rebecca Gruliow (New Brunswick, N.J., 1994); and Zhores A. Medvedev, *The Rise and Fall of T. D. Lysenko*, trans. I. Michael Lerner (New York, 1969). For more recent and sophisticated efforts that still place the bulk of their attention on the ideology of either the macropolitical or microbureaucratic varieties, see Nikolai Krementsov, *Stalinist Science* (Princeton, N.J., 1997); and Ethan Pollock, *Stalin and the Soviet Science Wars* (Princeton, N.J., 2006). For further discussion of this specific Lysenko emphasis, see Michael D. Gordin, "Was There Ever a 'Stalinist Science'?" *Kritika* (forthcoming).

[46] Loren R. Graham, *Science, Philosophy, and Human Behavior in the Soviet Union* (New York, 1987); David Joravsky, *Russian Psychology: A Critical History* (Oxford, 1989); M. G. Iaroshevskii, *Repressirovannaia nauka*, 2 vols. (Leningrad, 1991–94); David Holloway, "How the Bomb Saved Soviet Physics," *Bulletin of the Atomic Scientists* 50 (Nov. 1994): 46–55; and Nikolai Krementsov, *The Cure: A Story of Cancer and Politics from the Annals of the Cold War* (Chicago, 2002).

on hallmark episodes of the development of science in Russia: from the moment of its dramatic takeoff to just before the cold war "stagnation" (*zastoi*) of the Brezhnev years when several factors altered the configuration of Soviet science. This periodization also places the Bolshevik Revolution firmly in the center of our span, so we can openly explore the extent to which it was, in fact, a discontinuity in the history of science in Russia, rather than taking such a break for granted.

Place is a closely related reformulation. One of the supposedly distinctive features of the Stalinist Soviet Union was its autarky—its disconnect from the developments in Western Europe and North America. (The Lysenko Affair as modern Galileo Affair fits neatly into this narrative.) The essays in this volume, by contrast, constantly move between the physical space of Russia and the Soviet Union to the West—Paris, Zurich, Heidelberg, London, Houston—to show how the borders between Russian developments and those worldwide were more porous than has typically been realized. The history of science in Russia is thus an essential component of the history of science around the world and not just a cutoff boutique interest. To fully understand the impact of the autarkic periods (which certainly existed), one needs to explore how significant the interactions were at other times.

Our volume is divided into three sections, each of which explores a different traditional function of the various "intelligentsias" in Russian history through the lens of science and reformulates the traditional concepts in the process. The first section, "Intelligentsia as Social Organization," examines the institutions of the intelligentsia as forms of social cohesion and differentiation and how patterns that were characteristic of the more "mainstream" intelligentsia were both appropriated and adapted for the particular needs of science and technology.

Our story opens around 1860, when the intelligentsia found its obsession with science. Both science and the intelligentsia would change as a result of that relationship. A few years after 1855, when Tsar Alexander II assumed the throne, the greatly expanded universities filled with a cohort of students coming from the much wider, predominantly lower, ranks of gentry, clergy, and state officials, often poor and needing to work for material survival. The intelligentsia began to fragment into various strata, characterized by different tones: the younger set were typically more uncompromising and demanding, while the older generations—who continued to see themselves as intelligenty—characteristically combined free thinking with a confidence and leniency of aristocratic privilege. The new, more democratic radicalism demanded changes that even a reformist government was not able to satisfy, which quickly turned universities into an arena of chronic political protest.[47] The intelligentsia's generation of the 1860s added natural science as another major component to more traditional preoccupations with literature and social criticism. The transition was well marked by the arrival of the Russian translation of Darwin's *Origin of Species* (published 1859 in English; translated into Russian in 1864), which sparked multiple reactions—anticlerical, antihierarchical, anticapitalist, anticompetition, anti-Malthusian—and defies simple characterization.[48]

[47] On universities in this period, see Samuel D. Kassow, *Students, Professors, and the State in Tsarist Russia* (Berkeley, Calif., 1989); Susan K. Morrisey, *Heralds of Revolution: Russian Students and the Mythologies of Radicalism* (New York, 1998); and Allen Sinel, *The Classroom and the Chancellery: State Educational Reform in Russia under Count Dmitry Tolstoi* (Cambridge, Mass., 1973).

[48] On these various debates, see Todes, *Darwin without Malthus* (cit. n. 44); and Alexander Vucinich, *Darwin in Russian Thought* (Berkeley, Calif., 1988).

In time, there were tensions and competition between various activities that had previously been more or less unproblematically supported by the intelligentsia's commonly shared values. Many radical students struggled over the personal choice between literary and scholarly pursuits, involvement in social activism, and mainstream careers within the establishment. This volume's articles concentrate on those for whom science became a major preoccupation in life—the subspecies of scientific intelligentsia. Using as an example the case of chemistry, the leading science of the period, Michael Gordin's contribution analyzes the first major consequence of the shift: the training and professionalization of the Russian research community. For perhaps the first time, the intelligentsia's aspirations overlapped with the interests of the Russian state. Embarking upon a renewed effort of modernization "from above" (and later also as a conscious attempt to distract students from radical politics), the government supported research aspirations of able students with a program of state fellowships for study abroad. Gordin shows how this initiative eventually (and in this case unintentionally) enabled the organization of the professional Russian Chemical Society, which came about via a combination of various models of contemporary scientific research—largely taken in this case from the southern German states, in particular from the University of Heidelberg—and the native cultural model of *kruzhok*, or closely knit discussion circle, that had previously provided the basic structural social unit for the Russian intelligentsia.

The cult of science stood high enough for many other professional groups to long for the status of a scientific discipline. Andy Byford describes a revealing, if ultimately unsuccessful, attempt to establish a new science of pedology, or child studies, in the border area between academic psychology and high school teacher training. For experts in pedagogy, the creation of their own branch of science would have considerably raised the social prestige of the profession—and simultaneously enabled specialists in education to rightfully call themselves intelligenty. The first decade of the twentieth century opened an opportunity to achieve the goal of "scientific pedagogy" through nongovernmental academic and educational institutions, including professional training courses for teachers. Bypassing the state and establishing private and community-based institutions allowed a novel (for Russia) approach to the realization of these goals, as well as for many other controversial intelligentsia initiatives. Although the pedology movement lost much of its civil base after the 1917 revolution, its activists established themselves within the new organs of the Bolshevik state and carried the project further with intensified vigor, though only for the time being. In the 1930s, the changed political realities reversed the trend and delivered final victory to their opponents and to "practical" pedagogy.

One finds interesting parallels to this case from the end of our period, at the height of the Soviet Union's technological development. The Soviet nuclear power industry was born of twin origins: nuclear expertise, drawn from military research that had produced the first Soviet atomic bomb in 1949; and the experience of electrical engineers, who had occupied central positions in the technocratic intelligentsia since Lenin's massive electrification effort of the 1920s (GOELRO). Sonja Schmid explores the tension between these two groups of experts as the "designers" (*atomshchiki*) cast themselves as heirs of an older elite intelligentsia tradition, while the "operators" (*energetiki*) of the nuclear power plants continued as Stalinist technocrats. Taking this story of professional differentiation through to the 1986 Chernobyl reactor catastrophe, Schmid demonstrates how this long-standing competition of intelligentsia identities ran continuously through major transformations in the post-Stalin Soviet Union.

Indeed, throughout this volume, one can trace the constant tension between the Stalinist identification of "intelligentsia" with white-collar workers (such as scientists), and a protodissident identification of science as a radical democratic force analogous to the nineteenth-century traditions. Our second section, "Intelligentsia as Political Agent," explores the implications of this tension when the intelligenty sought to actively take part in statecraft and science policy.

Such political action, especially in the Soviet period, relied on an inflated notion of the power of science, shared by the Communist Party and the scientific intelligentsia, or "bourgeois experts" in Bolshevik parlance of the 1920s. The prominent role of nonparty experts in the creation of the new society and its institutions made the Soviet state in this period of transition from the October Revolution to Stalinism the closest real-life approximation of Plato's imaginary *Republic*. It also enabled a far-reaching transformation of science itself in an attempt to reach the impossible ideal. As Alexei Kojevnikov argues in his essay, the intelligentsia in power pioneered a series of important innovations that would change the social role of science in the course of the twentieth century. Those included a large-scale government involvement in supporting and directing research and development; recognition of science as a separate branch of civil service and its transformation into a mass profession; a broadening of the demographic base of science by a system of measures similar to those currently known as "affirmative action" in the American context; and last but not least, constructivist theoretical ideas about science in its relation to society.

Building a scientific state required cadres to work it, and building significant resources took a long time. Olga Valkova's contribution discusses the emergence of an important group of intelligentsia scientists—women scientists—from its origins in the 1860s until the onset of the Second World War. This *longue durée* approach emphasizes the continuities for which the intelligentsia category is so useful. For intelligentsia women in the nineteenth century, becoming professional scientists was extremely difficult, as in their case the government had long maintained legal barriers to university matriculation, the sine qua non for a scientific career. The feminist movement arose in the 1860s from young women's (and men's) struggle for access to higher education, which they attempted to achieve, at first, in Russia—and when that failed, abroad, by enrolling in western European universities.[49] But formal higher education was not the only possible route toward scientific activities—Valkova uncovers a variety of different cases and strategies employed by women in their quest to engage in scientific research. Even the removal of formal barriers by the Bolshevik government immediately after the revolution of 1917 and the establishment of legal equality in scientific education did not fully resolve the problem, as women continued to fight against more subtle forms of discrimination throughout their research careers. In this final period, women scientists occupied an interesting blend of contrary intelligentsia traditions: reformist technocratic politics from within, and oppositional resistance from a position of exclusion.

---

[49] In autocratic Russia, the demand for gender equality in educational opportunities remained much more important for feminism than the issue of voting rights, as men could not vote either. On the course of Russian feminism, see the classic but still useful Richard Stites, *The Women's Liberation Movement in Russia: Feminism, Nihilism, and Bolshevism, 1860–1930* (Princeton, N.J., 1978).

The most famous (and infamous) incident where this intelligentsia tension came to the fore was the well-known Lysenko case of 1948, or the decision to abandon Mendelian genetics in favor of a supposedly homegrown, neo-Lamarckian version of agricultural science, "Michurinist biology." Instead of primarily ideological or institutional explanations, Nils Roll-Hansen sees the root possibility of this move in a real and basic dilemma of science policy, which can be expressed by the famous Marxist dictum as "the unity of theory and practice." This postulate allowed the state to play the role of ultimate arbiter between the two competing approaches in science and also allowed rival communities of scientists to use and abuse the apparatus of the state for settling their own academic and institutional conflicts. Although the Lysenko case represents a somewhat extreme example, the underlying science policy dilemma is not unique to the Soviet Union and remains valid and meaningful today.

In the final essay in this section, Slava Gerovitch explores the specific identity of Soviet technical intelligentsia in his analysis of the rocketry and space programs in the 1950s and the 1960s. The chief designers of the Soviet space probes, Sergei Korolev and Valentin Glushko, survived arrests in the 1930s, worked their way into classified military research, and ultimately learned how to manipulate the system to win reluctant permission for the launch of *Sputnik* in 1957. All the while they remained loyal to the regime, yet also critical to a degree, lobbying for their own understanding of a space policy that might deviate considerably from the official one. The dual nature of this project—both classified military and openly propagandistic—relied on dual-use rocket technologies and developed dual identities and dual agendas within the space program, signifying the parting of ways between the Soviet state and its scientific elite.[50]

One of the consistent themes in this collection is that although the intelligentsia (or intelligentsias) consisted of real social structures and engaged in real politics, it also represented an ideal, an imaginary on which Russian scientists could project their aspirations. This is the central concern of our third section, "Intelligentsia as Utopia," borrowing the term from one of the classic historical surveys of the Soviet period, *Utopia in Power*.[51] Of course, this interest in the utopian was not confined to the Soviet period. In the first essay of this section, Kirill Rossiianov presents a novel analysis of Elie Metchnikov's landmark discovery of immune cells and the founding of immunology. As earlier research in the history of Russian science has established, many Russian authors tended to accept Darwinism without its Malthusian component, stressing as the chief mechanism of evolution the struggle of organisms with the forces of environment rather than against the members of the same species. It is also well known that Western scientific racism generally found little appeal and circulation in the nineteenth-century Russian empire, whose authors could view themselves (as Russians) simultaneously or interchangeably as both colonizers and colonized, privileged and oppressed, compared with other nations. Yet Metchnikov, one of the most distinguished Russian scientists of his era, did engage in a borrowed discourse about "primitive races." He managed, however, to combine it—in a counterintuitive and paradoxical way—with the Russian Populist (*narodnik*) moral and political ideal

---

[50] This terminology borrows from one of the classic studies of the formation of the Russian intelligentsia as an "unofficial opposition": Nicholas V. Riasanovsky, *A Parting of Ways: Government and the Educated Public in Russia, 1801–1855* (Oxford, 1976).

[51] Mikhail Heller and Aleksandr Nekrich, *Utopia in Power: The History of the Soviet Union from 1917 to the Present*, trans. Phyllis B. Carlos (New York, 1986).

of human equality, and to find a resolution to the inherent contradiction between the two, and a realization of his hopes for a harmonious society in his model of immunity in higher organisms.

In 1917, theoretical physics was still a fairly young subdiscipline even in central Europe, and few Russian physicists had embraced this ill-defined professional identity. Yet within the space of a few decades sizable cohorts of Soviet theorists had won worldwide acclaim, primarily for developing distinctive professional practices in quantum field theory and condensed matter theory. Along with mathematicians, however, these theorists were also seen by many as the natural scientists who had best managed to escape Lysenkoist politicization. Karl Hall describes the search for collective identity among theorists in this dystopian setting, focusing on the "school" as an important means for advancing the discipline in the Soviet period. The best-known school among theorists was that led by Lev Landau. Yet his peculiar brilliance has obscured the broader professional context for pursuing this mode of scientific community in the early decades of Soviet power. Hall locates some of these aspirations in the generational disruptions under the Bolsheviks but also finds many unacknowledged echoes of prerevolutionary intelligentsia concerns among Russian physicists. The article further demonstrates the international context for understanding these generational and disciplinary tensions within the young Soviet physics "community." Whereas the older generation could use its westward travels to establish its Soviet credentials under the new regime, the younger generation—often initially sympathetic to Bolshevik scorn for the classical intelligentsia—could, in turn, employ their Western exploits to fashion distinctive Soviet identities at the expense of their elders.

With the young supplanting the old, the primitive taming the civilized, and the quantum providing stability, it seems safe to claim that there was indeed something utopian going on in the Soviet Union. A revolution had unexpectedly overthrown a large and powerful empire. A small and marginal segment of the intelligentsia—the Bolsheviks, a socialist conspiracy in a predominantly peasant country—achieved supreme political power and embarked upon building a society of universal equality and prosperity. Asif Siddiqi looks at the postrevolutionary fate of a non-Bolshevik but equally utopian project, Russian Cosmism. Despite its deep roots in mysticism and religion, Cosmism had one aspect in common with the Bolshevik program: belief in the unlimited potential and transformative power of technology. Because of this, some of its representatives, in particular Konstantin Tsiolkovskii, received encouragement from the state, if not exactly munificent support. Though meant primarily as a public-relations campaign to inspire impressionable youth into the study of engineering, Cosmism's utopian project—or rather its technological component cut off from the mystical goal—also turned real forty years later, with the Soviet breakthrough into the cosmos.

What became of all these traditions—technocratic, oppositional, reformist, revolutionary, utopian—in the decades after *Sputnik*? While open dissidents were rare, the growing alienation from the regime eventually created a whole new generation of the oppositional-minded Soviet intelligentsia by the 1980s. Their renewed revolt during Gorbachev's perestroika against the system that was ultimately responsible for their privileged status in society would result in the third Russian revolution of the twentieth century, this time anticommunist as a successor to the liberal (1905) and Bolshevik (1917) revolutions. This time, some argued, the revolution also destroyed the very conditions that allowed intelligentsia to exert such influence in the first place, considerably undermining the group's prestige and making it disoriented and marginalized. Whether this is a terminal case or only a temporary development remains to be seen. The essays in this volume should add some perspective on the place of science in the developments that led to the rise of Russia's prominence in the intellectual sphere in the 1860s, and its possible—but debatable—decline since the 1960s. The Russian century may in fact be over, but the interrelation of the intelligentsia and science to form "intelligentsia science" proves enduring.

INTELLIGENTSIA AS SOCIAL ORGANIZATION

The Heidelberg Circle:
German Inflections on the Professionalization of Russian Chemistry in the 1860s

By Michael D. Gordin[*]

ABSTRACT

The success of the "second importation" of science to Russia during the Great Reforms of the 1860s is illustrated by examining the extended postdoctoral study of chemists in Heidelberg. While there, they adapted the Russian intelligentsia institution of the "circle," or *kruzhok*, to cope with their alienation from the German culture they were confronting. Upon their return to Russia, they felt the lack of the communicative network they had established while abroad and reimported the kruzhok to serve as a central model for the formation of the Russian Chemical Society in 1868.

INTRODUCTION

Science, as everyone knows, was not native to Russia. Although there were limited cosmological, medical, and metallurgical concepts and practices employed across the space now identified with Russia, it was not until the very late seventeenth century that large-scale imports of engineers from central and western Europe began to affect governance and the military. As for elite science—the collection of high-level theories, mathematics, experimental practices, and conceptual frameworks usually understood on the model of western European natural knowledge from the Renaissance onward—that had a very specific birth date in Russia. Tsar Peter the Great (r. 1689–1725), in one of his final decisions, acted upon a suggestion by the noted natural philosopher Gottfried Leibniz and created an Academy of Sciences in his new capital, St. Petersburg. Of course, Peter not only had to arrange for the institution but also had to provide the professionals qualified to staff it. He imported a collection of

---

[*] Department of History, 129 Dickinson Hall, Princeton University, Princeton, NJ, 08544; mgordin@princeton.edu.
  Abbreviations used in notes: ADIM—Arkhiv-Muzei D. I. Mendeleeva (D. I. Mendeleev Archive-Museum), St. Petersburg, Russia; BorP—A. P. Borodin, *Pis'ma: Polnoe sobranie, kriticheski sverennoe s podlinnymi tekstami*, ed. S. A. Dianin, 4 vols. (Moscow, 1927–50); DM-HS—Deutsches Museum, Handschriften, Munich, Germany; TsGIASPb—Tsentral'nyi Gosudarstvennyi Istoricheskii Arkhiv Sankt-Peterburga (Central State Historical Archive of St. Petersburg), St. Petersburg, Russia; ZhRFKhO—*Zhurnal Russkogo Fiziko-Khimicheskogo Obshchestva* (*Journal of the Russian Physico-Chemical Society*). Materials originating in Russia are dated according to the Old Style Julian calendar, which lagged twelve days behind the Gregorian New Style calendar in the nineteenth century. Materials originating in Germany are presented in New Style. Unless otherwise indicated, all translations are my own.

© 2008 by The History of Science Society. All rights reserved. 0369-7827/08/2008-0002$10.00

central European savants in various areas of the arts and sciences to be his first academicians.[1] Thus science was a foreign import.

However, science, as everyone knows, has been immensely successful in Russia. By whatever measure one chooses—numbers of scientists, rate of publication, important discoveries, peer recognition—Russian scientists have been at the forefront of international scientific developments for at least the last 150 years. So, although science was a foreign import, it was one that took exceptionally well to Russian soil.

Or did it? For over a century after the introduction of the main eighteenth-century institution of Western natural philosophy, the scientific academy, it is difficult to find any significant penetration of science or scientific institutions outside St. Petersburg. Most (although certainly not all) of the achievements of Russian science took place after the transformation of Russian governance during the so-called Great Reforms of the 1860s and 1870s, under the leadership of Tsar Alexander II (r. 1855–1881). Something very specific seems to have happened at the cusp of the 1860s that mobilized a scientific intelligentsia out of what had earlier been a shallow system that had relied on foreign talent. This essay will explore what those transformations were and how they altered the structures by which Russian science was organized.[2]

What happened to alter the fundamental structure of Russian science is fairly easy to map out schematically: upon the loss of the Crimean War (1853–1856), the Russian state realized that it would risk its future fiscal and military stability if it did not modify features of the Russian polity that made it, in contemporary Russians' terms, relatively "backward" with respect to western Europe. What came to be called the Great Reforms were initiated formally by the abolition of serfdom in February 1861, a reform that had actually been in the planning stages for some time.[3] Similar self-conscious "modernizing" reforms ensued in the areas of technical education and technical institutions. Instead of bringing the mountain to Muhammad, as they had done with the Academy of Sciences, Russian bureaucrats decide to send their talented graduate students and "postdocs"[4] abroad, largely to the German states, thereby taking Muhammad to the mountain. As this essay will argue, using the specific example of chemistry and chemical postdocs, this technical emigration (especially its reverse

---

[1] On the early eighteenth-century establishment of scientific institutions, see Michael D. Gordin, "The Importation of Being Earnest: The Early St. Petersburg Academy of Sciences," *Isis* 91 (2000): 1–31; and Iu. D. Kopelevich, *Osnovanie Peterburgskoi akademii nauk* (Leningrad, 1977).

[2] I will not address much of the content of scientific work during this time period; the organization and infrastructure were so complex and of such a striking nature that they need to be examined in detail before one can properly analyze the role of such a system in fostering specific intellectual developments.

[3] See W. Bruce Lincoln, *In the Vanguard of Reform: Russia's Enlightened Bureaucrats, 1825–1861* (DeKalb, Ill., 1982); Lincoln, *The Great Reforms: Autocracy, Bureaucracy, and the Politics of Change in Imperial Russia* (DeKalb, Ill., 1990); and Alfred J. Rieber, "Alexander II: A Revisionist View," *Journal of Modern History* 43 (1971): 42–58.

[4] A word of clarification about my use of the terms "postdoctoral" and "postdoc" in this essay: Strictly speaking, the terms are both anachronistic and inappropriate. The Russian educational system allows for three degrees of higher education: the candidate degree, the master's degree, and the doctoral degree. The candidate degree is very close to an American bachelor's degree. The master's degree is often considered close to a doctorate, but it is actually at a somewhat lower level than a full-fledged Western PhD. After working for several years, Russian scholars file for their actual *doktorskaia*, which is almost exactly analogous to a German *Habilitation*. When I use "postdoc" here, I mean students who have already received significant higher technical training, almost always past the master's degree. The system of transit mixes and matches them with contemporaries we would recognize as postdoctoral students from France, Britain, or the United States.

flow back to Petersburg) led to the creation of a specific form of professionalization of the sciences in postreform Russia, one that both drew from and reacted to the German milieu in which the Russians lived while abroad.

To the extent that this "German captivity" has been discussed with respect to Russian science by commentators, it has received mixed or negative reviews. Either the transformation in Russian institutions is seen as autochthonous, and essentially unrelated to the two- or three-year sojourns the Russians spent abroad, or the exposure to German institutions is seen as deleterious.[5] For those who hold to an essentialist vision of the Russian national character, to the extent that the returning Russians borrowed anything from the Germans, that borrowing was destructive and only served to hold back some form of authentic Russian science:

> The educational system was borrowed from Germany, its negative qualities were intensified while the most important positive qualities were partially or completely suppressed. The Russian national character was not taken into account by that system foreign to its spirit which was put as it were into a straight-jacket and had its wings clipped by the two most efficient tools in the hands of autocracy—censorship and espionage.[6]

This (clearly bigoted) quotation raises a series of intriguing questions: Was the system in fact borrowed from Germany? What features made it "German"? How was "Russian national character" understood, and how did it relate to the sciences? The only way to get at these questions is to back away from facile generalizations and really examine the local dynamics of what happened.

For those dynamics were very much *local* ones, as well as broader cultural transformations. The most important social group for the development of science in Russia— and then eventually the domestication and appropriation of that system of knowledge into *Russian* science—was the intelligentsia.[7] The intelligentsia, like any other social institution, had to organize itself somehow, and the basic structure of the Russian intelligentsia was subdivision into *kruzhkí* ("circles"; singular, *kruzhók*). (The accents are used here to clarify pronunciation.) In this essay, I argue that the organization of chemistry in Petersburg into a Russian Chemical Society in November 1868 was in large part mediated by the adaptation of the urban social institution of the kruzhok under the culturally adverse conditions of the German scientific emigration. Understanding the kruzhok, therefore, leads directly into the nexus between the formal structures of Western science and the social structures of the Russian intelligentsia. Although the kruzhok provides far less than a total explanation of the professionalization of scientists in Russia, its specific features do go a long way toward explaining the rapidity and vehemence of Russian national identification in the sciences, particularly in chemistry.

[5] For accounts that stress autonomous Russian developments, see Iu. I. Solov'ev, *Istoriia khimii v Rossii: Nauchnye tsentry i osnovnye napravleniia issledovanii* (Moscow, 1971); and also to a certain extent Nathan M. Brooks, "The Formation of a Community of Chemists in Russia: 1700–1870" (PhD diss., Columbia Univ., 1989).
[6] Alexander Petrunkevich, "Russia's Contribution to Science," *Transactions of the Connecticut Academy of Sciences* 23 (1920): 211–41, on 215.
[7] It has proven perpetually difficult to provide a clean and simple characterization for what exactly the "intelligentsia" was other than a heterogeneous grouping of educated individuals who conceived of themselves and organized themselves somewhat outside the social confines bounded by the autocratic state. As an introduction to the concept and its history, see the introduction to this volume and the essays in Richard Pipes, ed., *The Russian Intelligentsia* (New York, 1961).

It is quite difficult to formulate a precise definition of the kruzhok, which is somewhat of a cross between a concentrated, topical salon and an intellectual *Stammtisch*. Kruzhki were usually relatively small (fewer than twenty people) and had a defined membership; you could only become a member of a kruzhok if proposed by a standing member, and other members could blackball you if you were perceived as unreliable. Given the political stakes of a compromised kruzhok—Fyodor Dostoevsky was exiled to Siberia when a police mole reported on proscribed political discussions within the Petrashevskii kruzhok, of which the writer was a member—this insularity and exclusiveness were vital adaptations to a highly controlled political climate. They also offered remarkable communicability across increasingly divergent disciplines, as intellectuals and aristocrats tended to belong to several kruzhki at a time, carrying concerns from one into another. Most scholarship of kruzhki among Russian historians has focused on the two outstanding exemplars from the 1840s—the Westernizers and the Slavophiles—although the tradition extended earlier in time as well as later.[8] With emancipation in 1861, the historiography stresses these institutions as staples of student culture in the demographic boom of Moscow and Petersburg university populations, where they would eventually serve as kernels of Marxist, populist, and terrorist politics, or as circles of artists and literati.[9] Daniel Alexandrov—the only historian of science to take the kruzhok seriously as an organizing principle for Russian knowledge production—traces his genealogical line from these student kruzhki into the Soviet period, and his work gives a picture of the versatility of this institution in Soviet Russia (and abroad, in the case of the famous Kapitza Club at Cambridge).[10] I propose that the kruzhok's legacies were far richer than just this contribution to

---

[8] Roman Jakobson, "An Example of Migratory Terms and Institutional Models (On the Fiftieth Anniversary of the Moscow Linguistic Circle)," in *Selected Writings*, vol. 2, *Word and Language*, ed. Stephen Rudy (The Hague, 1971), 527–38; N. L. Brodskii, ed., *Literaturnye salony i kruzhki* (Moscow, 1930); Mark Aronson and Solomon Reiser, *Literaturnye kruzhki i salony* (St. Petersburg, 2001); and Frederick I. Kaplan, "Russian Fourierism of the 1840's: A Contrast to Herzen's Westernism," *American Slavic and East European Review* 17 (1958): 161–72.

[9] See, e.g., Allan K. Wildman, "The Russian Intelligentsia of the 1890's," *American Slavic and East European Review* 19 (1960): 157–79; Martin A. Miller, "Ideological Conflicts in Russian Populism: The Revolutionary Manifestoes of the Chaikovsky Circle, 1869–1874," *Slavic Review* 29 (1970): 1–21; Richard Pipes, "Russian Marxism and Its Populist Background: The Late Nineteenth Century," *Russian Review* 19 (1960): 316–37; Franco Venturi, *Roots of Revolution: A History of the Populist and Socialist Movements in Nineteenth-Century Russia*, trans. Francis Haskell (Chicago, 1960); Susan K. Morrissey, *Heralds of Revolution: Russian Students and the Mythologies of Radicalism* (Oxford, 1998); Barbara Walker, "*Kruzhok* Culture: The Meaning of Patronage in the Early Soviet Literary World," *Contemporary European History* 11 (2002): 107–23; and Walker, *Maximilian Voloshin and the Russian Literary Circle: Culture and Survival in Revolutionary Times* (Bloomington, Ind., 2005).

[10] D. A. Aleksandrov [Alexandrov], "Istoricheskaia antropologiia nauki v Rossii," *Voprosy istorii estestvoznaniia i tekhniki*, 1994, no. 4:3–22; and Alexandrov, "The Politics of Scientific 'Kruzhok': Study Circles in Russian Science and Their Transformation in the 1920s," in *Na perelome: Sovetskaia biologiia v 20–30'kh godakh*, ed. E. I. Kolchinskii (St. Petersburg, 1997), 255–67. A more sociological approach to the role of circles in Soviet science, but one that does not attempt the cultural connection to kruzhki, is Linda L. Lubrano, "The Hidden Structure of Soviet Science," *Science, Technology, and Human Values* 18 (1993): 147–75. It is important not to confuse the kruzhok, a culturally specific Russian institution, with the more generalized sociological concept of "social circle," which has been somewhat fruitful in analyzing scientific change. On social circle theory, see Charles Kadushin, "Power, Influence, and Social Circles: A New Methodology for Studying Opinion Makers," *American Sociological Review* 33 (1968): 685–99; and Belver C. Griffith and Nicholas C. Mullins, "Coherent Social Groups in Scientific Change," *Science* 177 (1972): 959–64. There is obviously a relation between the two, but they are not identical.

radical student culture: it was also a seedbed of the established, albeit idiosyncratic, professional culture.[11]

Before beginning with the origins of the emigration, a word of defense is required with respect to the choice of chemistry as the focus of this paper. Although the implications of the argument here for the formation of a professional scientific culture are of necessity broader than merely in one field, there are good reasons to focus on this specific science when exploring the "second importation" of the sciences to Russia. Chemistry was *the* dominant science in late imperial Russia, partially because of its utility to the state in the areas of mining, oil exploitation, agriculture, and munitions; partially because it was at the time the leading science internationally; and partially because it was simply the first science to cross the horizon of modernity by acquiring a professional society and official government recognition. The chemical community is vital to our historical understanding of the nature of professional organization in Russia because it subsequently served as a model for essentially all communities of scientists (and other technical experts) formed in the spaces of the Russian empire. Studying chemistry, then, provides the historian of science a tracer for the evolution of Russian nationalist conceptions precisely at a site where one would least expect it: at the heart of the most international and cosmopolitan physical science.[12] And these Russian developments could not have occurred—or would have occurred rather differently—without the midwifery of a specific German university town.

HEIDELBERG: THE CENTER OF THE CIRCLE

Generally speaking, there have been two strategies that states have followed for introducing Western science into a new cultural context, executed either individually or in combination: importing the scientists as retainers from foreign lands, thus outsourcing the *talent* (think of Peter the Great and his academy project); or sending one's countrymen abroad to receive their training, then bringing them back, thus outsourcing the *training*.[13] It was this latter strategy that became increasingly common among Russian institutions of higher education and tsarist bureaucrats in the mid-nineteenth century. This strategy was not completely new to Russia after the Crimean War. Medical doctors, for one, had been sent abroad for "improvement" since the mid-eighteenth century, and—after a brief hiatus from 1803 to 1817—continued to

---

[11] On professionalization in Russia and its divergence from the classic Anglo-American models, see, e.g., Christine Ruane, *Gender, Class, and the Professionalization of Russian City Teachers, 1860–1914* (Pittsburgh, Pa., 1994); and Harley D. Balzer, ed., *Russia's Missing Middle Class: The Professions in Russian History* (Armonk, N.Y., 1996). For helpful criticisms of the traditional models of professionalization for the history of science, see Jan Goldstein, "Foucault among the Sociologists: The 'Disciplines' and the History of the Professions," *History and Theory* 23 (1984): 170–92; Thomas Broman, "Rethinking Professionalization: Theory, Practice, and Professional Ideology in Eighteenth-Century German Medicine," *J. Mod. Hist.* 67 (1995): 835–72; and B. W. G. Holt, "Social Aspects in the Emergence of Chemistry as an Exact Science: The British Chemical Profession," *British Journal of Sociology* 21 (1970): 181–99.

[12] The counterpart would be a chauvinistic social science or discipline of the humanities. For these, more intuitive, cases, see, e.g., Alexander Vucinich, *Social Thought in Tsarist Russia: The Quest for a General Science of Society* (Chicago, 1976).

[13] These two strategies were undertaken simultaneously in the cases of Japan and China. See James R. Bartholomew, *The Formation of Science in Japan: Building a Research Tradition* (New Haven, Conn., 1989); Paula Harrell, *Sowing the Seeds of Change: Chinese Students, Japanese Teachers, 1895–1905* (Stanford, Calif., 1992); and Weili Ye, *Seeking Modernity in China's Name: Chinese Students in the United States, 1900–1927* (Stanford, Calif., 2001).

be so well into the nineteenth. Likewise, legal scholars had been sent abroad for many years, most notably under the sponsorship of the pivotal legal bureaucrat Mikhail Speranskii in the late 1820s.[14] Nevertheless, the scale of the effort that emerged in the late 1850s dwarfed these earlier qualified attempts to siphon some of the cream off the Western educational establishment.

Even at this point, however, the procedure was implemented on a trial basis. A trickle of scholars—almost always very talented graduate students who had already completed their *magisterskaia*, the second-highest academic degree—were sent abroad to various universities in central and western Europe. Almost none undertook the trip to England. A few ventured to Paris. The vast majority of those who went abroad chose to affiliate with an institution in one of the German states. (This includes medical students who by and large went to Vienna, which in the era before German unification made for a plausible German university.) When the program was instituted on a wider scale in the early 1860s, this trend only deepened.

This migration of students, especially in the early period, was not of a random character. Generally, postdocs in specific fields (legal scholars, chemists, physicians, classicists) tended to congregate at specific sites. For a variety of reasons, the small number of chemists—in the late 1850s not more than twenty—concentrated in Heidelberg, although they almost all traveled widely in Europe during their two- or three-year stay abroad. This core of Russian Heidelbergers proved to be a vital kernel for the institutionalization of a professionalized chemistry in St. Petersburg, and then in the Russian empire more broadly.

There were (as there still are) numerous attractions for the young student who opted to take his stipend in Heidelberg, at Heidelberg University. Founded in 1386, it was the third-oldest German university—and the oldest within the confines of present-day Germany (the older two being in Prague and Vienna). The somewhat tortured and lengthy history of the institution had reached a brighter passage by the early nineteenth century, as Heidelberg University began to transform itself into the very model of a modern research university.[15] The direct impetus for the reorganization of the university in 1803 was the establishment, by Napoleon, of new con-

---

[14] G. Skorichenko, "Imperatorskaia mediko-khirurgicheskaia akademiia v vedenii Ministerstva narodnago prosveshcheniia," in *Istoriia imperatorskoi Voenno-meditsinskoi (byvshei mediko-khirurgicheskoi) Akademii za sto let, 1798–1898*, ed. Ivanovskii (St. Petersburg, 1898), 155–238, 218; and P. Maikov, "Speranskii i studenty zakonovedeniia: Ocherk iz istorii russkago pravovedeniia," *Russkii Vestnik* 266, no. 8 (1899): 609–26; 267, no. 9 (1899): 239–56.

[15] Understandably, given the age and distinction of the university, its historiography is vast. Some of the best studies include: Eike Wolgast, *Die Universität Heidelberg, 1386–1986* (Berlin, 1986); Gerhard Hinz, "Die Geschichte der Universität Heidelberg: Überblick," in *Ruperto-Carola Sonderband: Aus der Geschichte der Universität Heidelberg und ihrer Fakultäten*, ed. Gerhard Hinz (Heidelberg, 1961), 20–39; Renate Klauser, "Aus der Geschichte der Philosophischen Fakultät Heidelberg," in ibid., 235–336; Helene Tompert, *Lebensformen und Denkweisen der akademischen Welt Heidelbergs im Wilhelminischen Zeitalter: Vornehmlich im Spiegel zeitgenössischer Selbstzeugnisse* (Lübeck, 1969); and Ludwig Schmeider, *Ruperto Carola: University of Heidelberg*, trans. D. Michael (Düsseldorf, 1931). On the reforms of the university in the early nineteenth century, see Georg Jellinek, ed., *Gesetze und Verordnungen für die Universität Heidelberg* (Heidelberg, 1908); Franz Schneider, *Geschichte der Universität Heidelberg im ersten Jahrzehnt nach der Reorganisation durch Karl Friedrich, 1803–1813* (Heidelberg, 1913); and Richard August Keller, *Geschichte der Universität Heidelberg im ersten Jahrzehnt nach der Reorganisation durch Karl Friedrich (1803–1813)* (Heidelberg, 1913). For the situation of the university before the reforms, see Gerhard Merkel, *Wirtschaftsgeschichte der Universität Heidelberg im 18. Jahrhundert* (Stuttgart, 1973).

fines for the grand duchy of Baden, in which Heidelberg is located. Baden was perhaps unique among the German states for maintaining a vibrant culture of liberalism, which was especially important in the period between the Congress of Vienna (1815) and the abortive revolutions of 1848. The attraction for German students from various *Länder* (states or administrative regions) was quite strong, and Heidelberg offered a counterpart to the famous exaltation of *Wissenschaft* and academic freedom in the Prussian north, but in a more congenial political environment.[16] As a result, the student population—undergraduate and graduate—and the professoriate boomed.

Those populations did not boom haphazardly. As Peter Borscheid argued in a seminal monograph thirty years ago, after 1848 a distinct emphasis was placed on the natural sciences, then a part of the philosophy faculty. (This narrative runs counter to the typical presentation of Heidelberg as a seat of German Romanticism—which, of course, it also was.[17]) According to Borscheid, southern German states perceived the revolutions to be at their roots agricultural disturbances caused by instability in crop production. Justus von Liebig (1803–1873), at the time the doyen of German chemistry from his post in tiny Giessen, argued that substantial development of chemistry in a university context would foster future political stability among the lower classes in two ways: advanced agricultural chemistry would guarantee greater crop stability across harvests, and the development of a cadre of technical experts within universities would encourage the maturation of the chemical industry, which could absorb an impoverished proletariat.[18] Liebig was heavily courted by Baden to take charge of the development of the sciences in Heidelberg, but he accepted an offer from Bavaria instead and shortly thereafter moved to Munich. The second choice in the southern German bidding war was Robert Wilhelm Bunsen (1811–1899). Bunsen had taught in many different German universities throughout his career before moving to Heidelberg in 1854, where he remained until his death.

To the extent that chemistry in Heidelberg has attracted any interest at all, it has been because of Bunsen. Bunsen was one of the towering figures of nineteenth-century chemistry, although most of his major contributions to the field occupied the earlier, pre-Heidelberg, part of his career.[19] Bunsen at various points in time was interested in

---

[16] On the fortunes of Baden's political liberalism from the Napoleonic period to German unification, see Lloyd E. Lee, *The Politics of Harmony: Civil Service, Liberalism, and Social Reform in Baden, 1800–1850* (Newark, Del., 1980); and Lothar Gall, *Der Liberalismus als regierende Partei: Das Grossherzogtum Baden zwischen Restauration und Reichsgründung* (Wiesbaden, 1968).

[17] Cyrus Hamlin, "Heidelberg im Zeitalter der Romantik: Die Entdeckung des geschichtlichen Bewußtseins," in *Heidelberg—Stadt und Universität*, ed. Cyrus Hamlin et al. (Heidelberg, 1997): 173–92; and Michael Buselmeier, "Mythos Heidelberg," in *Auch eine Geschichte der Universität Heidelberg*, ed. Karin Buselmeier, Dietrich Harth, and Christian Jansen (Mannheim, 1985), 491–500.

[18] Peter Borscheid, *Naturwissenschaft, Staat, und Industrie in Baden, 1848–1914* (Stuttgart, 1976). See also the additional figures and observations in Arleen Marcia Tuchman, *Science, Medicine, and the State in Germany: The Case of Baden, 1815–1871* (New York, 1993).

[19] On Bunsen, see Georg Lockemann, *Robert Wilhelm Bunsen: Lebensbild eines deutschen Naturforschers* (Stuttgart, 1949); Henry Roscoe, "Bunsen Memorial Lecture," *Journal of the Chemical Society* 77 (1900): 513–54; Heinrich Debus, *Erinnerungen an Robert Wilhelm Bunsen und seine wissenschaftlichen Leistungen* (Kassel, 1901); Siegfried Lotze, "Die Chemie in Kurhessen vor 150 Jahren: Robert Wilhelm Bunsens 175. Geburtstag," *Zeitschrift des Vereins für Hessische Geschichte und Landeskunde* 91 (1986): 105–31; Fritz Krafft, "Das Reisen ist des Chemikers Lust—auf den Spuren Robert Bunsens: Zu Robert Wilhelm Bunsens 100. Todestag," *Berichte zur Wissenschaftsgeschichte* 22 (1999): 217–38; Margot Becke-Goehring, Ekkehard Fluck, Herbert Grünewald, Karl Rumpf, and Günther Wilke, "Betrachtungen zur Chemie in Heidelberg," in *Semper Apertus: Sechshundert Jahre Ruprecht-Karls-Universität Heidelberg, 1386–1986*, 6 vols. (Berlin, 1985): 2:332–60; and

almost every aspect of the physical sciences (including geology, the subject of a celebrated trip to Iceland), but his most famous discovery, before the Heidelberg-based discovery of spectral analysis in 1859, was his work on the chemistry of cacodyls—highly toxic organic arsenic compounds—which he had completed in Kassel, as well as the laboratory gas burner that bears his name. When Bunsen moved to Heidelberg, he altered his research profile substantially. He gradually ceased to train graduate students and shifted his attention to the first- and second-year laboratory courses for undergraduates, and he reoriented from organic chemistry to inorganic chemistry. In addition, as a former student, Henry Roscoe, recalled, Bunsen tended to shy away from theoretical issues in chemistry: "Bunsen did not enlarge in his lectures on theoretical questions; indeed to discuss points of theory was not his habit and not much to his liking. His mind was eminently practical; he often used to say that one chemical fact properly established was worth more than all the theories one could invent."[20] In the late 1850s and 1860s, a time of some of the most exciting theoretical developments in chemistry, this would only serve to further alienate graduate students. Bunsen also insisted on significant training in physics and is often quoted (probably erroneously) as saying: "Ein Chemiker, der kein Physiker ist, ist gar nichts."[21]

The attraction for undergraduate students was unquestioned, and they flocked to attend his lectures. His international reputation also appealed to foreign students, especially those who had to justify to state institutions the expense of an education abroad.[22] (This was the case for Russian students, although the experience of working with Bunsen, addressed below, did not turn out very happily.) However, perhaps the most significant draw to study with Bunsen in Heidelberg was the extensive laboratory facilities. As a contemporary noted: "Bunsen's newly built laboratory was for this time, as that of Liebig's in Gießen had been earlier, a gathering place of young chemists from near and far. Men from almost all countries were found here."[23] Bunsen had negotiated this laboratory as part of his offer, and construction began on it in 1854. It was overcrowded almost from the start, and in fall 1859, Bunsen had to get new spaces because more than sixty people were working there during the roughly 100 hours he taught experimental chemistry courses each semester.[24]

Bunsen also managed, as part of his offer from Heidelberg, to arrange for the hires of several major figures in German physical sciences, perhaps his most significant contribution to the transformation of the institution. The first such hire was Gustav

---

Bunseniana: Eine Sammlung von humoristischen Geschichten aus dem Leben von Robert Bunsen nebst einem Anhang von pfälzischen Lyceums-Anekdoten (Heidelberg, 1904), a collection of humorous anecdotes about Bunsen.

[20] Roscoe, "Bunsen Memorial Lecture" (cit. n. 19), 550.

[21] "A chemist who is not a physicist is nothing." Quoted in Roscoe, "Bunsen Memorial Lecture" (cit. n. 19), 554. Debus questioned the attribution: "*It is very unlikely that Bunsen said something like this*. He would have said: 'A chemist without physical knowledge is nothing.'" Debus, *Erinnerungen* (cit. n. 19), 148–9 (emphasis in original).

[22] Helmut Neubauer, "Chemiker und Musikant: Alexander Borodins Heidelberger Jahre (1859–1862)," *Heidelberger Jahrbücher* 24 (1980): 81–94, 87.

[23] Edvard Hjelt, "Friedrich Konrad Beilstein," *Berichte der Deutschen Chemischen Gesellschaft* 40 (1907): 5041–78, on 5043.

[24] Theodor Curtius and Johannes Rissom, *Geschichte des Chemischen Universitäts-Laboratoriums zu Heidelberg seit der Gründung durch Bunsen* (Heidelberg, 1908); and August Bernthsen, "Die Heidelberger chemischen Laboratorien für den Universitätsunterricht in den letzten hundert Jahren," *Zeitschrift für angewandte Chemie* 42 (1929): 382–4.

Kirchhoff (1824–1887), who had been Bunsen's colleague in Breslau and who in 1859 in Heidelberg discovered, together with Bunsen, spectral analysis.[25] But the most famous new recruit was Hermann von Helmholtz (1821–1894), who, besides serving as rector, published his important works on optics and acoustics at Heidelberg, straddling the border between physiology and physics.[26] The reason to expand the faculty in this particular way was not just for utility—drawing more students in specific areas that needed strengthening—but also to adorn Heidelberg University as an institution and Baden as a state with a cultural ornament.[27] For roughly a decade—from the arrival of the final of the three scholars that made up the so-called *Dreigestirn* (literally "three [guiding] stars")—until German unification, Heidelberg University was perhaps *the* center of physical sciences education in the German states. Upon unification, Kirchhoff and Helmholtz departed (separately) for Berlin, the new capital, and Bunsen was left to soldier on alone. Heidelberg's scientific supremacy faded somewhat in consequence.

Heidelberg's scientific decade corresponds almost exactly to the height of the Russian postdoctoral emigration there, which proved most fortunate for the fate of science in Russia. Aside from the presence of the Dreigestirn—which was a particular attraction for science students—there were other factors at Heidelberg that made it uniquely appropriate for Russian students. Since the early nineteenth century, Russian nobility had traveled to Heidelberg during long European sojourns to take advantage of the mild southern German climate and the convenient gambling at nearby Baden-Baden. As a result, there was a fairly constant circulation of Russians through the town, and a Russian reading room was formally established there in 1862. It proved (relatively) easy to obtain Russian "thick journals" (*tolstye zhurnaly*) and thus stay abreast of the news back home, and the companionship of local Russians proved a relief, as we shall see, from what the students perceived as the baleful influence of the native Germans.[28] Heidelberg University was also one of the last universities not to require an Abitur, making registration fairly easy, and it did not demand a written dissertation.[29] Those who were exiled from Russia for political (read: revolutionary)

---

[25] For biographical information, see Friedrich Pockels, "Gustav Robert Kirchhoff," in *Heidelberger Professoren aus dem 19. Jahrhundert*, ed. Universität Heidelberg, 2 vols. (Heidelberg, 1903), 2:243–63.

[26] Franz Werner, *Hermann Helmholtz' Heidelberger Jahre, 1858–1871* (Berlin, 1997).

[27] Frank A. J. L. James, "Science as a Cultural Ornament: Bunsen, Kirchhoff, and Helmholtz in Mid-Nineteenth-Century Baden," *Ambix* 42 (1995): 1–9. See also Arleen Marcia Tuchman, "Experimental Physiology, Medical Reform, and the Politics of Education at the University of Heidelberg: A Case Study," *Bulletin of the History of Medicine* 61 (1987): 203–15.

[28] Willy Birkenmaier, *Das russische Heidelberg: Zur Geschichte der deutsch-russischen Beziehungen im 19. Jahrhundert* (Heidelberg, 1995), especially chap. 3 on the scientists; Birkenmaier, ed., *Russische Stimmen aus Heidelberg* (Heidelberg, 1991); Otto Krätz, "Iwan Turgenjew und die russischen Chemiker in Heidelberg," *Chemie in unserer Zeit* 21 (1987): 89–99; Sergei Sviatikov, "Russkie studenty v Geidel'berge (K 50-letiiu russkoi chital'ni v Geidel'berge)," *Novyi zhurnal dlia vsekh*, Dec. 1912, 69–82. Russo-German antipathy already had a long and fairly specific history, largely a consequence of significant immigration of German specialists and German residents of the borderlands to St. Petersburg and Moscow. On some of the particulars, see Ludmila Thomas and Dietmar Wulff, eds., *Deutsch-russische Beziehungen: Ihre welthistorischen Dimensionen vom 18. Jahrhundert bis 1917* (Berlin, 1992); and Dagmar Herrmann and Alexander L. Ospovat, eds., *Deutsche und Deutschland aus russischer Sicht: 19. Jahrhundert: Von der Jahrhundertwende bis zu den Reformen Alexanders II* (Munich, 1998).

[29] Krätz, "Iwan Turgenjew" (cit. n. 28), 91.

reasons also found Heidelberg a hospitable outpost (in addition to London, Zurich, and other metropoles of the Russian Revolution abroad).[30]

From July 1862 to 1865, the attraction to Heidelberg was further cemented as it became the quasi-official center of the Russian postdoctoral export program. Noted physician N. I. Pirogov (1810–1881), who was seminal in the reconstruction of Russian pedagogy in the middle of the nineteenth century, was deputed to the German states in 1862 to serve as the liaison for the Russians abroad. Each Russian student had to visit Pirogov periodically to have his program of study approved, and Pirogov chose to settle in Heidelberg. As a result, Heidelberg had a constant traffic of Russian expatriate students, some enrolled in the university, others not. This Pirogov era, however, falls outside of the temporal scope of this essay as the apex of the Russian chemical emigration was already over by the time Pirogov had settled on the Neckar River.[31]

And so the Russian students came, particularly the chemists, expecting that they would work with Bunsen, Kirchhoff, and Helmholtz. Kirchhoff proved reasonably amenable. Bunsen, however, would have very little to do with graduate students, especially those in organic chemistry, and Helmholtz did not care for Russians very much and usually declined to work with them.[32] Bunsen tended to farm out his organic teaching—especially theoretical lectures or advanced organic laboratory instruction—to one of the chemical privatdocents he had hired. When Russians started to arrive, August Kekulé (1829–1896), soon to be the luminary of chemical structure theory, had just left this position, ceding it to Emil Erlenmeyer (1825–1909), a pharmacist turned chemist.[33] The Russian chemists worked almost exclusively with Erlenmeyer; his surviving attendance books for his classes show that in the fall and spring semesters of academic year 1861/62, for example, half of the students in his laboratory practicum were Russians.[34] (In 1865, Erlenmeyer received the Order of St. Anna from the tsar for being so supportive of his émigré students.)[35] The Russians did not just flock to Erlenmeyer because of subject matter and lack of alternative; they also genuinely seemed to like him. Yet working with Erlenmeyer to the exclusion of almost any other faculty contact only served to make the Russian postdocs more insular, which shaped their social organization in unexpected ways.

---

[30] Birkenmaier, *Das russische Heidelberg* (cit. n. 28), 9, 53, and chap. 7.

[31] Ibid., 8; and Aleksandr Brezhnev, *Pirogov* (Moscow, 1990), 375–84. On the longer-term trends in the foreign emigration of students from Russia, see the data presented in A. E. Ivanov, "Rossiiskoe studencheskoe zarubezh'e: Konets XIX–nachalo XX vv.," *Voprosy istorii estestvoznaniia i tekhniki*, 1998, no. 1:91–120.

[32] Enrollment data makes this fairly clear, although the hagiography of Helmholtz tends to attribute to the master's influence any discovery by a Russian whose time in Heidelberg coincided with that of Helmholtz or who met Helmholtz. See, e.g., Annette Vogt, "Hermann von Helmholtz' Beziehungen zu russischen Gelehrten," in *Universalgenie Helmholtz: Rückblick nach 100 Jahren*, ed. Lorenz Krüger (Berlin, 1994), 66–86.

[33] W. H. Perkin, "Emil Erlenmeyer," *Journal of the Chemical Society* 99 (1911): 1649–51; Richard Meyer, "Emil Erlenmeyer," *Chemiker-Zeitung* 23 (13 Feb. 1909): 161–2; Otto Krätz, "Emil Erlenmeyer, 1825–1909," *Chemie in unserer Zeit* 6 (1972): 52–8; and M. Conrad, "Emil Erlenmeyer," *Berichte der Deutschen Chemischen Gesellschaft* 43 (1910): 3645–64.

[34] As Erlenmeyer wrote to Aleksandr Butlerov on May 4, 1862: "I have about 10 Russians in the laboratory." Reproduced in G. W. Bykow and L. M. Bekassowa, "Beiträge zur Geschichte der Chemie der 60-er Jahre des XIX. Jahrhunderts: I. Briefwechsel zwischen E. Erlenmeyer und A. M. Butlerow (von 1862 bis 1876)," *Physis* 8 (1966): 185–98, 189. Complete attendance lists can be found in Erlenmeyer's papers, DM-HS 1968–589/3, especially, 5–7.

[35] Conrad, "Emil Erlenmeyer" (cit. n. 33), 3647.

THE CIRCLE: INNOCENTS ABROAD

One of the distinctive features of kruzhki is that they are both formal and informal: they possess a rather defined social form but at the same time are not "organized" in a classical Weberian sense. The informality of the Heidelberg chemical kruzhok, in particular its flexible membership, which would change as people rotated in and out of Heidelberg for postdoctoral stints, means that it is impossible to compose a complete inventory of participants; any enumeration would of necessity be partial and incomplete. At the same time, however, given that the interactions among kruzhok members were both intellectual *and* social, certain elements of the Heidelberg circle can be elucidated through individual interactions among members. Instead of providing a futile effort at an exhaustive inventory of individuals and their involvement, I will illustrate the simultaneous diversity and homogeneity of the Heidelberg kruzhok by following the paths of four specific individuals, selected for me by the photograph below (Figure 1), for which they sat in 1860: Aleksandr P. Borodin, Dmitrii I. Mendeleev, Nikolai Zhitinskii, and Ladislaus Olevinsky.

These four do not represent a random choice of individuals, as they happened to be captured in this moment, which they all must have agreed to (and agreed to pay for). But nonetheless these four individuals represent a single age cohort, with the same interests (organic chemistry), and each went on to a different career path. Those paths, in sum, prove emblematic of the choices confronting many of these scientific postdocs once they returned to Russia.

For purposes of exposition, I shall begin with the man standing in the background, Aleksandr Porfir'evich Borodin (1833–1887), whose reputation today stems entirely from his musical composition of the operatic masterpiece *Prince Igor* and fewer than twenty other symphonic, vocal, and chamber works.[36] Borodin's life was colorfully atypical from the moment of his birth. He was born on October 31, 1833, in St. Petersburg, a royal bastard. His father, Luka Stepanovich Gedianov (1772–1843), was an Imeretian prince originally from Transcaucasia, suitably Russified and living in the center of Petersburg in the 1830s. Gedianov sired young Aleksandr with Avdot'ia Konstantinovna Antonova, a soldier's daughter from Narva, who was twenty-four at the time. To resolve the problem of legitimacy, Borodin was registered as the son of Gedianov's valet, Porfirii Ionovich Borodin, and his wife, Tat'iana Girgor'evna Borodina—which technically meant that the boy was a serf. His biological mother—whom he called "auntie" (*tetushka*) for the rest of her life—took charge of his education at home, having him tutored in German (by Fräulein Luischen, a housekeeper), in French (by Monsieur Béguin, who taught at the Lycee), and in English (by John Roper, who served as a governor at a commercial school).[37] Aleksandr was registered as a free peasant on November 3, 1849, and the following year, when he turned seventeen, his mother attempted to register him as a student at St. Petersburg University. This proved abortive, but she managed to enroll him as a student at the Medico-Surgical Academy on the Vyborg Side, in the northeast of Petersburg—largely because her current beau, F. A. Fedorov, knew the inspector, Il'inskii, who directed

---

[36] Michael D. Gordin, "Facing the Music: How Original Was Borodin's Chemistry?" *Journal of Chemical Education* 83 (April 2006): 561–5.

[37] This information is emphasized by prominent critic Vladimir Stasov in the canonical obituary that became the basis for all future biographical studies: "Aleksandr Porfir'evich Borodin," *Istoricheskii Vestnik* 28 (1887): 137–68, 138–9.

Figure 1. Four members of the Heidelberg chemical kruzhok. The individuals are, from left to right: Nikolai Zhitinskii, Aleksandr P. Borodin, Dmitrii I. Mendeleev, and Ladislaus Olevinsky. (Source: R. B. Dobrotin et al., Letopis' zhizni i deiatel'nosti D. I. Mendeleeva [Leningrad, 1984], 65.)

admissions there. Antonova, who had had several affairs, some of which produced half-siblings for Borodin, and had married retired physician Kh. I. Kleineke in spring 1839, moved her family to Aleksandr's new neighborhood.[38]

The most significant move Borodin made at the academy was to approach his

[38] The biographical particulars here and in what follows are drawn from the most reliable Soviet-era biographies of Borodin: N. A. Figurovskii and Yu. I. Solov'ev, *Aleksandr Porfir'evich Borodin: A Chemist's Biography*, trans. Charlene Steinberg and George B. Kauffman (New York, 1988); S. A. Dianin, *Borodin: Zhizneopisanie, materialy i dokumenty* (Moscow, 1960); and A. P. Zorina, *Aleksandr Porfir'evich Borodin* (Moscow, 1987).

chemistry professor, Nikolai N. Zinin, and ask to perform experiments in his laboratory as training for a career in chemistry (as opposed to medicine). Zinin's imprint on the young man—from taking him in, directing his specialization in precisely the same areas of experimental organic chemistry he himself studied, and sending him off to Heidelberg for further study, and even apparently controlling issues of his personal toilet—would be hard to overemphasize.[39] The role of Zinin was so prominent that several psychobiographers have cast Zinin as the first of a series of father figures by which Borodin sought to replace his biological father—absent by illegitimacy, his mother's dominance, and his father's death.[40] (We shall return in the conclusion to the historiographical ramifications of Zinin's position in the narrative of Russian chemistry.)

Borodin graduated on March 25, 1856, and served briefly as a physician at the Second Infantry Hospital, but he preferred to pursue a career in chemistry, not medicine. Borodin returned to work with Zinin and on May 3, 1858, defended his dissertation (the first in the history of the academy written and defended in Russian, not Latin), "On the Analogy of Arsenous Acid with Phosphoric Acid in Its Chemical and Toxicological Relations." (He later would obtain a master's in chemistry from St. Petersburg University while working in Zinin's lab at the Medico-Surgical Academy.)[41] Borodin had already been abroad once, escorting the distinguished oculist Ivan Ivanovich Kabat to an international ophthalmological congress in Brussels, during which he visited chemistry laboratories (such as Marcellin Berthelot's) in Paris.[42] Zinin believed that a postdoctoral trip to study abroad would be beneficial for the development of young Borodin's chemical career and arranged for him to embark on a subsidized three-year stay in Heidelberg (and incidentally also in Paris and Pisa).[43]

Borodin's letters from Heidelberg to his biological mother provide some of our most vivid sources concerning the life of the chemistry postdocs there. His first missives

---

[39] On the micromanagement of Borodin's personal life, see A. P. Dianin, "Aleksandr Porfir'evich Borodin: Biograficheskii ocherk i vospominanii," *ZhRFKhO* 20, khim. ch. (1888): 367–79, 369.

[40] Other such father figures were Milii Balakirev and Franz Liszt. See A. Sokhor, *Aleksandr Porfir'evich Borodin: Zhizn', deiatel'nost', muzykal'noe tvorchestvo* (Moscow, 1965), 45; Bärbel Zaddach-Dudek, "A. P. Borodin—russischer Musiker und Naturwissenschaftler im 19. Jahrhundert," in *Wissenschaftsgeschichte in Osteuropa: Europa literarum artiumque scientiam communicans*, ed. Aloys Henning and Jutta Petersdorf (Wiesbaden, 1998), 87–100; and R. P. LaCombe's theory as reported in George B. Kauffman and Kathryn Bumpass, "An Apparent Conflict between Art and Science: The Case of Aleksandr Porfir'evich Borodin (1833–1887)," *Leonardo* 21 (1988): 429–36, 434. Stasov stoked these flames: "[Zinin] considered [Borodin] his spiritual son, and Borodin from his side considered him a second father." Stasov, "Aleksandr Porfir'evich Borodin" (cit. n. 37), 149. Much of this is derived from Dobroslavin's reminiscences solicited by Stasov: A. P. Dobroslavin, "Vospominaniia o A. P. Borodine," *Muzykal'noe nasledstvo* 3 (1970): 261–5, 261.

[41] See the correspondence regarding permission to take his master's exam: Borodin to Rector of St. Petersburg University Aleksandr Pletnev, St. Petersburg, 23 March 1859, f. 14, op. 1, d. 5983, l. 1, TsGIASPb; Pletener to Dean of Physico-Mathematical Faculty of St. Petersburg University Emilian Khristianovich Lenz, St. Petersburg, 29 May 1859, f. 14, op. 3, d. 14709, ll. 51-51ob, TsGIASPb.

[42] Borodin to his mother (Avdot'ia Konstantinovna Kleineke), 15 Aug. 1857, Heidelberg, *BorP*, 1:27–8; Dianin, *Borodin* (cit. n. 38), 40.

[43] The subsidy in Borodin's case came from the Ministry of War, which connected it with potential future service in the Medico-Surgical Academy, which was under its jurisdiction. Most other subsidies came from the Ministry of Popular Enlightenment, which was in charge of the university system. Occasionally, the view has been voiced that Zinin sent Borodin to Heidelberg so Mendeleev—who was already there—could straighten him out from his musical leanings and make him serious about chemistry, as for example in Victor I. Seroff, *The Mighty Five: The Cradle of Russian National Music* (1948; repr., Freeport, N.Y., 1970), 71. There is not a single trace of evidence for this claim.

reprise the usual aesthetic awe one finds among nineteenth-century visitors to Heidelberg, as in this letter of November 25, 1859:

> Heidelberg is a very nice and clean town—clean to the point that there is no need for galoshes. On Saturdays awkward German women wash not only the sidewalks but also the streets. The location of the city is unusually picturesque: on the one side are mountains (on one of them the wonderful ruins of a castle, overgrown with ivy), on the other side a charming river. The view from my windows is unparalleled—directly before my windows begins an enormous mountain—the *Kanzel* with a tower at its peak.[44]

Even before reflecting on his surroundings, however, Borodin was quick to connect with the society of Russians in the town, especially those he had already met as students in St. Petersburg. (The link of higher education in St. Petersburg, or some connection with the institutions of that city, was one of the most common features of members of the Russian chemical kruzhok.) Almost his first stop, as he reported on November 5, 1859, was to the Russian watering hole of the Badischer Hof: "Having stopped at the Badischer Hof we immediately happened upon the Hôtel where all our Russians living in Heidelberg dine. At the table d'hôte I saw Mendeleev[,] Sechenov and many others. After dinner we all headed to Mendeleev's place; he has a very nice laboratory, clean and even supplied with gas."[45] Already here we can see several features of importance: the conglomeration of Russians, the specific locations where they gathered, and the importance of proximity to a laboratory. Borodin even couched his move to Karpfengasse 2, a few months later, in terms of proximity to his research: "The chief advantage is that I live next to the laboratory—that's why I moved."[46] Erlenmeyer's laboratory was two doors down.

The hotel and restaurant Badischer Hof was only one of several regular meeting sites, which can be determined easily from Borodin's letters. The most obvious was Erlenmeyer's facility at Karpfengasse 6, but there were others. Among the most often invoked was the pension run by Karl Hofmann (1811–1877), a former professor of Greek literature at Moscow University (appointed in 1835), who had been forced to leave Russia for Heidelberg in 1849 as a direct result of Tsar Nicholas I's reaction to the German revolutions of the previous year. He set up his establishment with his Russian wife, Sof'ia Petrovna, and he taught as a privatdocent at Heidelberg University from 1850 onward. His wife in particular turned the establishment at Bergheimer Straße 14 into a lively center for the émigré Russians. As Borodin's future wife, whom he met in Heidelberg while she was on a rest cure for tuberculosis, recalled: "We arrived in Heidelberg and stayed in the pension of Hofmann, a former professor of Moscow University. All Russians then would stay at Hofmann's, as if in memory of his former relationships in Moscow."[47] Borodin himself returned while on a scientific visit to Heidelberg in 1877, hoping to see Hofmann again, but found only his widow and some fond memories.[48]

Perhaps most the most important feature one finds in Borodin's letters, however,

---

[44] Borodin to his mother, 25 [13, OS] Nov. 1859, Heidelberg, *BorP*, 1:36.
[45] Borodin to his mother, 5 Nov. 1859, Heidelberg, *BorP*, 1:33.
[46] Borodin to his mother, [3/31 or 4/1] 1860, Heidelberg, *BorP*, 1:39.
[47] E. S. Borodina, "Vospominaniia ob A. P. Borodine, zapisannye S. N. Kruglikovym," *Muzykal'noe nasledstvo* 3 (1970): 241–52, on 246.
[48] Borodin to his wife, 30 [18, OS] July 1877, Heidelberg, *BorP*, 2:163. On the details of Hofmann and the pension, see Birkenmaier, *Das russische Heidelberg* (cit. n. 28), 158. For a lively account

was not so much Russians visiting each other in defined spaces, or the way socializing and chemistry were interlinked, but the social institution in which these phenomena took place. Borodin was very explicit that the major form was the kruzhok:

> The society of foreigners forms here its own circles [kruzhki] and does not get acquainted with Germans. There are a lot of Russians here; among them even two literary women—Marko-Vovchok and another lady of some sort who writes articles. There are even Russian literary evenings.
>
> The Russians divide into two groups: those who do nothing, i.e., the aristocratic Golitsyns, Olsuf'evs, etc., and those who do something, i.e., students. These latter all gather together and go for dinner and for evenings out. I have come out, in short—of course with Mendeleev and Sechenov—as a perfect gentleman, extraordinarily simple and very active. The society of Germans is unbearable in the extreme, [with its] primness [and] horrible gossips. . . . And the women here! Simply horrible!—what mugs. The society of German students is yet more repulsive: terrible schoolboyish behavior—downright infants. Imagine that they are all divided into parties, each of which has its own boss, a seigneur. The students of the different parties differ by dress and colors; some have yellow caps, others red ones, a third group white ones and so on. Besides this each student has a silk strap across his shoulders; the seigneur has a tricorn hat. The fashion of the caps is the most curious. Add to this still enormous jackboots of the strangest form, and you will begin to have a sense of the dress of a German student. On Sundays the students get drunk and it is a rare week that goes by without a duel. . . . These duels, on the other hand, are always restricted to trivialities: one has his forehead cut open, another his face—and that's it. All of their ventures are conducted with a heap of the most ridiculous formalities, which are however always carried out to the letter.[49]

The antipathy to Germans as people (often on aesthetic grounds) and the mocking of the tradition of dueling are not idiosyncratic features of Borodin's account; they are tropes in foreigner lore. One of the most distinctive features of Heidelberg as a college town in the middle of the nineteenth century was the relatively prominent position of regional societies (*Burschenschaften*) among the student body. Roughly a quarter of the students in the first half of the century belonged to one of these organizations, with membership usually linked either to one's home region or to the notion of a potentially unified German nation (as in the Teutonia society).[50] In addition to providing social cohesion for *Landsmänner* who were far away from their home states, the Burschenschaften also served as a forum for manly rites of passage, such as the ritual dueling that left scars on the cheeks of the loser. Dueling, of course, was a noted aristocratic (and even nonaristocratic) activity in many countries in the nineteenth century, Russia not least of them; but the nonlethal and corporate nature of German university students' contests as a distinctive social form sparked particular comment from both Germans and foreigners.[51]

---

of evenings at the pension during the late 1850s, see A. V. Romanovich-Slavatinskii, "Moia zhizn' i akademicheskaia deiatel'nost', 1832–1884 gg.," *Vestnik Evropy* 38 (1903): no. 1:138–197; no. 2:606–50; no. 3:168–214; no. 4:527–66; no. 5:181–205; no. 6:499–508.

[49] Borodin to his mother, 25[13, OS] Nov. 1859, Heidelberg, *BorP*, 1:36–7.

[50] Peter Classen and Eike Wolgast, *Kleine Geschichte der Universität Heidelberg* (Berlin, 1983), 44.

[51] Ute Frevert, *Men of Honour: A Social and Cultural History of the Duel*, trans. Anthony Williams (Cambridge, UK, 1995), chap. 4; Frevert, "Honour and Middle-Class Culture: The History of the Duel in England and Germany," in *Bourgeois Society in Nineteenth-Century Europe*, ed. Jürgen Kocka and Allen Mitchell (Oxford, 1993): 207–40; and Herman Haupt, ed., *Handbuch für den Deutschen Burschenschaftler* (Frankfurt am Main, 1922). For a humorous American interpretation of the duels, see the wonderful account in Mark Twain, *A Tramp Abroad* (1880; repr., New York, 1997), chaps. 5–6.

Perhaps most noteworthy about Borodin's (and other Russians') hostility to the Burschenschaften is what is left unsaid: these were organizations created primarily to provide some sense of territorial identity and cohesion for Germans studying in Baden, away from their families and homesteads.[52] This has two implications. First, the Russian postdocs were *by definition* excluded from participation in this incredibly important socialization ritual, and thus the Burschenschaften reinforced Russians' isolation from their German peers and encouraged them even further to seek the society of their compatriots. Second, the omnipresence of this kind of semistructured social formation—in the case of the Burschenschaften, quite structured, with uniforms and insignia—performed a function analogous to that of the kruzhok and must have been an additional spur to the mostly unconscious importation of that Russian institution to Heidelberg.

Of the individuals in our photograph, Borodin may have been the most prolific and astute reporter of that institution, but he was not its dominant figure. That role was assumed, perhaps obviously from the image, by the young man sitting in the center: Dmitrii Ivanovich Mendeleev (1834–1907).[53] Today Mendeleev happens to be the most well-known nineteenth-century Russian chemist, a fame built entirely on his 1869 formulation of the periodic system of chemical elements, a feat still nine years in the future for him at the time of the photograph. During his two-year stay in Heidelberg, from 1859 to 1861, Mendeleev was not yet a prominent inorganic chemist and central figure in the Petersburg community. What force he had in the group was due to his apparent cleverness and breadth in chemical knowledge, his willingness (as documented above by Borodin) to secede from formal laboratory structures and strike out in organic chemistry on his own in his apartment laboratory, and his lively personality. A satellite member of the kruzhok, physiologist Ivan M. Sechenov, later noted in his autobiography: "Mendeleev made himself, of course, the center of the kruzhok, moreover since, regardless of his young age (he is years younger than I), he was already a prepared chemist, and we [others] were students."[54] T. P. Passek, the sister of the prominent Russian socialist thinker and activist Aleksandr Herzen, joined other travelers in similar accolades for Mendeleev's personality, entirely divorced from (because ignorant of) any chemical acumen.[55] The kruzhok met frequently at Mendeleev's apartment at Schulgasse 2, which lay at some remove from Erlenmeyer's chemical laboratory and Borodin's apartment on Karpfengasse. Of course, Mendeleev did not spend all of his time on chemistry or even in Heidelberg. Of the twenty-two months Mendeleev spent abroad on his postdoctoral stay, only five months and twenty days were spent in Heidelberg—the rest was on the road, visiting Italian, French, and German university towns to gather materials, equipment, and literature for the return trip to St. Petersburg, where such supplies were rare.[56] And, like Borodin, Mendeleev

---

[52] Frevert, *Men of Honour* (cit. n. 51), 87; and Rudolf Sperling-Sueviae, *Der Ausschuss der Heidelberger Studentenschaft* (Heidelberg, 1911).

[53] For biographical details on Mendeleev, see Michael D. Gordin, *A Well-Ordered Thing: Dmitrii Mendeleev and the Shadow of the Periodic Table* (New York, 2004).

[54] Ivan Mikhailovich Sechenov, *Avtobiograficheskie Zapiski Ivana Mikhailovicha Sechenova* (Moscow, 1945), 96.

[55] T. P. Passek, "Vospominaniia T. P. Passek [Chapter 32]," *Russkaia Starina* 20 (1877): 277–300, 294. See also the memoirs of one of Passek's visitors at this time, E. F. Iunge, *Vospominaniia (1843–1860 g.g.)* ([Moscow?], [1914]), 285–8.

[56] M. D. Mendeleeva, "Novye materialy o zhizni i tvorchestve D. I. Mendeleeva v nachale 60-kh godov," *Nauchnoe Nasledstvo* 2 (1951): 85–94, 92.

clearly felt his experience in Heidelberg was a unique one while it was happening, for he kept a diary of his time abroad, a practice he discontinued shortly after his return to St. Petersburg.[57] Mendeleev and the kruzhok also read contemporary Russian literature—notably I. A. Goncharov's seriocomic satire of Russian apathy *Oblomov*[58]—and he tore himself away from his peers long enough to father an illegitimate child with an actress.[59]

To return to our photograph, the man sitting on the left is Nikolai Zhitinskii, and he appeared numerous times in correspondence concerning Mendeleev. Mendeleev was asked on occasion by his friends in St. Petersburg to pass regards to Zhitinskii, and Borodin referred to him twice in letters home as "a certain Zhitinskii," always in the context of Mendeleev.[60] And this is all we know about him today. He published no articles deriving from his chemical research in Heidelberg, and when he returned to St. Petersburg, he vanished without leaving a documentary trace. In this sense, he was a far more typical Russian Heidelberger than either Borodin and Mendeleev, who achieved contemporary fame in music and science, respectively. Most of those who went abroad came back, failed (or failed to try) to find academic jobs, and were lost to history. Zhitinskii serves thus as a marker for both the majority experience of foreign study and for what Russians such as Mendeleev and Borodin feared would happen to them.

The case of the man standing on the right in the photograph is more sobering. His name was Ladislaus Olevinsky, and he was, technically speaking, not a Russian but a Pole. (Poland, of course, had ceased to exist as an independent state since its final partition in 1795, and Olevinsky hailed from the Russian empire.) While most of the Russian postdocs engaged in experimental organic chemical research, Olevinsky was almost unique in having high ambitions of participating in the transformation of organic theory then being effected (separately) by August Kekulé and Aleksandr Butlerov. Although he did publish a few articles while in Heidelberg based on work in Erlenmeyer's laboratory, he was also engaged in formulating a theory of chemical affinity that he believed would transform chemistry. Borodin, for his part, thought the theory was not as interesting as Olevinsky expected it would be and considered most of it heavily derivative of Butlerov's work. Indeed, Olevinsky had spent significant time shadowing Butlerov in Paris, and Borodin was certain that both the ambition and the results stemmed from that proximity.[61]

Olevinsky descended into paranoia. He was convinced that Butlerov had been stealing ideas from him, and he also commented on several occasions that Erlenmeyer had been appropriating his work. A friend sadly commented in a letter to Mendeleev

---

[57] For this illuminating document, see the helpful publication D. I. Mendeleev, "Dnevnik 1861 g.," *Nauchnoe nasledstvo* 2 (1951): 111–212.

[58] Sechenov, *Avtobiograficheskie Zapiski* (cit. n. 54), 97. Sechenov incorrectly remembers this book as *Obryv*, also by Goncharov, an impossible choice due to publication dates. Mendeleev also recalls *Oblomov* specifically in his diary.

[59] Although known to Mendeleev's peers, this fact was kept quiet until recently. For details on Agnes Voigtmann, Mendeleev's mistress, see Birkenmaier, *Das russische Heidelberg* (cit. n. 28), 111; and Annette Nolte, *D. I. Mendeleev in Heidelberg*, *Russica Palatina*, vol. 22 (Heidelberg, 1993), 79.

[60] Borodin to his mother, [3/31 or 4/1] 1860, Heidelberg, *BorP*, 1:38; and Borodin to Ivan Maksimovich Sorokin, 8 April 1860, Heidelberg, *BorP*, 1:41.

[61] Borodin to Mendeleev, 19 [7, OS] Jan. 1862, Pisa, *BorP*, 4:253. For Olevinsky's research, see, e.g., Ladislaus Olewinsky, "Ueber das chemische Verhalten der Metall-Aldehydate," *Zeitschrift für Chemie und Pharmacie* 4 (1861): 360–2; and Olewinsky, "Ueber die Wirkung des Benzoylchlorürs auf Natriumbenzyladehydrat," *Zeitschrift für Chemie und Pharmacie* 4 (1861): 625–6.

(already back in St. Petersburg) that Olevinsky's antics stemmed from his having "lost faith in his capability as a chemist."[62] Mendeleev could hardly have been surprised by this. A visitor to the Heidelberg kruzhok had noted that a similar relationship of envy and persecution existed between Mendeleev and Olevinsky, analogizing them to Aleksandr Pushkin's depiction of Mozart and Salieri.[63] The level of alienation grew, especially after Mendeleev left Heidelberg in early 1861, and Olevinsky started to distance himself from the only support network he had left: the kruzhok. Poles in Heidelberg—who, because of quotas in tsarist universities, constituted the largest foreign population in Heidelberg—refused to sit with Olevinsky at dinner on one occasion (the reason is unclear from the sources, but it seemed related to debates within Polish politics that would eventually erupt in the abortive 1863 Polish uprising against the Russian empire). This only reinforced Olevinsky's feeling of persecution. As Borodin wrote to Mendeleev shortly afterward: "The next week they found [Olevinsky] dead. He dispatched himself with potassium cyanide. What drove him to this is unclear, and the papers he left behind him don't explain anything. He only wrote that the reason was fear of the government, that he already over a week ago burned his diary, memoirs, etc., and bought CyK, but wavered and hadn't decided to take it. It's a blow."[64]

These four individuals, then, represented four paths for Russian Heidelbergers. Some, such as Borodin, returned to Russia, worked as chemists professionally, but devoted most of their time not to research but to nonscientific activities. Others, such as Mendeleev, continued on the path of independent research—although understandably with less spectacular success than Mendeleev achieved. And others went nowhere, either in terms of their careers, or, tragically, because the strains and tensions of being young and on their own proved too much.

Although most chemists we have record of did not descend to an abyss as extreme as Olevinsky's, all felt some apprehension on leaving Heidelberg to return to Russia. Borodin, for example, wrote home after borrowing some money from Erlenmeyer and heading off to Rome with Mendeleev in October 1860:

> I admit I am a little sad to part with Heidelberg, where I so peacefully and well spent almost an entire year. True, besides Erlenmeyer, I got to know almost no Germans, and only recently was I in one very nice English group. Thus our Russian *kruzhok* lived here truly as friends, amiably, each lending another mutually what he could. It is unlikely that one will find such a tight and friendly kruzhok in another place.[65]

Not even in St. Petersburg, their home.

---

[62] N. Il'in to Mendeleev, 5 Feb. 1862, Heidelberg, ADIM I-V-55-1-28. Olevinsky's surviving two letters to Butlerov, dated 1861, both explicate his "theory of limits" in organic chemistry and accuse Butlerov of stealing it. See the reproductions in G. V. Bykov, ed., "Pis'ma russkikh khimikov k A. M. Butlerovu," *Nauchnoe nasledstvo* 4 (1961), 294–301.

[63] Romanovich-Slavatinskii, "Moia zhizn' i akademicheskaia deiatel'nost'," (cit. n. 48), 541.

[64] Borodin to Mendeleev, 19 [7, OS] Jan. 1862, Pisa, *BorP*, 4:253–4. The suicide has proven embarrassing to certain historians, who choose instead to elide the circumstances of the sudden death, as in Figurovskii and Solov'ev, *Aleksandr Porfir'evich Borodin* (cit. n. 38), 41.

[65] Borodin to his mother, 28 [16, OS] Oct. 1860, Rome, *BorP*, 1:53–4. For another example, see D. Woeikof to Erlenmeyer, 3 March 1879, San Marco, Italy, DM-HS 1968-478/3, 2.

SQUARING THE CIRCLE: HOMECOMINGS

Why were these Heidelbergers so apprehensive about returning to St. Petersburg? It turns out that they had good reason to worry. While in Heidelberg, as Borodin wrote in early 1861 to the now repatriated Mendeleev, he tried not to think of what awaited him upon his return:

> You know, I am beginning to feel that I have already lived through the greater half of my life abroad; I am somehow beginning to look at myself as a guest abroad. In the first half of my stay abroad this wasn't true; the date of my return seemed to me unusually far off. I didn't even think of Russia. Now it's different; I am beginning to think of Petersburg, of setting up a laboratory (it's funny, I know) and other miscellany, relating to everyday life in Petersburg.[66]

The miscellany were not a surprise to the Russians who returned. As the fact that Borodin and Mendeleev were in epistolary contact shows, the correspondence among young chemists in St. Petersburg and abroad served a vital function of filling in students on political and (especially) academic gossip about their potential for being hired or equipping a laboratory when they returned.[67]

Just as Mendeleev served as one of Borodin's informants, Mendeleev's primary source about the academic scene in St. Petersburg was one of his close friends, N. P. Il'in, who kept him updated about the fate of positions at St. Petersburg University—the plum jobs in the capital—and speculations about his future from Aleksandr Voskresenskii, their mutual mentor. A lot of gossip went back and forth.[68] At first, the prospects for a position seemed to be good if Mendeleev managed to get enough accomplished while abroad. Il'in reported: "Voskresenskii sort of said that if you finished a significant work, then maybe you might hope to receive an adjunct post; that wouldn't be bad for you; after all a docent position doesn't count for much."[69] It was so important to get the work done while away because there was little opportunity to set aside enough time for research in Russia, given the need to teach a large number of hours to support oneself. Time had to be scrounged where it could be. Il'in again: "The whole shame is that there isn't enough free time; I am counting heavily on Holy Week, when I'll be completely free and hope to finish everything I had planned."[70]

---

[66] Borodin to Mendeleev, 5 March [21 Feb., OS] 1861, Paris, *BorP*, 4:246.

[67] As Birkenmaier notes: "Letters often had the additional function of private reporting from abroad, since news from abroad was not then so omnipresent as today." Birkenmaier, *Das russische Heidelberg* (cit. n. 28), 44. Most of the conventionally political gossip concerned serf emancipation, as in this letter from Sechenov to Borodin and Mendeleev, 29 June 1860: "They say the emancipation matter is entirely done. It will be promulgated at the end of the harvest, i.e., in October or November. God make it true. Then you would return to a free Russia." Reproduced in T. Volkova, "Perepiska I. M. Sechenova s D. I. Mendeleevym," *Priroda*, 1940, no. 2:86–92, on 90. Emancipation did not occur until February 1861, two weeks after Mendeleev's return to Petersburg.

[68] For example, Mendeleev in this period was continually very competitive with Nikolai N. Sokolov, and he received regular reports from Il'in about Sokolov's dissertation defense (which was fairly brutal) and his maneuvers at the university to become Voskresenskii's successor. See N. P. Il'in to Mendeleev, 3 Jan. 1860, St. Petersburg, ADIM I-V-55-1-27; and Il'in to Mendeleev, 21 June 1859 [O.S.], St. Petersburg, ADIM I-V-55-1-27; Il'in to Mendeleev, 22 Feb. 1860, St. Petersburg, ADIM I-V-55-1-30.

[69] Il'in to Mendeleev, 22 Feb. 1860, St. Petersburg, ADIM I-V-55-1-30.

[70] Ibid. And, in another letter a month and a half earlier: "Thus you see how it isn't easy, and doubtless this is one of the reasons for our not-too-fruitful researches." Il'in to Mendeleev, 3 Jan. 1860, St. Petersburg, ADIM I-V-55-1-27.

The problem was so widely recognized that it even appeared in published articles, such as one by Dmitrii Voeikov, about his analysis of iodine performed in Erlenmeyer's laboratory in Heidelberg. Voeikov lamented in print: "Sadly it was shortly before my departure for Russia that I was able to make the determination of the iodine content and the spec. weight. I reserve the right to carry this research to its end in Russia."[71] Despite Il'in's encouragement, then, Mendeleev expected that he had not much more in his immediate future than hack science-writing jobs and laborious and financially unremunerative docentships, as he stated in his diary: "What I will do now for money? I'll do something. Well, in Russia I'll have to get a little poorer—not a big misfortune, on the other hand."[72]

If the first foreseeable problem was lack of time to conduct appropriate research, the second—and in many senses equally problematic—difficulty was lack of adequate laboratory space and resources to conduct interesting research. The difficulties were diverse and unlikely to be resolved by any one individual, even assuming one had a decent paying professorship lined up, as Borodin did at the Medico-Surgical Academy. Recalling Mendeleev's apartment-laboratory in Heidelberg, Borodin wrote to St. Petersburg to see whether there were gas lines set up by the city on the Vyborg Side, so he could equip his laboratory with the most elementary of burners: "But, on the other hand, it is unlikely that it is possible to set this up in Russia."[73] It would be years before the urban infrastructure caught up to Borodin's fairly modest demands.

There was also a recent history of organizational dysfunction that could not but signal to the St. Petersburg chemists that there would be significant hurdles to organizing chemists at home and seeking forums for publication. In the late 1850s, in the capital of St. Petersburg, the epicenter of the educational system of tsarist Russia, there were long-standing places where chemistry could be studied (institutions such as the Technological Institute, St. Petersburg University, and the Academy of Sciences). After the Crimean War, these sites continued to produce a small number of specialists who mostly went into teaching, training a meager number of pharmacists and industrial chemists. From 1857 to 1860, an attempt to provide a forum for chemists to organize emerged that was designed to be not merely an educational opportunity for further training but a stepping-stone to a fully functioning chemical community: the private laboratory and journal of Nikolai N. Sokolov (1826–1877) and Aleksandr N. Engel'gardt (1832–1893). Both these chemists were ambitious and talented and undertook strategies *exactly* like those that D. I. Mendeleev and A. M. Butlerov would later employ to such excellent effect—Sokolov in speculative theoretical chemistry like Mendeleev, and Engel'gardt in what would become Butlerov's area of experimental organic chemistry. And yet their professional strategies came to naught. These efforts provide a background of failure to institutionalize and organize that contrasts vividly with the rapid entrenchment of a national chemical community based in St. Petersburg only a decade later. It is only by reflecting on the contrast between the states before and after that the importance of the kruzhok and postdoctoral study in Heidelberg can be appreciated.

[71] D. Woiekoff [Voeikov], "Ueber die Einwirkung von Zinknatrium auf Jodallyl," *Zeitschrift für Chemie und Pharmacie* 6 (1863): 537–9.
[72] Diary entry of 11 Jan. 1861, Mendeleev, "Dnevnik 1861 g." (cit. n. 57), 115. For more detail on Mendeleev's dire financial situation at the time, see Gordin, *Well-Ordered Thing* (cit. n. 53), 19–22.
[73] Borodin to Mendeleev, 5 March [21 Feb., OS] 1861, Paris, *BorP*, 4:247.

At first, probably using Engel'gardt's funds from his patrimonial estate, the two chemists put together a private laboratory—explicitly modeled on Justus von Liebig's Giessen laboratory—that would be open to chemists in the St. Petersburg area.[74] The idea was to provide a place for individuals to advance chemical knowledge while waiting for an appointment at an institution that could provide them with more permanent laboratory space, generating a chemical network as a byproduct. As a complement to this effort, in 1859 Sokolov and Engel'gardt set up the other sine qua non of professionalized midcentury chemistry: a chemical journal. This periodical, *Sokolov and Engel'gardt's Chemical Journal*, came out in only four volumes over two years. The journal intended to offer an outlet for Russian chemical works published in their native language, but besides publishing the dissertations of the editors and a few incidental original pieces, the journal quickly devolved into publishing translated abstracts of important Western articles. As it was unable to sustain itself in this derivative format—most chemists in Russia could read the German, French, and English originals—the journal went under in 1860.[75] The laboratory closed that same year, mostly because Sokolov received a privatdocent post at St. Petersburg University and simply donated the laboratory to the university.[76] In 1869, Engel'gardt was exiled from St. Petersburg for his populist agricultural writings and confined to his rural estate, thus terminating his scientific career.[77]

Sokolov and Engel'gardt did not significantly participate in the burgeoning professionalization of Russian chemistry in the 1860s. The fault (not their own) was that they were born too early; when they reached intellectual maturity, the state was not yet willing to encourage science. There are two additional points to make about this brief venture: First, although there was some demand for both the laboratory and the journal, neither had enough demand to make them going ventures financially. Second, when Sokolov obtained a better post, he simply disbanded the laboratory and moved it to the state institution that employed him. There was neither any sense of loyalty to the project of an autonomous professional community on his part nor enough corporate sensibility among other chemists to resist him. The point of the laboratory may have originally been to focus the network of chemists, but a community had not congealed yet. Nor was it the only failure. N. P. Il'in wrote to Mendeleev in Heidelberg on December 22, 1860 (that is, after the collapse of the Sokolov/Engel'gardt venture):

> [L. N.] Shishkov is agitating here for the establishment of a Chemical Society, and the affair moves forward somewhat murkily because he proposed printing the minutes of

---

[74] On the importance of Liebig for the transformation of chemical pedagogy via laboratory instruction, see W. H. Brock, *Justus von Liebig: The Chemical Gatekeeper* (Cambridge, UK, 1997); and J. B. Morrell, "The Chemist Breeders: The Research Schools of Liebig and Thomas Thomson," *Ambix* 19 (1972): 1–46.

[75] Nathan M. Brooks, "Russian Chemistry in the 1850s: A Failed Attempt at Institutionalization," *Annals of Science* 52 (1995): 577–89; and Iu. S. Musabekov, "Pervyi russkii khimicheskii zhurnal i ego osnovateli," in *Materialy po istorii otechestvennoi khimii*, ed. N. A. Figurovskii et al. (Moscow, 1953), 288–302.

[76] N. Menshutkin, "Pamiati N. N. Sokolova," *ZhRFKhO* 10 (1878): 8–15; P. Lachinov, "Moi vospominaniia ob N. N. Sokolove," *ZhRFKhO* 10 (1878): 15–9.

[77] The writings in question are translated as *Aleksandr Nikolaevich Engelgardt's Letters from the Country, 1872–1887*, trans. and ed. Cathy A. Frierson (New York, 1993). See also N. S. Kozlov, "Nauchnaia i obshchestvennaia deiatel'nost' A. N. Engel'gardta," *Trudy Instituta istorii estestvoznaniia i tekhniki* 30 (1960): 111–34.

meetings, which requires money and material. The first we can and could have in the necessary quantity, but the second would be, probably, rather little.[78]

The idea (in this case modeled on the Société Chimique in Paris) went nowhere, and the chemists remained disorganized. Every single attempt to organize in the late 1850s and early 1860s simply floundered. Some kind of catalyst to corporate cohesion appeared to be missing.

The failure to coordinate was evident to the Heidelberg postdocs. Borodin wrote Mendeleev sarcastically from Paris about the continued misperception abroad that the situation was better than it was: "Maréchal Vaillant gave a speech and said, among other things, that 'the founding of the Société des amis des sciences cannot remain without influence on other nations and serves as a spur for the foundation of similar societies in England and Russia.'—Where?—I somehow haven't heard of such a society at home."[79] However, Heidelberg showed the expatriate chemists exactly what the potential of group organization could be. Heidelberg's Natural Historical and Medical Association (*Verein*) was reopened in 1856 (it had closed in 1848) and comprised forty-eight initial members, including some of the most prominent scientists in Heidelberg (such as Bunsen, Kirchhoff, and Erlenmeyer). This was a model the Russians saw up close, one that functioned essentially without official endorsement.[80]

Alongside these various foreign formal models for the organization of chemists as a professionalized entity, there was also an informal model for organization that the Russians could use when they returned to St. Petersburg: the kruzhok, which they had exported to Heidelberg initially to deal with the cultural alienation they felt. Back in St. Petersburg, it was a natural solution—albeit relatively untried in the sciences—to the organizational failures of Russian academia, to Engel'gardt and Sokolov's abortive ventures, to Shishkov's deflated effort, and to the newly acquired feeling of isolation induced by contrast to the closeness they had felt in Heidelberg.

OPEN CIRCLES: THE FORMATION OF THE RUSSIAN CHEMICAL SOCIETY

The transformation of the local kruzhok into an ersatz chemical society happened on two fronts: in Heidelberg and in St. Petersburg. The Heidelberg transformation was one largely of perception. The kruzhok already existed, and chemistry was already its focal point. What the local Russian chemistry students started to do in the early 1860s was to *think* of their informal circle as a budding chemical society. Borodin noted this connection in mid-1861 in a letter from Heidelberg to chemist Petr Alekseev: "Here we are forming (at first, it stands to reason, only in our own kruzhok) a chemical society—all in the family, for now."[81] He stated it more bluntly for the just-departed Mendeleev: "A chemical society is being formed here."[82] Of course, this was more wishful thinking than a reflection of true organization and professionalization; the kruzhok had merely acquired a conceptual patina.

In St. Petersburg, by contrast, the kruzhok was reinvigorated by its former Heidel-

---

[78] Quoted in V. V. Kozlov, *Vsesoiuznoe khimicheskoe obshchestvo imeni D. I. Mendeleeva, 1868–1968* (Moscow, 1971), 11.

[79] Borodin to Mendeleev, 3 April [22 March, OS] 1861, Paris, *BorP*, 4:248.

[80] Wolfram Schmitt, "Struktur und Funktion des Naturhistorisch-Medizinischen Vereins zu Heidelberg im 19. Jahrhundert," *Heidelberger Jahrbücher* 22 (1978): 71–92.

[81] Borodin to P. P. Alekseev, 24 [12, OS] May 1861, Heidelberg, *BorP*, 4:251.

[82] Borodin to Mendeleev, 28 [16, OS] May 1861, Heidelberg, *BorP*, 4:251.

berg participants who now found themselves back home with no one to talk to. This new social forum allowed for a continuation of the intimacy attested to in Heidelberg, and since it took place in their new permanent home, they would never have to leave it.[83] It would soon grow into a more formal organization due to two factors: demographic growth and the reforms of the 1860s.

The demographic growth was straightforward. Heidelberg proved a common destination for Russian chemical postdocs, but it was still a relatively exclusive club. Not every graduate student got to go abroad, and not everyone went to Baden, so the population attending the kruzhok stayed within certain confines. Upon their return to Russia, the chemists felt an even stronger pull to settle in St. Petersburg. This was the growing center of the chemical world in Russia (competing with and then displacing Kazan University, and before the growth of Moscow University's chemistry laboratories during the 1870s) and was thus the most likely place to gain employment, either in academia or in the nascent chemical industry. "Chemical evenings" became very common, as one can find from traces in much of the correspondence. Borodin jumped right into the fray when he returned from Baden, even choosing the chemists' kruzhok over the musical kruzhok at the apartment of his composition mentor, Milii Balakirev.[84] (This other kruzhok eventually grew into the "mighty little heap" [*moguchaia kuchka*] of Balakirev, Borodin, César Cui, Modest Musorgskii, and Nikolai Rimskii-Korsakov—the epicenter of Russian nationalist music.)[85] Borodin's vigor soon flagged, and in 1864 he wrote to Alekseev: "I haven't been to the chemical meetings for a long time. I have gotten lazy and there is little time."[86] They continued without him, however, including papers read by P. P. Alekseev, F. K. Beilstein, N. N. Beketov, Voskresenskii, Nikolai Zinin, N. P. Il'in, M. D. L'vov, G. V. Struve, A. I. Skinder, A. S. Famintsyn, A. I. Khodnev, and L. N. Shishkov—at least half of whom had participated in the Heidelberg kruzhok.[87] Mendeleev was still attending them as late as October 1868, a month before the formation of the Russian Chemical Society.[88] The sheer number of chemists in the capital and the corporate interests in intradisciplinary communication and publication provided a strong internal impetus to the creation of just such a *professional* body. It was a kruzhok grown too big for its britches.

There was simultaneously external pressure from the modernizing bureaucracy and the autocracy of tsarist Petersburg in the 1860s. As noted earlier, the tsarist state expressed a prominent need to "modernize" in terms of emulating the structures of western Europe. Although bureaucrats obviously recognized the importance of adapting certain structures from educational institutions—witness the establishment of the postdoctoral emigration in the first place—a broader cultural movement agitated

---

[83] P. Trifonov, "Aleksandr Porfir'evich Borodin: Biograficheskii ocherk," *Vestnik Evropy* 23 (1888): no. 10:747–79, 755; no. 11:46–81; Iu. V. Ionov and A. Iu. Ionov, "A. P. Borodin—vrach, khimik, pedagog (K 150-letiiu so dnia rozhdeniia)," *Sovetskoe zdravookhranenie*, 1984, no. 1:61–4, 62; and Dianin, "Aleksandr Porfir'evich Borodin" (cit. n. 39), 370.

[84] Borodin to Balakirev, 8 Dec. 1863, St. Petersburg, *BorP*, 1:59.

[85] On the kuchka, see Francis Maes, *A History of Russian Music: From Kamarinskaya to Babi Yar*, trans. Arnold J. Pomerans and Erica Pomerans (1996; repr., Berkeley, Calif., 2002); and Richard Taruskin, *Opera and Drama in Russia: As Preached and Practiced in the 1860s* (1981; repr., Rochester, N.Y., 1993).

[86] Borodin to P. P. Alekseev, 12 March 1864, St. Petersburg, *BorP*, 4:264.

[87] See Kozlov, *Vsesoiuznoe khimicheskoe obshchestvo imeni D. I. Mendeleeva* (cit. n. 78), 11.

[88] Iu. F. Fritzsche to Mendeleev, 12 Oct. 1868, St. Petersburg, ADIM I-V-7-1-24.

for the erection of the decentralized structures of Western professional life. Voluntary associations assumed a new importance in the Russia of the Great Reforms, as aspects of rural and urban governance were simultaneously being devolved to the newly emergent category of "citizen." Before the Great Reforms, Moscow only had a few voluntary societies, the majority of them charities. The autocracy really began to spur the emergence of such societies by allowing, even encouraging, *zemstva* (rural land councils) and other local organizations to encroach on its public role. Scientific societies were a particular form of this activity after the reforms, and almost every significant scholar was to some degree part of a social grouping, picking up on the tradition of the Russian Geographical Society, which had been a seedbed of the reforms in the first place.[89]

Nevertheless, no social movement on this scale moves monotonically or without resistance. Although parts of the autocracy clearly endorsed this new civic culture, other segments (particularly the security apparatus) were concerned about these new destabilizing forces. Professional societies in the sciences sold themselves as part of the more conservative modernization trend, claiming to bolster scientific progress and thus fiscal stability. In August 1861 the announcement of an organization of Moscow doctors lamented the scarcity of similar organizations: "We have few scientific societies and still less that have in mind the practical side of knowledge which comes into contact with the daily activity of the country. The absence of such associations serves as one of the chief reasons of the extremely slow dissemination of useful information to the people."[90] The chemists similarly encouraged this rhetoric of their own with a plea for a chemical organization in the St. Petersburg newspaper *Russkii Invalid*:

> A chemical society, in our opinion, is entirely possible in Petersburg. There live our most famous chemists, Messrs. Voskresenskii, Zinin, Mendeleev, Sokolov, Shishkov, Khodnev, and Engel'gardt—and in general in Petersburg many young people occupy themselves by studying chemistry. Why shouldn't our scientists gather around themselves an entire society?
>
> We consider it unnecessary to discuss the utility of such a society. Under the society there could be a public laboratory, which there isn't in Petersburg at this time. The University laboratory is too small and serves only for University students. . . . It is too hard to get access to the Academy laboratory. . . . The establishment of a physicochemical society could enable the publication of a "Chemical Journal," in which a division could also be opened for physics.[91]

These calls did not result directly in the creation of a chemical society, of course. There were two potential obstacles: one internal to the chemical community, and one external. In Mendeleev's papers, one finds the remnants of a poll taken at one of the

---

[89] Joseph Bradley, "Voluntary Associations, Civic Culture, and *Obshchestvennost'* in Moscow," in *Between Tsar and People: Educated Society and the Quest for Public Identity in Late Imperial Russia*, ed. Edith W. Clowes, Samuel D. Kassow, and James L. West (Princeton, N.J., 1991), 131–48; and E. V. Soboleva, *Organizatsiia nauki v poreformennoi Rossii* (Leningrad, 1983). On the Geographical Society, see Lincoln, *In the Vanguard of Reform* (cit. n. 3), 98–100.

[90] *Russkii Vestnik—Sovremennaia Letopis'*, no. 32 (Aug. 1861): 23, quoted in Kozlov, *Vsesoiuznoe khimicheskoe obshchestvo imeni D. I. Mendeleeva* (cit. n. 78), 13.

[91] "Vnutrenniia Izvestiia," *Russkii Invalid*, 17 Aug. 1861, 733. This anonymous piece was almost certainly penned by Mendeleev—as indicated by the cheekiness of including his postdoctoral name among the actual intellectual and institutional leaders of Petersburg chemistry and the insistence on a cooperation between chemistry and physics.

meetings of the chemical kruzhok in St. Petersburg about whether they should petition the Ministry of Popular Enlightenment for official recognition of a chemical society. The vote was seventeen in favor, four opposed. Those four were A. R. Shuliachenko, A. N. Engel'gardt, P. A. Lachinov, and G. G. Gustavson. None of them was a Heidelberger. An early petition from August 17, 1865, was left "without action" on the desk of a ministry functionary.[92] In January 1868, the Chemical Section of the first Russian Congress of Natural Scientists and Physicians, meeting in St. Petersburg—a conference established by the ministry to increase communication among Russian naturalists—petitioned for the authorization of a chemical society. At this point, with the massive proliferation of such specialist scientific societies both in Russia and abroad, the ministry began to relent and approved the plea of October 26. The Chemical Society, under president N. N. Zinin, began meeting shortly afterward.[93] The kruzhok dissolved as a matter of course.

CONCLUSION

Even though it was only one component in a synergy of factors that led to the professionalization of science in Russia, the implications of the role of the Heidelberg kruzhok in the creation of the Russian Chemical Society are potentially far reaching for understanding some features of the "second importation" of science into Russia. Even across the traumas of two revolutions, a civil war, the cataclysm of the Second World War, and the collapse of Communism, science has maintained a strong foothold in Russian culture. How does one explain the vitality from the middle of the nineteenth century forward, given the comparative lethargy of the pre–Great Reforms Academy of Sciences? I would argue that it was precisely in this hybridization of foreign professional organizations with the domestic institution of the kruzhok that allowed for such permanence.

This has everything to do with decentralization. The lesson of the first importation of Russian science into the academy was not lost on either the bureaucrats or the scientists themselves. It was simply not possible to import a science complete from abroad, hoping that institutions would emerge without the cultural backdrop to make sense of this new form of life. Instead, a blend between the forms of science already developed in western Europe with specific Russian forms helped make science something that *fit* with the environment posed by Russia. The fact that these institutions were quasi-formal, with a strong remnant flavor of the underground kruzhok, helps account for the stability of these institutions (the scientific societies, the scientific community) across violent ruptures in other formal structures. It is precisely this enduring stability of science in Russia despite shocks such as Lysenkoism and the collapse of the Soviet Union that requires attention to the enduring organization of the intelligentsia.

The irony of all this is that the vigor of Russian sciences was accomplished through the midwifery of alienation within Germany. And the contribution of Germany was not all negative and reactive; it was also positive in providing a forum in which different kinds of hybridizations could be experimented upon. After the establishment of

---

[92] See Kozlov, *Vsesoiuznoe khimicheskoe obshchestvo imeni D. I. Mendeleeva* (cit. n. 78), 13–4.
[93] For details, see Brooks, "Formation of a Community of Chemists in Russia" (cit. n. 5).

the Russian Chemical Society, however, the German contributions, both positive and negative, were elided or ignored completely. Whereas correspondence with Heidelberg—with Erlenmeyer in particular—continued up to the formation of the society, it dropped off strongly afterward, and communication with the West happened through the formal reports submitted to foreign scientific societies of the doings in St. Petersburg. One way to illustrate this reinvention and domestication of the origin story is through the "founder myth" of Russian chemistry.

The founder represented is Nikolai Zinin, the first president of the Russian Chemical Society. Upon his death in 1880, an image of Zinin was presented to the chemical community by his former students from both Kazan University and the Medico-Surgical Academy in St. Petersburg as entirely *native*, without foreign influence. Consider, for example, Zinin's obituary in the *Journal of the Russian Chemical Society*, lovingly penned by his former students Aleksandr Butlerov and Aleksandr Borodin:

> With his scientific and pedagogical activity is joined the emergence of a Russian chemical school; Russian chemistry is obligated to him preeminently for its entry into autonomous life; his works first persuaded the scientists of Western Europe to allot a distinguished place to Russian chemistry....
>
> He appeared as one of the few gifted, purely Russian scientists who had real scientific achievements to his name. The German element was then very strong in Russia's scientific estates, and in public one was hardly able to seriously distinguish chemistry from apothecaries' affairs, the laboratory from the pharmacy, and, having become used to seeing Germans as pharmacists, one all the more paid attention to a leading chemist among the Russians.[94]

It was not enough that Zinin be seen as somehow sprung out of whole cloth from Russian soil—despite the fact that he had also studied abroad in his youth—but he was given by Borodin, in a funeral oration, full credit for emancipating Russian chemistry from the German yoke:

> The deceased became a professor in those distant times, when science in Russia was a phenomenon imported from the West. If one came across rare instances of autonomous developments of it here, then it was the exclusive achievement of only a few privileged scientists—professors, academics—and of that larger group of foreigners or, at least, of those who received special education abroad. Science, thus, was obtained ready-made from the West and was considered some kind of sacred object, which only the high priests of science dared deal with, guarding it from the hands of the unordained. In Russia people bowed to it, but they did not serve it; they taught it, but they did not cultivate it.
>
> Having broad knowledge, a deep, bright mind, and a fiery love for Russia, N. N. [Zinin] understood before many others that science will not be at home with us until young forces, in the heart of the fatherland itself, are drawn to working on it, until the foundation for an autonomous Russian school is laid. And there, in Kazan, he first laid the basis of the Russian school of chemistry.[95]

For the chemists who survived Zinin's death, his image was used as a way of erasing all traces of Heidelberg and the kruzhok on the development of what had now

---

[94] A. P. Borodin and A. M. Butlerov, "Nikolai Nikolaevich Zinin: Vospominaniia o nem i biograficheskii ocherk," *ZhRFKhO* 12, khim. ch. (1880): 215–254, on 215, 226.

[95] Borodin, funeral oration of N. N. Zinin, 9 Feb. 1880, published in *Zdorov'e*, 1 May 1880, 183–4; reprinted in *BorP*, 3:86.

become *Russian* chemistry. A usable past had been found that eliminated the entire history told in this essay.

And this points to perhaps the most important implication of the German origins of Russian *national* identification in the sciences. It was certainly not the case that chemistry in Russia was imported from Germany: there were chemists working in the Russian empire before the 1850s, and even metallurgists and the like before Peter the Great. There was long a tradition of science in Russia. But the experience of living abroad, emulating foreign institutions, and then figuring out self-perpetuating ways of adapting them—*this* promoted a sense of there being a national character to the science, a Russian *national style*. That style was theoretical, bold, impulsive, and stridently argumentative. It was the style of D. I. Mendeleev and V. V. Markovnikov. It was also the style of Emil Erlenmeyer.

Turning Pedagogy into a Science:
Teachers and Psychologists in Late Imperial Russia (1897–1917)

By Andy Byford*

ABSTRACT

The article explores the Russian teachers' tortuous campaign at the beginning of the twentieth century to rise above the status of "semiprofessionals" by rooting the legitimacy of their professional expertise, training institutions, and working practices in the authority of "science." This involved a radical reshaping of traditional pedagogy and its fusion with new, controversial approaches to child psychology. It also led to a proliferation of teacher-training courses and conferences devoted to "pedagogical psychology," "experimental pedagogy," and "pedology." The article analyzes how the teachers' professional aspirations interacted with the conflicting agendas of rival groups of psychologists, who were themselves engaged in bitter squabbles over the legitimate identity of psychology as a scientific discipline.

INTRODUCTION

The first decades of the twentieth century were a decisive period in the history of Russian education, marked by the awakening of public and professional initiative and a gradual dismantling of the authoritarian, bureaucratic formalism that had emblematized tsarist schooling since the 1870s.[1] The death in 1897 of the education minister Ivan Delianov—best known for his "cook's children" circular,[2] the most outrageous example of the tsarist policy of blatantly restricting educational access to the lower classes—marked the start of a provisional "thaw" in official circles as well, leading to a series of intermittent ministerial reviews, with potentially wide-ranging reforms in mind.[3] The partial democratization of Russian society following the 1905 revolu-

---

* Wolfson College, University of Oxford, Linton Road, Oxford, OX2 6UD, UK; a_byford@hotmail.com.
  Research for this article has been funded by the Leverhulme Trust (F/87306/A), under the auspices of the project "Childhood in Russia, 1890–1991: A Social and Cultural History," directed by Professor Catriona Kelly (New College, Oxford). I thank (in alphabetical order) Vitaly Bezrogov, Michael Gordin, Karl Hall, Catriona Kelly, Alexei Kojevnikov, and Kathy Olesko, for their valuable input and suggestions. I also wish to express my gratitude to the staff at the Helsinki University Library, especially Irina Lukka of its Slavonic Library, for their help in providing the bulk of my sources.
  [1] Patrick L. Alston, *Education and the State in Tsarist Russia* (Stanford, Calif., 1969), 87–104, 115–29, 140–53.
  [2] On the affair surrounding this circular, see P. A. Zaionchkovskii, *Rossiiskoe samoderzhavie v kontse XX stoletiia* (Moscow, 1970), 347–54; and G. I. Shchetinina, *Universitety v Rossii i ustav 1884 goda* (Moscow, 1976), 198–204.
  [3] Alston, *Education and the State* (cit. n. 1), 153–65.

© 2008 by The History of Science Society. All rights reserved. 0369-7827/08/2008-0003$10.00

tion opened up further scope for civic enterprise, leading to expansion and diversification through a burst of alternative, privately funded projects in various domains of schooling and training.[4]

One major issue in these transformations was the problem of the social standing of teachers, including their professional autonomy, financial situation, working conditions, and corporate identity.[5] Russian educators were heterogeneous in terms of qualifications, social background, gender, earnings, and social aspirations.[6] There was a vast gulf between those occupying posts in elite classical gymnasiums in the capitals and the lowly village teachers, who could barely make ends meet in the provincial outback. However, toward the end of the nineteenth century, the profession, irrespective of category, seemed bound by a common sense of dissatisfaction and low self-esteem. This led, between 1900 and 1907, to the formation of an increasingly politicized "teachers' movement" that not only called upon government and society for improvements to be made to their professional standing and working conditions but also called for the profession to be properly recognized as performing a key social function, one of vital significance to the people and the empire.[7]

Indeed, the teachers' aspirations were not just narrowly trade unionist.[8] Among the challenges deemed essential for realizing their self-perception as legitimate members of the nation's intelligentsia was the construction of a suitable form of *expertise*, one that would raise educators to a level similar to that of other, more reputable, professionals (such as doctors, lawyers, and academics) and would provide them with the required legitimacy both within the school and in society at large.[9] The most visible result of this drive was the proliferation during the first two decades of the twentieth century of new kinds of courses, conferences, and societies that emphasized the grounding of pedagogy and teacher training in the *scientific study of children*,[10] especially their psychology—a movement that went by a number of interrelated labels, such as "scientific pedagogy" (*nauchnaia pedagogika*), "pedagogical psychology" (*pedagogicheskaia psikhologiia*), "experimental pedagogy" (*eksperimental'naia pedagogika*), and "pedology" (*pedologiia*).

The "child-centrism" of this movement was inseparable from the formation of a new expertise. Insistence on the specificity of the child was a way of problematizing

---

[4] Ibid., 119–220; J. C. McClelland, *Autocrats and Academics: Education, Culture, and Society in Tsarist Russia* (Chicago, 1979), 29–55.

[5] On rural teachers, see Scott J. Seregny, "Professional Activism and Association among Russian Teachers, 1864–1905," in *Russia's Missing Middle Class: The Professions in Russian History*, ed. Harley D. Balzer (Armonk, N.Y., 1996), 169–95. On city teachers, see Christine Rouane, *Gender, Class, and the Professionalization of Russian City Teachers, 1860–1914* (Pittsburgh, Pa., 1994). For professionalization efforts after 1905, see Alston, *Education and the State* (cit. n. 1), 228–42. See also V. R. Leikina-Svirskaia, *Intelligentsiia v Rossii vo vtoroi polovine XIX veka* (Moscow, 1971), 147–73.

[6] Christine Ruane and Ben Eklof, "Cultural Pioneers and Professionals: The Teacher in Society," in *Between Tsar and People: Educated Society and the Quest for Public Identity in Late Imperial Russia*, ed. Edith W. Clowes, Samuel D. Kassow, and James L. West (Princeton, N.J., 1991), 199–211.

[7] Ibid., 206–9. The height of this movement's politicization was during the 1905 revolution; that politicization declined in the repressions that followed.

[8] For more on the political side of teacher professionalization, see Rouane, *Gender, Class* (cit. n. 5), 87–184.

[9] On the special role of "academic knowledge" in professional formation, see Andrew Abbott, *The System of Professions: An Essay on the Division of Expert Labor* (Chicago, 1988), 52–8.

[10] Alston, *Education and the State* (cit. n. 1), 228–42; E. A. Kniazev, *Genezis vyshego pedagogicheskogo obrazovaniia v Rossii XVIII–nachala XX veka: Smena paradigm* (Moscow, 2001), 187–238.

the authority that adults automatically assumed over the process of children's upbringing. It presumed the need for an intermediary—the "expert in child development."[11] The first such experts came from the ranks of doctors, psychiatrists, and psychologists.[12] However, these professionals were relatively few and far between and lacked the necessary institutional support to gain social relevance and public recognition for their still fragile and newly developing knowledge. In seeking such support, they targeted especially the masses of teachers.[13] By fusing the idea of an "expertise in children" with the teachers' supposed "expertise in education," they assumed the role of "educators of educators," promising to empower teachers with "science."

The present article is a discussion of the complexities involved in this process. The focus of analysis is on the Russian teachers' drive to rise above the level of "semi-professionals,"[14] specifically through reforms of pedagogy (*pedagogika* or sometimes *pedagogiia*), as their specialist discipline. Although attempts to improve the educators' skills and expertise are discussed in most accounts of teacher professionalization in late imperial Russia,[15] the question of pedagogy's problematic status as an academic discipline or "science" (*nauka*)[16] and the controversies surrounding its institutional supports and disciplinary reconceptualizations have not been explored in any detail. Yet these issues lay at the heart of the educators' struggle for social recognition and played a key role in attempts to transform their professional and public identity at this historical juncture.

Calls to root pedagogy in science date back to the late 1850s and 1860s. During the liberalizing reforms of Tsar Alexander II, Russian teachers—inspired by figures such as the physician and educational administrator Nikolai Pirogov (1810–1881) and the educational theorist Konstantin Ushinskii (1824–1870)—first started to organize themselves into a distinct professional corps, whose grand social mission was to bring knowledge and enlightenment to the Russian people.[17] The role of teachers

---

[11] See I. A. Sikorskii, "O postanovke prepodavaniia i vospitaniia soobrazno estestvennomu khodu umstvennogo razvitiia," *Russkaia shkola* (hereafter cited as *RS*), 1900, no. 2:32–46.

[12] To name but a few Russian pioneers: the hygienist A. S. Virenius (see E. A. Pokrovskii, "Doktor med. Aleksandr Samoilovich Virenius [Po povodu 30-letii ego sluzheniia shkol'nym vrachem]," *Vestnik vospitaniia* [hereafter cited as *VV*], 1894, "Khronika," no. 4:270–80); the psychiatrist I. A. Sikorskii (see "Prof. Ivan Alekseevich Sikorskii [biograficheskii ocherk]," VV, 1895, "Khronika," no. 3:291–5); and the psychologist P. F. Kaptereva," *Voprosy psikhologii* [hereafter cited as *VP*], 1978, no. 2:150–6). For a general discussion, see A. A. Nikol'skaia, *Vozrastnaia i pedagogicheskaia psikhologiia v dorevoliutsionnoi Rossii* (Dubna, 1995).

[13] The other key target group were the parents, especially in so-called parent circles (*roditel'skie kruzhki*). See N. Arep'ev, "'Roditel'skii kruzhok' v Peterburge," VV, 1897, no. 1:181–97; no. 3:158–69; no. 5:188–97. Particularly important also was the series *Entsiklopediia semeinogo vospitaniia i obucheniia*, which started coming out in 1898. On conferences that focused on family education, see *Trudy pervogo vserossiiskogo s"ezda po semeinomu vospitaniiu*, 2 vols. (St. Petersburg, 1914).

[14] See Amitai Etzioni, ed., *The Semi-Professions and Their Organization: Teachers, Nurses, Social Workers* (New York, 1969).

[15] Seregny, "Professional Activism" (cit. n. 5), 184–8; Rouane, *Gender, Class* (cit. n. 5), 21–41; Alston, *Education and the State* (cit. n. 1), 228–42.

[16] In what follows, the term "science" will be used as the Russian term *nauka*, which is closer to the German *Wissenschaft* and includes more than just the natural sciences, implying academic knowledge in a more general way.

[17] Nicholas Hans, *The Russian Tradition in Education* (London, 1963), 45–64; N. I. Pirogov, "Voprosy zhizni," *Morskoi sbornik*, 1856, no. 9:559–97; K. D. Ushinskii, *Chelovek kak predmet vospitaniia: Opyt pedagogicheskoi antropologii*, 2 vols. (St. Petersburg, 1868–1869). In addition to his work in pedagogical theory, Ushinskii is valued especially as a campaigner for the democratization of education in Russia and its development on national foundations.

in Russian society was, in fact, meant to emblematize precisely that ideal combination of values expected of the nation's intelligentsia *as a whole*: on the one hand, firm roots in rational and systematic, scientifically based knowledge; on the other, a direct and active (civic or professional) involvement in the material and spiritual emancipation of the nation. More specifically, the ideal of the education profession was conceived as the direct harnessing of the emergent anthropological sciences to a practical technology of human betterment and social progress.

However, at this point, the idea of making pedagogy itself truly scientific had little basis in institutional or intellectual reality, and the whole idea was mainly deployed to mobilize young university-educated talent to join the ranks of a much-needed and seemingly ennobled profession. Yet the 1870s school reforms of education minister Dmitrii Tolstoi created no room for pedagogy as an academic discipline, relegating it instead to purely practical-professional functions, especially in the training of lower-level, primary school educators.[18] It was not until the 1890s that the upper, secondary school echelons of the profession—especially via a new generation of professional journals, such as *Russkaia shkola* and *Vestnik vospitaniia* (both founded in 1890)—again started to campaign for the transformation of pedagogy into an academically legitimate and socially acclaimed form of corporate expertise, one that could legitimate the teachers' inclusion among the ranks of Russia's expanding professional intelligentsia.[19]

In what follows, I shall first sketch the professional "identity crisis" that Russian teachers were experiencing just before and around 1900, in order to locate the profession's main social motivations for rebuilding pedagogy into a scientific discipline. I shall then examine what concrete forms the movement for "scientific pedagogy" took, on two overlapping levels—that of *institutional development* and that of *disciplinary redefinition*. First, I shall chart the contradictions surrounding the challenge of creating suitable institutional supports for the production and transmission of a scientific form of pedagogy. Second, I shall analyze the controversial conceptual strategies used to reshape pedagogy into something resembling science, above all through a juxtaposition of the functions of education with the study of child psychology.

A PROFESSION WITHOUT A SCIENCE

Although the problem of making pedagogy scientific was important to Russia's teaching profession as a whole, it was those working in secondary education—the classical gymnasiums and the realschulen—to whom this discipline's academic status and identification with nauka mattered the most.[20] At the turn of the twentieth century, Russian secondary school teachers were trapped in a limbo between the

---

[18] On the abolition of the university "pedagogical courses" by Dmitrii Tolstoi in 1867–1868 (on the grounds that they were unsuitable for an institution of nauka), see *Polnoe sobranie zakonov Rossiiskoi Imperii*, 2nd ser., vol. 42 (1866), statute 44767.

[19] Much of the impetus for this came from professionals other than teachers—especially from doctors. See Andy Byford, "Professional Cross-Dressing: Doctors in Education in Late Imperial Russia (1881–1917)," *Russian Review* 65 (2006): 586–616.

[20] The idea of making pedagogy scientific was also promoted to primary school teachers but in a simplified way, as secondhand popularization. See N. V. Tulupov and P. M. Shestakov, eds., *Prakticheskaia shkol'naia entsiklopediia: Nastol'naia kniga dlia narodnykh uchitelei i drugikh blizhaishikh deiatelei v oblasti narodnogo obrazovaniia* (Moscow, 1912); and N. E. Rumiantsev, *Lektsii po pedagogicheskoi psikhologii dlia narodnykh uchitelei* (Moscow, 1913).

academy, identified with "science," and primary school teaching, viewed as pure "technique." Those working in secondary schools, in contrast to primary school teachers, all had higher, mostly university, education, which was emphatically academic rather than practical-professional, and they clearly saw themselves as part of the nation's intelligentsia, in the broad sense of "educated elite." Yet while refusing, for this reason, to be reduced to the level of "mere" schoolteachers, emblematized by the lower level of primary educators (*narodnye uchitelia*), they were not in a position to emulate the kind of respectability identified with university-based scholars and scientists (*uchenye*).

Secondary school teachers tended to perceive themselves as subject specialists, and privileged their academic knowledge over any general teaching expertise. On the whole, they received little or no explicit teacher training. Yet at the same time, their teaching was seen as contributing to the students' "general education," rather than "science," and in this context, the role that they were identified with was that of a "pedagogue" (*pedagog*), with the discipline of "pedagogy" featuring as their supposed professional specialism. However, pedagogy was not taught at Russian universities because it was not deemed a proper scholarly discipline,[21] being often denounced, especially by academics, as merely an outdated collection of scholastic recipes, based on questionable subjective values and traditional norms, rather than anything resembling nauka.[22]

This vacuum of academic legitimacy was recognized as one major factor in secondary schools' being pervaded, through and through, by bureaucratic rather than properly professional power relations.[23] Pedagogy's lack of academic authority seemed to make it easier for educational practice to be governed by extrinsic ministerial instructions and programs and by mechanical and arbitrary evaluations of the teachers' achievement through students' school marks and exam results, enforced by the bureaucratic authority of school directors and inspectors.[24]

Such subordination of teachers internal to the school translated into a sense of overall social inadequacy and entailed the incorporation of a veritable corporate inferiority complex.[25] In the 1890s, teachers were very concerned about their public image, characterized by "lifeless" formalism, routine, and mediocrity. They decried, yet also admitted the symbolic truth of, popular literary and journalistic representations of the

---

[21] A university chair in pedagogy existed in the 1850s when it replaced the chair in philosophy (which was repressed following the 1848 revolutions in Europe), but the pedagogy chair was then abolished in the 1863 university charter as unworthy of independent scientific status. Some theoretical aspects of pedagogy were occasionally taught in philosophy lectures. See V. Ivanovskii, "O prepodavanii pedagogiki v universitetakh," *VV*, 1906, no. 7:109–35, 118–21. Chairs in pedagogy existed at ecclesiastical academies, and for this reason, graduates from these establishments were sometimes given preference for posts as school inspectors (see E. A. Bobrov, "O zhelatel'nosti uchrezhdeniia kafedr pedagogiki v Russkikh universitetakh," *Voprosy pedagogiki*, 1912, no. 1:16–20, 17). However, pedagogy taught in ecclesiastical institutions was deemed inadequate for purposes of modern "rational" education (see N. Ticher, "Diletantizm i prizvanie v pedagogicheskom dele" [Iz inostrannoi pedagogicheskoi literatury], *Narodnoe obrazovanie*, 1905, "Knizhnoe i zhurnal'noe obozrenie," no. 12:555–76, 556).

[22] R. Iu. Vipper, "Spetsial'naia podgotovka prepodavatelia srednei shkoly ili podniatie ego polozheniia?" *VV*, 1898, no. 6:52–74.

[23] Alston, *Education and the State* (cit. n. 1), 144–53; Rouane, *Gender, Class* (cit. n. 5), 42–61. See also A. Divil'kovskii, "Avtoritet i svoboda—vospitatelei," *VV*, 1915, no. 5:56–95.

[24] G. R[okov?], "Eshche o polozhenii uchitelei srednei shkoly," *VV*, 1900, no. 5:66–97.

[25] Ibid., 68.

teacher as an automaton encased in a shell of bureaucratic routine;[26] or a petty, bitter pedant engrossed in "gloomy dissatisfaction";[27] or an embarrassed recluse, ashamed to mention his job title when introduced to a stranger on a train.[28]

Teachers complained in their periodicals about being unjustly targeted by society and under fire not only from the schools' bureaucratic authorities (*nachal'stvo*) but also from students, parents, and the press. They were especially resentful that much of their bad reputation came from the poor opinion their students were spreading about them,[29] although they believed that children assimilated this disrespectful attitude especially from their parents, who refused to recognize teaching as implying any expert authority. They complained that anyone seemed to think himself qualified to judge their methods and quibble with their educational decisions or student evaluations.[30] Teachers also strongly felt their inferiority in relation to other professions. They were particularly sensitive to doctors and university professors' looking down on them, claiming that they knew far better than teachers did how children should be educated and schools should be managed.[31]

Many teachers considered overcoming excessive bureaucratization to be an essential step toward enhancing their social status. However, they also recognized that this would not in itself overcome public contempt and condescension. In fact, many of them also warned that it would be dangerous to abandon all "control" and allow complete "freedom of teaching," as this was likely to lead to anarchy.[32] Because the problem was ultimately that of building public trust, it was argued that if individual teachers happened to abuse this trust even minimally, it could bring irreparable damage to the profession as a whole. As one commentator put it, because the stakes were so high "in this matter, which concerned the fates of living individuals, namely children, and the vital interests of their parents and of the whole of society," one could simply not rely solely on the teachers' "only human" judgment.[33]

Thus, teachers still needed to be controlled somehow, although not bureaucratically, as had been the case throughout the last quarter of the nineteenth century. The problem with this bureaucratic regime was that the teachers' own bureaucratic clout was weak within the school hierarchy (in relation to the nachal'stvo), while appearing arbitrary and unconvincing outside it (in relation to the parents). The educators' motivation behind overturning this particular structure of power was not just the idealistic drive to democratize education and rebuild it on rational and humane grounds, or to emancipate their own profession, even if all of these happened to be their sincerely stated aims. A key drive behind the campaign to replace a (negatively connoted) "bureaucratic" order with a (positively connoted) "scientific" one was to reinforce the

---

[26] Ibid., 66. Usually with specific reference to Anton Chekhov's story "Man in a Case," whose central character is Belikov, a provincial teacher of classical Greek.

[27] G. Rokov, "O pedagogicheskoi professii i ee predstaviteliakh," *VV*, 1897, no. 1:79–118, on 82.

[28] S. Zenchenko, "O podgotovke prepodavatelei srednikh uchebnykh zavedenii k pedagogicheskoi deiatel'nosti," *VV*, 1898, no. 4:60–96, 95.

[29] On the problem of the teachers' authority in class, see also G. R[okov?], "Avtoritet lichnosti i avtoritet vlasti v dele vospitaniia," *VV*, 1905, no. 2:22–44; and P. L. Kovan'ko, *Pedagogicheskii avtoritet* (Kiev, 1905).

[30] V. Vagner, "Universitet i sredneobrazovantel'naia shkola," *Russkaia mysl'*, 1898, no. 2:136–46.

[31] A. S. Virenius, "Soiuz medikov s pedagogami," *Vestnik obrazovaniia* (hereafter cited as *VO*), 1891, no. 3:65–89; Virenius, "Gigiena, kak predmet prepodavaniia v shkole," *RS*, 1897, no. 11:132–43; no. 12:110–25.

[32] R[okov?], "Eshche o polozhenii" (cit. n. 24), 81–2.

[33] Ibid., 81.

symbolic foundation of the teachers' *authority*—a foundation that needed to exist not just in abstract terms but also as an indispensable support in concrete professional actions and decisions (for example, those related to student evaluations as exemplary of the teachers' power over children's destinies).

The required control over the profession was supposed to come from a system of "expert knowledge"—a system that would also exist as a communication network, interrelating educational professionals and organizing them into a unified corps with an appropriate hierarchical structure based on different levels of expertise. The idea was to transform the partially extrinsic bureaucratic authorities (directors and inspectors) into pedagogical mentors—a special rank of teachers whose role would be to supervise and instruct less experienced colleagues in matters of pedagogy.[34]

In principle, novice teachers were already supposed to be initiated on the job by the school director, but in practice the amount of supervision provided was minimal and arbitrary.[35] As early as the 1860s, the Ministry of Education had tried to organize systematic teacher training precisely as a form of mentorship, with university students observing "exemplary lessons" taught by specially chosen "excellent teachers."[36] However, these exemplary teachers were selected either from among those whose students all achieved high grades (which was the result more often of strict mental dressage than of the teacher's pedagogical talent); or those who managed to impress the bureaucrats with some random, supposedly original, pedagogical trick; or those whose textbooks enjoyed particular popularity. Indeed, a simple recasting of figures of authority was not enough because, as the teachers themselves readily admitted, pedagogy was an underdeveloped and ambiguous discipline, providing no definitive consensus about what good teaching really was.

Educational theory mainly relied on quotations from revered classics of educational philosophy, such as the works of Herbart, Komenský, Locke, Pestalozzi, Rousseau, and Ushinskii, but there was no authoritative academic institution in charge of defining the criteria of teaching quality in unequivocal, scientific terms.[37] Thus, since the teachers' scientific clout was no greater than their bureaucratic powers, the crucial task facing the profession seemed to be that of developing the academic standing of their discipline. In the first instance, this presumed the formation of training establishments in which pedagogy would at last be treated in serious fashion and become a respectable nauka.

THE FAILURE OF PEDAGOGY AT THE UNIVERSITIES

Toward the end of the nineteenth century, imperial Russia had very few higher-educational establishments offering specialist professional training for those intending to teach in secondary education.[38] Teaching in secondary schools was something

---

[34] See also N. V. Chekhov, "Pedagogicheskii nadzor i pedagogicheskaia ekspertiza," *RS*, 1914, nos. 7–8:32–6.

[35] M. Miropiev, "O pedagogicheskoi podgotovke uchitelei dlia srednikh uchebnykh zavedenii," *RS*, 1899, no. 1:106–22, 106.

[36] On what follows, see Zenchenko, "O podgotovke prepodavatelei," (cit. n. 28), 80–1.

[37] R[okov?], "Eshche o polozhenii" (cit. n. 24), 72.

[38] On qualifications required to occupy a teaching post in secondary education, see S. Stepanov, "K voprosu o pedagogicheskoi podgotovke prepodavatelei srednei shkoly v Rossii: Istoricheskii ocherk," *Zhurnal ministerstva narodnogo prosveshcheniia* (hereafter cited as *ZhMNP*), 1909, "Sovremennaia letopis'," no. 3:1–35, 12–5.

that those with higher education did as a matter of course, often while waiting for a better job in the civil service or sometimes in academia, typically combining it with other intellectual work. Of course, many in the end made a career in education, usually with the hope of progressing to posts as educational bureaucrats.

Contemporary critical surveys of government failures to devise an efficient provision of teacher training for secondary schools regularly highlighted the fact that, throughout the nineteenth century, the Ministry of Education was creating structures in which there was too much emphasis on specialized subject knowledge at the expense of the (academic and practical) development of pedagogy as professional expertise.[39] The education profession never had a faculty of its own, as the legal and the medical ones did, and secondary school teachers usually came from the ranks of those graduating in the natural sciences and the humanities. These two faculties had little room for teacher training in their already overstretched degree structures, and they were too absorbed in forging their own academic legitimacy to pay much attention to educational problems. Attempts to append extracurricular teacher training to the university degree structure remained marginal and inadequate.[40] The fact that pedagogy referred to the hierarchically inferior level in the same system of knowledge production and dissemination (secondary as opposed to higher education) meant that, at the university, pedagogical questions would usually be addressed with condescension and postponed until the "purely" scientific ones had been resolved first.

This sidelining of pedagogy in relation to academic knowledge doomed the discipline even in those structures that the state had set up specifically to prepare future secondary school teachers, such as the Main Pedagogical Institute in St. Petersburg (1828–1858) and the Historical-Philological Institutes in St. Petersburg (founded in 1867) and Nezhin (founded in 1874).[41] These teacher-training establishments doubled as higher-educational institutions, and their academic standards emulated university learning, even though they did not have the credentials of sites of "pure" science. "Pedagogy and didactics" was a recognized chair at the Historical-Philological Institutes, but this discipline occupied a marginal role in the curriculum. It was training in specialist subjects (the classics, Russian language and literature, history, and geography), rather than anything to do with general pedagogy, that ultimately distinguished their graduates. In other words, even here, the cult of academic knowledge (albeit tailor-made for the bureaucratized, classics-dominated Tolstoian school, named after the education minister Dmitrii Tolstoi) overshadowed any attempt to develop a profession-specific expertise.

When the drive to insert nauka into pedagogy reemerged in the 1890s, the most obvious idea was finally to admit pedagogy to the "temple of science," that is, to

---

[39] Zenchenko, "O podgotovke prepodavatelei" (cit. n. 28), 68–81; Ivanovskii, "O prepodavanii pedagogiki v universitetakh" (cit. n. 21); Stepanov, "K voprosu o pedagogicheskoi podgotovke" (cit. n. 38); V. F. Kagan, "O podgotovlenii prepodavatelei matematiki dlia srednikh uchebnykh zavedenii," in *Trudy 1-go vserossiiskogo s"ezda propodavatelei matematiki: 27-go dekabria 1911 g.–3-go ianvaria 1912 g.*, vol. 1 (St. Petersburg, 1913), 479–554.

[40] On the 1860s extracurricular "pedagogical courses," see Zenchenko, "O podgotovke prepodavatelei" (cit. n. 28), 70–1. On the role of early forms of the university seminar as teacher-training frameworks, see Andy Byford, "Initiation to Scholarship: The University Seminar in Late Imperial Russia," *Russian Review* 64 (2005): 299–323. Moscow University had a Pedagogical Society, but its activities were limited. See *Proekt ustava sostoiashchego pri Moskovskom universitete Pedagogicheskogo obshchestva: Ob" iasnitel'naia zapiska* (Moscow, 1897).

[41] V. A. Zmeev, "Vysshaia shkola Rossii: Ot reform 60-kh k reforme 80-kh godov XIX v.," *Sotsial'no-politicheskii zhurnal*, 1998, no. 5:164–79, 166–9.

incorporate it into the university degree structure as an academic discipline in its own right. This was something that Ushinskii had already been vainly calling for in the 1860s, yet the matter was still proving to be far from straightforward at the turn of the century.

A feverish debate on the issue flared up in 1898–1899, launched by the zoologist Vladimir Vagner in an article in *Russkaia mysl'*.[42] Vagner argued that Russian universities produced two types of teachers: strong specialists in a particular subject who knew little about how to convey that knowledge to schoolchildren, or individuals who, after obtaining a degree, and without a clear occupation in mind, ended up teaching a subject they might not have studied at the university, supposedly in a "general educational" way. For Vagner, the only way to produce proper "general educators" was to provide specialized teacher training, and his solution was to establish a separate university chair or department in pedagogy, closely following the American example.[43]

Some educators were worried that this would not in itself prevent the traditional marginalization of pedagogy within the Russian academy. Writing in *Vestnik vospitaniia*, the teacher Sergei Zenchenko (who also lectured in courses for primary school teachers in the provinces)[44] argued controversially that the university, in fact, went beyond its sphere of competence when it accorded someone the right to teach simply by awarding him a university degree.[45] Zenchenko called for the creation of a two-year *postgraduate* teacher-training program, which would still be attached to the universities but would be taught by specialists in pedagogy. Zenchenko speculated that the first professors in this specialty should be *not* university academics but talented "autodidact pedagogues" (who might need some additional training abroad). He insisted that the purpose of such a program would be very different from what the university traditionally provided, insofar as the aim would be to transform particular "academic disciplines" into "school subjects."

Zenchenko believed that after attending such a program, the novice secondary school teacher would cease to look up with envy at his academic counterpart—the university privatdocent (untenured assistant professor)—and would no longer feel ashamed for not knowing some minor factual detail in their common discipline.[46] In other words, the teacher would stop seeking (in vain) to be compared to a scholar but would form a different identity as someone in possession of knowledge that was, in fact, *not* in the domain of subject-specific scholarly expertise.[47] Zenchenko was keen to ensure not just that a careful distinction be made between the teaching profession and the academic one but also that the two professions should become equal in status,

---

[42] Vagner, "Universitet i sredneobrazovantel'naia shkola" (cit. n. 30).

[43] Ibid., 138–41. At this time, there were twenty-six universities with education departments in the United States. There was little unity in their organization, but American practice indicated some of the main content requirements (psychology, physiology, history of education, school management, school hygiene, subject-specific teaching methods) and forms of training (seminar presentations and discussions, observations of lessons, trial lessons).

[44] See N. Bunakov, "O pedagogicheskikh kursakh dlia uchitelei i uchitel'nits tserkovno-prikhodskikh shkol penzenskoi eparkhii letom 1902 g. (Zametki rukovoditelia)," *Narodnoe obrazovanie*, 1902, no. 10:267–81, 268.

[45] Zenchenko, "O podgotovke prepodavatelei" (cit. n. 28), 61.

[46] Most privatdocents while lecturing at universities also taught in secondary schools to supplement their low and irregular income. Nonetheless, they presented themselves as "scholars" first and foremost, with an image quite distinct from that of "pedagogues."

[47] Zenchenko, "O podgotovke prepodavatelei" (cit. n. 28), 96.

insofar as pedagogical expertise would institutionally lie at the same level as professional nauka—above the university degree.

Both Vagner's and Zenchenko's schemes were challenged by more conservative academics. The professor of world history Robert Vipper resented the portrayal of the academy as some sort of villain that inhibited the true vocation of practicing teachers.[48] In his view, the problem lay not in the teachers' alleged lack of professional expertise but in their poor pay and working conditions. Vipper insisted on preserving the identification of secondary school teachers' professional self-value with the value of academic learning and wanted to see teachers further improve their ability to orientate themselves in the latest scholarly research in their specialist subjects. In counterattack, Vipper mocked pedagogy itself. He argued that, in reality, pedagogy existed only as a secondary school subject in the final grades of the female gymnasiums, where it included only very basic elements of physiology and hygiene, tidbits of psychology and ethics, and random quotations of famous educational philosophers, suitable only for the training of future primary educators.[49] Vipper argued that instead of vainly trying to make something scientific out of this, one would do better to rely on the existing university program as the only program adequate for training secondary school tutors.

Vipper's scathing comments about pedagogy angered much of the educational community, evoking a flurry of heated responses.[50] One of them came from Aleksei Ostrogorskii, who complained in *Pedagogicheskii sbornik* that Vipper had deliberately ignored the failure of the universities to engage properly with problems of education.[51] Ostrogorskii wrote that, although the university should remain dedicated to science, it ought to devote far more attention to "applied" matters and therefore work toward forging a scientific form of pedagogy. Ostrogorskii believed that the way forward was a collaboration of existing academic specialists and secondary school educators. In particular, he suggested that the theoretical expertise of professors in *psychology* (who taught in philosophy departments) needed to be combined with the practicing teachers' knowledge of schoolchildren and educational realities.[52] However, Ostrogorskii's understanding of the division of labor and the hierarchy of authority in such collaboration remained unclear.

The debate peaked at the end of 1898 when a ministerial circular (16 Nov 1898, No. 27971) called upon the curators of educational districts to carry out regional surveys and prepare for a discussion of the impending reforms of teacher training for secondary education.[53] The press quickly responded to the circular: some articles offered

---

[48] Vipper, "Spetsial'naia podgotovka" (cit. n. 22).

[49] On the pedagogy class in female gymnasiums, see, e.g., S. Brailovskii, "Zametka ob organizatsii zaniatii v pedagogicheskom (VIII-m) klasse zhenskikh gimnazii," *RS*, 1892, no. 2:91–102; A. P. Nechaev, "O postanovke prepodavaniia pedagogiki v zhenskikh gimnaziiakh," *RS*, 1899, no. 1:145–63; K. El'nitskii, "Prepodavanie obshchei pedagogiki v zhenskoi gimnazii," *RS*, 1905, no. 3:85–108.

[50] See editorial note to Vipper, "Spetsial'naia podgotovka" (cit. n. 22), 52; and S. Zenchenko, "I podniatie polozheniia prepodavatelia srednei shkoly i spetsial'naia ego podgotovka (Otvet professoru R. Vipperu)," *VV*, 1898, no. 8:83–103.

[51] A. Ostrogorskii, "I spetsial'naia podgotovka prepodavatelia srednei shkoly, i uluchshenie ego polozheniia (Po povodu st. prof. Vippera)," *Pedagogicheskii sbornik*, 1899, no. 1:60–78.

[52] Ibid., 73–4.

[53] "Tsirkuliarnoe predlozhenie g. popechiteliu uchebnogo okruga," *ZhMNP*, 1898, "Pravitel'stvennye rasporiazheniia," no. 12:57–9.

specific proposals,[54] while others provided lessons from the past;[55] some gave examples of teacher-training arrangements of other ministries (e.g., the Department of Military Education),[56] while others politicized the issue by arguing that one had to focus on creating a civic-minded identity for educators, rather than bureaucratize professional training, which they suspected was the government's underlying aim.[57]

The responses revealed a general lack of consensus about the ideal solution. They presented a number of different and often incompatible, if overlapping, formats. Some of the responses from teachers were especially critical of the academic environment and proposed a radical dissociation of teacher training from the universities by setting up district pedagogical seminaries under the supervision of local administrations, with an affiliation to local schools.[58] However, the majority of proposals favored a strong input from university professors as essential to boosting pedagogy's academic legitimacy. A summary of opinions was published in *Zhurnal ministerstva narodnogo prosveshcheniia* at the end of 1899.[59]

The dominant view, formulated especially by former curator of the Caucasus Educational District Kirill Ianovskii, was to organize teacher training as a postgraduate course in a new form of establishment—district pedagogical institutes, based mainly in university towns, where the theoretical courses would be run by university professors, while practical training would be provided in local schools. However, those who responded to the ministerial circular disagreed on most of the concrete details. Among the unresolved issues was the question of whether these institutes should be formally affiliated with a university or a secondary school; whether there should be professional exams at the end, and how the candidates would be evaluated; what format the tutoring and assignments would take; how the candidates would be selected, and who would be ultimately responsible for them (university professors, some special director, or teachers who mentored them in schools). The diversity of available options, together with persistent doubts as to whether the whole endeavor was really worth the trouble and the money, did little to boost the Ministry of Education's desire to invest its scarce resources into reorganizing secondary school teacher training on a grand scale.

Nevertheless, given the increasing public pressure to improve matters in this domain, the government set up, between 1900 and 1905, several commissions and subcommissions to deal with the problem; these formulated a variety of mutually contra-

[54] Miropiev, "O pedagogicheskoi podgotovke uchitelei dlia srednikh uchebnykh zavedenii" (cit. n. 35); A. Anastasiev, "K voprosu ob organizatsii pedagogicheskoi podgotovki prepodavatelei dlia srednikh uchebnykh zavedenii," *RS*, 1899, no. 3:108–22; A. S. Virenius, "K voprosu o podgotovlenii uchitelei (pedagogov) dlia srednikh uchebnykh zavedenii," *RS*, 1899, no. 10:105–18.

[55] N. Arep'ev, "K voprosu o podgotovlenii prepodavatelei (Istoricheskaia spravka)," *RS*, 1899, no. 1:123–8.

[56] A. S., "K voprosu o pedagogicheskoi podgotovke ofitserov-vospitatelei kadetskikh korpusov," *Pedagogicheskii sbornik,* 1899, "Chast' neoffitsial'naia," no. 6:586–90.

[57] "Polozhenie prepodavatelei srednikh uchebnykh zavedenii," *VV*, 1899, no. 3:16–40.

[58] See Anastasiev, "K voprosu ob organizatsii pedagogicheskoi podgotovki prepodavatelei" (cit. n. 54); Ia. Gurevich, "K voprosu o pedagogicheskoi podgotovke prepodavatelei dlia sredneuchebnykh zavedenii i ob uluchshenii ikh material'nogo polozheniia," *RS*, 1899, no. 4:77–93; N. S., "Po voprosu o spetsial'noi podgotovke prepodavatelei v srednei shkoly," *Pedagogicheskii sbornik,* 1899, "Chast' neoffitsial'naia," no. 8:128–36.

[59] "K voprosu o pedagogicheskoi podgotovke uchitelei dlia srednikh uchebnykh zavedenii i ob uluchshenii material'nogo ikh polozheniia," *ZhMNP*, 1899, "Sovremennaia letopis'," no. 10:11–42; no. 11:1–20; no. 12:49–78.

dictory conclusions about the desired way forward.[60] Apart from the issue of funding, which was always in short supply, the greatest tensions surrounded the questions of whether and how the universities should incorporate pedagogy and how exactly academics should be involved in teacher training.[61] The ideal director of the proposed institute was supposed to be an "experienced practicing pedagogue who must also possess serious scholarly competence,"[62] yet this seemed a blend impossible to achieve in an environment in which such sharp divisions reigned between the images of "serious scholars" and "practicing pedagogues."

In the absence of clear solutions, intermediary measures were proposed, namely the forming of temporary university chairs in pedagogy and the resurrection of university-based extracurricular pedagogical courses, with the aim of eventually transferring these to the proposed institutes.[63] However, proper chairs in pedagogy failed to materialize. Instead, in 1904, the Ministry of Education instructed philosophy departments to organize special series of lectures in pedagogy and provided some funding for the privatdocents who would be running them.[64] Yet such courses were only facultative, and the lecturing remained unsystematic, something blamed especially on the absence of full professorships in the discipline.[65]

This did not prevent students at some universities from organizing active "pedagogical circles," especially in Warsaw (beginning in 1910, with support from the philosophy professor Evegenii Bobrov)[66] and in Moscow (from 1912, again with strong support from some members of staff, especially Moisei Rubinshtein, a privatdocent in the Philosophy Department).[67] Both Bobrov and Rubinshtein campaigned for the formation of full professorial chairs in pedagogy (ideally attached to philosophy departments) and were critical of plans to create separate institutes, arguing that these were too expensive and simply duplicated much of what was already taught at the universities. Moreover, Rubinshtein claimed that without the universities' exercising full control over scientific pedagogy, there was a danger of the discipline turning into amateurish "trash" (*khlam*).[68]

However, nothing practical came of any of the above proposals, and pedagogy was not granted either a chair or a faculty at the universities. Instead, the curators of educational districts (first in Odessa, in 1909, then a year or two later in Kiev, Moscow, and St. Petersburg) organized temporary, annual pedagogical courses for future teachers, with professors doing most of the lecturing, but without a clear overarching program and with only a very loose connection between these courses and the university itself.

---

[60] See Kniazev, *Genezis vyshego pedagogicheskogo obrazovaniia* (cit. n. 10), 191–8.
[61] See Stepanov, "K voprosu o pedagogicheskoi podgotovke" (cit. n. 38), 21–33.
[62] Ibid., 23.
[63] Ibid., 35.
[64] Bobrov, "O zhelatel'nosti uchrezhdeniia kafedr pedagogiki" (cit. n. 21).
[65] Ibid. See also M. M. Rubinshtein, "Pedagogi bez pedagogiki (O polozhenii pedagogiki v nashikh universitetakh)," *VV*, 1913, no. 6:114–27; and I. K., "Zametka po povodu stat'i M. M. Rubinshteina 'Pedagogi bez pedagogiki,'" *VV*, 1913, no. 9:195–9.
[66] See *Pedagogicheskii kruzhok studentov pri imperatorskim varshavskom universitete* for 1910 and 1911—supplements to *Voprosy pedagogiki*, 1912, no. 1, and 1913, no. 2, respectively.
[67] "Studencheskii nauchno-pedagogicheskii kruzhok pri Moskovskom universitete," *VV*, 1915, no. 6:153–62. On Rubinshtein, see A. A. Nikol'skaia, "Nekotorye aspekty nauchnogo tvorchestva M. M. Rubinshteina," *VP*, 1978, no. 5:137–45.
[68] Rubinshtein, "Pedagogi bez pedagogiki" (cit. n. 65), 125.

Since each district authority organized its courses in a different way,[69] none created anything resembling an authoritative form of university-based pedagogy.

NEW INSTITUTIONS OF "SCIENTIFIC PEDAGOGY"

As pedagogy kept waiting before the gates of the "temple of science," independent, locally organized teacher-training initiatives were booming. Civic efforts led to the creation of a large variety of privately funded training courses and establishments, which mushroomed especially after the 1905 revolution.[70] They ranged from occasional lecture series (organized by local authorities and pedagogical societies of various persuasions),[71] to full-blown pedagogical institutes, some of which, though privately financed, had the government's special blessing (e.g., the postgraduate Shelaputin Institute in Moscow,[72] opened in 1911, based largely on the above-mentioned 1899 proposal by Ianovskii; and the Women's Pedagogical Institute in St. Petersburg,[73] founded in 1903 as the main establishment for the training of women staffing the female gymnasiums). The new private teacher-training establishments also included, for example, courses for kindergarten tutors and childminders organized by the St. Petersburg Froebel Society,[74] P. F. Lesgaft's Free Higher School in St. Petersburg, which specialized in physical education,[75] and the pedagogical courses of A. L. Shaniavskii's People's University in Moscow, which targeted mainly primary school teachers.[76]

All of these training programs sought in one way or another to enhance and to systematize the skills and knowledge of different categories of teachers, and they were all committed to adhering to what they understood as the most authoritative pedagogical principles, based on modern ideas and research, adapted to the needs of their different clientele. However, they were mostly practical-professional, focusing on matters of subject-specific teaching methods (*metodika*), didactics, and the popularization of educational psychology, school hygiene, and school management (*shkolo-*

---

[69] See Kagan, "O podgotovlenii prepodavatelei matematiki dlia srednikh uchebnykh zavedenii" (cit. n. 39) and "Khronika," *VV*, 1911, no. 5:87–99. In Kiev and Odessa, lectures focused on subject-specific knowledge and teaching. In St. Petersburg, the courses were geared toward practical training (trial lessons, school excursions). In Moscow, the courses consisted of lectures in the history of pedagogy, psychology, logic, and school hygiene. See also "Khronika," *VV*, 1909, no. 7:72–4.

[70] For a typology of different institutions, see Kniazev, *Genezis vyshego pedagogicheskogo obrazovaniia* (cit. n. 10), 187–238.

[71] See P. M. Fedotova, "K voprosu o znachenii provintsial'nykh pedagogicheskikh kruzhkov," *Voprosy psikhologii, kriminal'noi antropologii i gipnotizma* (hereafter cited as *VPKAG*), 1904, no. 8:662–7. On the courses of the St. Petersburg Pedagogical Society, see *Trudy kursov dlia uchitelei srednei shkoly (II god): 5–25 iunia 1907 g.* (St. Petersburg, 1908). For an example of active programs in the provinces, see O. V. Petrov, "Psikhologicheskoe prosveshenie v Samarskom krae v nachale XX stoletiia," *VP*, 1998, no. 1:99–107. On courses organized by the group around the journal *Obnovlenie shkoly*, see "Peredvizhnye pedagogicheskie kursy," *Obnovlenie shkoly*, 1911, no. 11:79–80.

[72] The Shelaputin Institute was answerable to the Ministry of Education and was built thanks to the donation of the millionaire P. G. Shelaputin, who insisted that only male Russians of Orthodox faith be accepted as students. See A. I. Iasinskii, ed., *Izvestiia pedagogicheskogo instituta imeni P. G. Sheliaputina v g. Moskve*, vol. 1 (Moscow, 1912–13); and Kniazev, *Genezis vyshego pedagogicheskogo obrazovaniia* (cit. n. 10), 208–15.

[73] Zhenskii Pedagogicheskii Institut emerged out of Zhenskie Pedagogicheskie Kursy and was subordinated to Vedomstvo uchrezhdenii Imperatritsy Marii. See Kniazev, *Genezis vyshego pedagogicheskogo obrazovaniia* (cit. n. 10), 215–20.

[74] See "Pedagogicheskie kursy frebelevskogo obshchestva," *RS*, 1907, "Pedagogicheskaia khronika," nos. 5–6:81–3.

[75] See Kniazev, *Genezis vyshego pedagogicheskogo obrazovaniia* (cit. n. 10), 150–7.

[76] Ibid., 166–77.

vedenie). Although they promoted and enhanced the idea of pedagogy as a science, they did not position themselves as establishments of *nauka* as such.

However, as the traditional seats of scientific authority—the universities—were failing to provide proper academic backing, the Russian educational field continued to crave it. Many commentators worried about dilettantism and confusion emerging from such a variety of relatively fragmented teacher-training efforts,[77] and they called for the creation of institutions of pedagogical science proper, on the model of those emerging in Belgium, France, Germany, Switzerland, and the United States.[78]

Toward the end of the 1900s, two privately funded establishments were created in St. Petersburg precisely as centers of educational *science* rather than merely practical teacher training. These were the Pedagogical Academy of the League of Education and the Pedagogical Faculty of Vladimir Bekhterev's Psycho-Neurological Institute, both founded at roughly the same time, during 1907–1908, and both seeking to assume a position at the very top of the field of educators, as groundbreakers in scientific pedagogy.[79]

The origins of the Pedagogical Academy lay in the initiatives of the Department of Military Education and the work of its Pedagogical Museum in St. Petersburg.[80] In 1891, General Apollon Makarov was appointed as head of the latter,[81] and he, in particular, began campaigning for a more systematic development of pedagogy on a "rational" basis, calling upon progressive educational theorists, doctors, and psychologists to take part in regular meetings and to introduce Russian parents and teachers to cutting-edge educational research and ideas coming from the West. Among Makarov's initiatives was the setting up in 1899 of teacher-training courses for officer-teachers working in schools run by the military.[82] These courses were generously funded by the Ministry of the Army, and as part of the project, Makarov sponsored the setting up, in 1901, of a fully equipped psychological laboratory devoted to educational problems, the first of its kind in Russia, organized on the initiative of the young psychology tutor Aleksandr Nechaev.[83]

---

[77] Ticher, "Diletantizm i prizvanie v pedagogicheskom dele"; Ivanovskii, "O prepodavanii pedagogiki v universitetakh." (Both cit. n. 21.)

[78] V. Ivanovskii, "Kafedra pedagogiki v Sorbonne," *VV*, 1902, "Referaty i melkie soobshcheniia," no. 1:71–7; "Izuchenie psikhologii detskogo vozrasta v amerikanskikh uchitel'skikh seminariiakh (Normal'nykh shkolakh)," *VV*, 1902), "Referaty i melkie soobshcheniia," no. 4:91–7; Ticher, "Diletantizm i prizvanie v pedagogicheskom dele" (cit. n. 21), focusing on the U.S. example. N. E. Rumiantsev, "K voprosu o pedagogicheskikh akademiiakh," *Obnovlenie shkoly*, 1911–12, no. 5:67–8, focusing on German efforts. "Novaia pedagogicheskaia vysshaia shkola v Zheneve (Institut Zhan-Zhaka Russo)," *Svobodnoe Vospitanie*, 1911–12, no. 7:73–8; "Institut nauk o vospitanii (Institut J. J. Rousseau)," *VV*, 1917, nos. 8–9:116–39.

[79] There was also a failed attempt by a group of Moscow professors to create something similar in Moscow at about the same time, although they hoped to organize it as a "pedagogical faculty" affiliated with Moscow University. See T. D. Faddeev, "O pedagogicheskom fakul'tete" and "Prilozhenie," *RS*, 1907, no. 2:149–56,157–66.

[80] On pedagogical museums, see N. Flerov, "O pedagogicheskom muzee," *Pedagogicheskii vestnik Moskovskogo uchebnogo okruga*, 1913, nos. 4–5:88–111; *Kratkii obzor deiatel'nosti Pedagogicheskogo muzeia voenno-uchebnykh zavedenii za 1894–95 g. (Dvadtsat piatyi obzor)* (St. Petersburg, 1895)—supplement to *Pedagogicheskii sbornik*, 1895, no. 11.

[81] For more on Makarov, see I. M., "Iubilei A. N. Makarova," *VV*, 1906, "Khronika," no. 6:168–71.

[82] See A. N. Makarov, *Pedagogicheskie kursy dlia podgotovleniia ofitserov k vospitatel'noi deiatel'nosti v kadetskikh korpusakh*, vol. 1 (St. Petersburg, 1902); *Pedagogicheskie kursy vedomstva voenno-uchebnykh zavedenii, 1900–1910*, vol. 3 (St. Petersburg, 1911).

[83] For more on Nechaev, see A. A. Romanov, *Opytno-eksperimental'naia pedagogika pervoi treti XX veka* (Moscow, 1997), 40–119; V. V. Anshakova, *Vklad A. P. Nechaeva v stanovlenie i razvitie vozrastnoi i pedagogicheskoi psikhologii* (Astrakhan, 2002); V. V. Bol'shakova, "Eksperimental'noe

Nechaev started off as a privatdocent at the Philosophy Department of St. Petersburg University in the late 1890s. But after having his master's thesis in educational psychology (which relied on controversial new mental testing techniques) rejected by his own mentor—the neo-Kantian philosopher Professor Aleksandr Vvedenskii—he ended up effectively barred from a university career and was instead recruited by Makarov to teach at the Pedagogical Museum.[84] Although the standard of psychology in Nechaev's course for officer-teachers had to be relatively low (given the fact that the trainees had not previously been to the university), Nechaev was able to market his psychology laboratory as a unique research establishment devoted to problems of experimental educational psychology, and he positioned it as Russia's (at that time) one and only center dedicated to the forging of a new, truly scientific basis for pedagogy, inspired by similar efforts taking place in the West.

The successful promotion of this laboratory prompted the setting up in 1904 of the so-called Pedology Section of the Pedagogical Museum, conceived as the kernel of a future institute devoted to all-round "child study" or "pedology" (from the Greek *pais* meaning child), emulating the establishment set up by Granville Stanley Hall at Clark University in Worcester, Massachusetts, while also claiming to be realizing Ushinskii's (i.e., a native Russian's) dream of a broad-based, anthropologically geared "pedagogical faculty" devoted to the comprehensive study of "man as an object of education."[85] The section organized its own annual courses, open to educators employed by all ministries.[86]

Although these courses targeted practicing teachers, Makarov, in his inaugural speech, insisted that they were *not* practical-professional but involved the development of a "science of education" (*nauka vospitaniia*). To assert the higher credentials of these courses, Makarov had to *split* scientific pedagogy from day-to-day "teaching practice."[87] In fact, he first stated that many teachers seemed to "look down with con-

---

izuchenie psikhologii shkol'nikov v trudakh A. P. Nechaeva," in *Istoriia stanovleniia i razvitiia eksperimental'no-psikhologicheskikh issledovanii v Rossii*, ed. B. F. Lomov et al. (Moscow, 1990), 189–99; A. P. Nechaev, "Zapiski psikhologa," in ibid., 201–13; V. M. Kadnevskii, "A. P. Nechaev i stanovlenie eksperimental'noi pedagogiki v Rossii," *Pedagogika*, 2005, no. 1:71–8.

[84] A. P. Nechaev, *Sovremennaia eksperimental'naia psikhologiia v ee otnosheniiakh k voprosam shkol'nogo obucheniia* (St. Petersburg, 1901). See also the review of this book by Iu. Aikhenval'd in "Kritika i bibliografiia," *Voprosy filosofii i psikhologii* (hereafter cited as *VFP*) 59 (1901): 405–16; A. P. Nechaev, "Otvet g. Aikhenval'du," and Aikhenval'd, "Neskol'ko slov po povodu stat'i g. Nechaeva," *VFP* 60 (1901), "Polemika": 504–26; Nechaev, "Vtoroi otvet g. aikhenval'du," *VFP* 61 (1902), "Polemika": 625–32; the review of Nechaev's book by A. I. Vvedenskii in "Kritika i bibliografiia," *ZhMNP*, 1901, no. 12:444–84 (also published as Vvedenskii, *Ob eksperimental'noi didaktike A. P. Nechaeva* [St. Petersburg,1902]); Nechaev, "Otvet g. prof. A. I. Vvedenskomu," *ZhMNP*, 1902, "Kritika i bibliografiia," no. 1:203–28.

[85] K. Dobropistsev, "Otkrytie pedologicheskogo otdela pri pedagogicheskom muzee voenno uchebnykh zavedenii," *VPKAG*, 1904, no. 4:268–71.

[86] M. K[onor]-ov, "Otkrytie lektsii i prakticheskikh zaniatii, organizuemykh pedologicheskim otdelom imeni K. D. Ushinskogo," *RS*, 1904, "Pedagogicheskaia khronika," nos. 10–11:94–8; Konorov, "Otkrytie pedologicheskikh kursov," *Pedagogicheskii sborniku*, 1904, no. 12:548–53; I. M., "Pervoe publichnoe zasedanie Pedologicheskogo otdela imeni K. D. Ushinskogo," *VV*, 1904, "Khronika," no. 4:109–15; I. M., "Pedologicheskie kursy," *VPKAG,* 1905, no. 1:58–62; I. M., "Pedologicheskie kursy imeni K. D. Ushinskogo," *VV*, 1905, "Khronika," no. 6:138–40.

[87] Makarov had to insist on the separation of the "science of education" from "teaching" because his target audience was a profession whose supposed expertise—pedagogy—had no scientific credentials but was invariably dismissed as "art," "technique," or "practice." Moreover, the institution he was inaugurating aspired to the authority of nauka yet did not possess an automatic and unequivocal association with it, as did the universities.

descension upon the science of education,"[88] refusing to recognize that one could advance educational practice through scientific research.[89] Yet in truth, he then argued, teachers would acquire the kind of professional status that they desired and deserved only once "the foundations of education [were] developed scientifically and not by the arbitrary views of practitioners."[90] In the end, Makarov urged "practicing pedagogues" (*pedagogi praktiki*) to *look up* to the new category of experts that he dubbed "representatives of the science of education" (*predstaviteli nauki vospitaniia*). Indeed, all those giving keynote speeches at the inaugural ceremony, and embodying the scientific authority of this new type of institution, were not educators but psychologists, psychiatrists, physicians, and criminologists.[91]

However, the rhetorical move that followed this separation of the science of education from teaching was, on the contrary, one of *fusing* scientists in education with the teachers themselves. Speaking after Makarov, the psychologist Nechaev, as one of the "representatives of the science of education," explained that the courses of the Pedology Section were, after all, intended for "representatives of living school practice" (*predstaviteli zhivogo shkol'nogo dela*). However, they were specifically designed to engage educational professionals in "free scientific work," to train them as *researchers* in "child study." Insistence on the scientific nature of these courses was incorporated into actual training methods, with practical classes focusing not on teaching skills but on research projects, with the achievement of a published research paper replacing any exam or diploma.

Aiming to promote this new paradigm of scientific pedagogy to a wider audience of educational practitioners nationwide, Nechaev and his colleagues also organized Russia's first major conference in "pedagogical psychology" in 1906 and followed it up with one-month courses for teachers on the above scientific model in the summers of 1906 and 1907.[92] These courses offered something genuinely different from those in most standard teacher-training programs. There were ninety hours of lectures and a considerable number of practical classes in experimental psychology, as well as excursions to local museums, laboratories, crèches, schools, and sanatoriums, followed by evening discussions on topical issues in education and psychology. There were lectures in anatomy, physiology, educational psychology, experimental psychology, psychophysiology of sensory organs, pathological pedagogy, school hygiene, speech therapy, children's games, and school management. The program was intensive—twenty-five days filled with nonstop lectures, practical classes, and public discussions, from 9 a.m. to 10–11 p.m. Lectures and discussions were accompanied by practical demonstrations using magic lantern displays, psychological instruments, diagrams and charts, anatomical displays, and physiological experiments on frogs and pigeons.

In the course of 1907–1908, the courses of the Pedology Section were transformed

---

[88] Note Makarov's provisional inversion of power relations here in presenting practicing teachers as *looking down on* the experts. This was prompted partly by the fact that the experts were here trespassing on the teachers' territory. The inversion was, of course, rhetorical and ironic—the device is one of setting up a provisional structure to be *transcended* or *reversed* by the orator's antithetical argument.

[89] K[onor]-ov, "Otkrytie lektsii i prakticheskikh zaniatii" (cit. n. 86), 94.

[90] Ibid., 94–5.

[91] They included, for example, the psychologist A. P. Nechaev, the pediatrician N. P. Gundobin, the psychiatrist L. N. Blumenau, and the criminologist D. A. Dril'.

[92] N. E. Rumiantsev, "Letnie pedologicheskie kursy dlia uchitelei v S.-Peterburge s 20 iulia po 18 avgusta 1907 g. otchet.," *VPKAG*, 1907, no. 4:153–86.

into a more permanent establishment—the Pedagogical Academy.[93] This academy was devised as a postgraduate institution that awarded diplomas after a two-year course, but only to those who had already obtained a first degree at some other higher-educational institution (a university or Higher Women's Courses, for instance). It was founded under the auspices of the League of Education[94] and became officially subordinated to the Ministry of Education, which ratified its charter. Yet the ministry provided no financial assistance and actually denied the academy the right to award doctorates in pedagogy. This was a serious blow to its ambitions to become the establishment creating future professors of pedagogy—those missing "pedagogue-scholars" who would be hired to occupy the hypothetical university chairs in the discipline or directorships of the envisaged pedagogical institutes.

In the end, the Pedagogical Academy had to express its official role somewhat vaguely and unambitiously as providing "specialist education to persons who wished to devote themselves to pedagogical work."[95] In practice, though, there was considerable ambiguity about what "pedagogical work" actually meant and whether the academy provided top-up training for already established teachers, or prepared future educational "managers," or if it trained researchers and lecturers in educational psychology, school hygiene, and pedagogical pathology.

Most of the tutoring at the academy was done through lab demonstrations and seminars, with an emphasis on group research projects and trial lessons. The student teachers were not novices but already experienced "practicing pedagogues" and were given a considerable amount of initiative and freedom in the way they organized their studies. The majority came from the provinces and were mature individuals, predominantly in their thirties and usually with a number of years spent in teaching. The course was not meant to provide them with teaching qualifications but to turn them into a special new form of scientifically minded "teacher-researchers"—an ideal, however, that was difficult to fit anywhere in the existing educators' job market and professional hierarchy.[96]

A similar approach was characteristic of the Pedagogical Faculty of Bekhterev's Psycho-Neurological Institute, with the difference being that it offered a first degree and hence assumed a broader higher-educational program, which is why it also had a much larger number of students than the Pedagogical Academy did.[97] The institute's

---

[93] On what follows, see "Ustav pedagogicheskoi akademii ligi obrazovaniia," *VPKAG*, 1907, no. 4:187–90; M. K[onorov?], "Pedagogicheskaia akademiia," *RS*, 1907, no. 12:83–5; A. P. Nechaev, "Pervye shagi pedagogicheskoi akademii," *Ezhegodnik eksperimental'noi pedagogiki* (hereafter cited as *EEP*) 3 (1910): 10–21; *Trudy s.-peterburgskoi pedagogicheskoi akademii*, vol. 1, *Nachalo dela: Vozniknovenie pedagogicheskoi akademii, obshchye osnovy ee organizatsii i pervye raboty slushatelei* (St. Petersburg, 1910); *Trudy s.-peterburgskoi pedagogicheskoi akademii*, vol. 2, *1910–11/1911–12* (St. Petersburg, 1913).

[94] On Liga obrazovaniia (a loose grouping of educational societies), see "Ustav ligi obrazovaniia," *VPKAG*, 1907, no. 4:190–4.

[95] "Ustav pedagogicheskoi akademii" (cit. n. 93), 187.

[96] Some believed that the academy's graduates ought to be particularly well suited for posts of school directors. See review of *Trudy s.-peterburgskoi pedagogicheskoi akademii*, vol. 2, *1910–11/1911–12* (cit. n. 93), by R. G., in *VV*, 1914, "Kritika i bibliografiia," no. 5:41–5. Other options included recommendations for the creation of a rank similar to the German *Gymnasialprofessor*, which would be awarded to teachers who made special scholarly contributions to pedagogy. See V. Polovtsov, "O pedagogicheskoi podgotovke uchitel'skogo personala srednei shkoly," *ZhMNP*, 1916, "Otdel po narodnomu obrazovaniiu," no. 1:97–9.

[97] For more on the Psycho-Neurological Institute in this period, see M. A. Akimenko and A. M. Shereshevskii, *Istoriia instituta imeni V. M. Bekhtereva na dokumental'nykh materialakh*, vol. 1

Pedagogical Faculty existed alongside the Faculties of Law and Medicine, thereby ensuring that teachers had formal rights analogous to those of other, better-established and higher-ranked professions. Despite this orientation toward creating professionals, the ethos of the Pedagogical Faculty was rigorously scientific, with an emphasis on initiation in science through the prolonged study of "fundamental disciplines," followed by research projects. (These included work in the institute's psychology lab and visits to different types of schools, sanatoriums for abnormal children, and the Pedological Institute—an affiliated small research center devoted to the study of early child development.)

As regards training in practical teaching technique, the faculty had little to offer apart from matter-of-course lectures in subject-specific *metodiki*, which meant that students had to rely on practice alone to gain teaching skills as such. Instead, the "practical" dimension of the course was conceived largely as a form of "applied science." A particularly good example of the fusion of "education" and "science" in this program would be the projects supervised by the psychology tutor Aleksandr Lazurskii (formerly Bekhterev's student and privatdocent at the St. Petersburg Military Medical Academy). Lazurskii devised a method he called the "natural experiment," which involved making use of concrete school lessons and educational activities as experimental frameworks for performing personality tests.[98] A student teacher in his course would sit at the back of a class and observe the behavior of one specific child in a series of lessons. These lessons would be specially designed to appear "natural," while in fact assessing specific cognitive abilities and personality traits. The researcher would take notes on the child's behavior during the lesson, based on a pre-prepared program that specified exactly which psychological characteristic was being "measured" in which school activity. In the end, a detailed verbal profile (*kharakteristika*) of the child's personality and a provisional diagram of cognitive abilities were produced and used for research-scientific as well as practical-educational purposes.[99]

Both the Pedagogical Academy and the Psycho-Neurological Institute hoped to embody an objective and neutral form of scientific authority in the field of teaching and education. However, the novelty of their institutional positioning, the originality of their programs, the lack of financial and moral support from the state, and the fact that they tended to attract as students and lecturers especially those who were not admitted to the academic mainstream (women, Jews, the "politically unreliable," and so on) meant that, ultimately, their public image was that of "radicals" in pedagogy, partisans of new and as yet untested methods. Their ideas and practices had exciting potential but were at the same time controversial, provoking skepticism and opposition from a great many teachers, bureaucrats, and academics of a more conservative or simply cautious bent. Although they were often praised as independent, civic institutions and although it was usually acknowledged that they were the leaders in

---

(St. Petersburg, 2002); and Kniazev, *Genezis vyshego pedagogicheskogo obrazovaniia* (cit. n. 10), 158–66.

[98] A. F. Lazurskii, ed., *Estestvennyi eksperiment i ego shkol'noe primenenie* (Petrograd, 1918); Lazurskii, *Shkol'nye kharakteristiki* (St. Petersburg, 1908); L. N. Val'vat'eva [Filosofova], "Eksperimental'nye uroki po prirodovedeniiu," *VV*, 1913, no. 5:44–82; Lazurskii and Filosofova, "Estestveno-eksperimental'nye skhemi lichnosti uchashchikhsia," *VV*, 1916, no. 6:1–51.

[99] Similar sorts of observations were being carried out in the experimental school affiliated with the Pedagogical Academy. See A. P. Nechaev, "Ob eksperimental'noi shkole pri pedagogicheskoi akademii," *VV*, 1911, no. 3:1–20.

the drive to further pedagogy's credentials as a science, the Pedagogical Academy and the Psycho-Neurological Institute also had reputations for pursuing this goal a bit too hastily, for uncritically falling for the latest "fads" in psychology, for leaning too much toward "materialist" extremism, and for not observing the more measured approach to science building that the traditional academy was renowned for and on which nauka's authority in Russian society depended.

This is why a considerable number of teachers, craving a stabler and less controversial scientific foundation for their profession, would have preferred to see the scientific form of pedagogy based on a more traditional form of nauka—the one embodied by the universities. This way, in the words of a teacher, one would avoid not just the "stagnant old belief" (*kosnoe staroobriadnichestvo*) that still reigned in the outdated programs of lower-level teachers' seminaries for primary educators but also the "spurious gimmicks" (*bezpochvennye novshestva*) of radical new establishments such as the Pedagogical Academy and the Psycho-Neurological Institute.[100]

In addition to sheer legitimacy, the problem that plagued the latter two institutions was that, unlike the universities, they had to position themselves strategically *between* the hierarchically opposed levels of "science" and "education," and they defined as their ideal the inextricable, yet ambiguous, *fusion* of the two. They courted practicing teachers as their main following, but at the same time they sought to forge and embody a hierarchically superior form of scientific authority that would reign *over and above* educational practitioners. Such positioning was contradictory as it left open the question of whether teachers could indeed become the owners of what was being devised, institutionalized, and sold to them as *their* science, or whether, in order to be empowered by the latter, they had to become a different kind of professional altogether.

THE TEACHER AS PSYCHOLOGIST

The movement for scientific pedagogy had indeed started precisely as an attempt to redefine teaching in terms of a rather different kind of practice—namely an "applied" form of child and developmental *psychology*.[101] Teaching seemed so inferior to the work of other professions that some of those campaigning to improve the teachers' image thought it was essential to transform the very nature of the teachers' work,

[100] S. Liubomudrov, "Psikhologiia i pedagogika," *ZhMNP*, 1910, "Sovremennaia letopis'," no. 8:65–84, on 83–4.

[101] Pedagogy's close relation to psychology was traditional, dating back at least to the early nineteenth-century German philosopher Johann Friedrich Herbart. It was based on the understanding of education as the development of a child's moral character and spiritual forces. In Russia, this disciplinary connection was popularized in the 1860s by Ushinskii. On this period, see A. A. Nikol'skaia, "Russkaia detskaia i pedagogicheskaia psikhologiia v 60-e gg. XIX v.," *VP*, 1978, no. 1:137–46; Nikol'skaia, "Rol' N. I. Pirogova v russkoi pedagogicheskoi psikhologii," *VP*, 1982, no. 1:127–33; Nikol'skaia, "Psikhologiia obucheniia v trudakh K. D. Ushinskogo (k 160-letiiu so dnia rozhdeniia)," *VP*, 1983, no. 6:92–103. On the role of psychology in education in late nineteenth-century Russia, see Nikol'skaia, *Vozrastnaia i pedagogicheskaia psikhologiia* (cit. n. 12); Nikol'skaia, "Zadachi razrabotki istorii detskoi i pedagogicheskoi psikhologii," *VP*, 1984, no. 4:134–40; Nikol'skaia, "Obshchii obzor literatury po detskoi i pedagogicheskoi psikhologii v dorevoliutsionnoi Rossii (vo 2-oi polovine XX v.)," *VP*, 1973, no. 6:115–23; Nikol'skaia, "Obshchii obzor literatury po detskoi i pedagogicheskoi psikhologii v dorevoliutsionnoi Rossii (nachala XX veka do oktiabr'skoi revoliutsii)," *VP*, 1974, no. 2:156–67; Nikol'skaia, "K istorii vozniknoveniia i pervonachal'nogo razvitiia detskoi i pedagogicheskoi psikhologii," *VP*, 1976, no. 1:143–55; Nikol'skaia, "Osnovnye etapy razvitiia pedagogicheskoi psikhologii v dorevoliutsionnoi Rossii," *VP*, 1987, no. 4:109–18.

so that one could not possibly identify it with the monotonous, formalist drill that the majority, including the teachers themselves, associated with the profession.[102] A particularly widespread solution was to construct the new, progressive teacher as effectively a *child psychologist* or, more specifically, a "developer" of schoolchildren's "souls."

In addition to making the teachers' tasks more interesting, important, and responsible, the hypothetical turning of teachers into applied child psychologists was used especially as a way of distinguishing the pedagogue's expertise from that of the scholar. (This was crucial, as we have seen, to the self-image of secondary school teachers, who, as subject specialists, occupied a role dependent on and subordinate to academics.) It was argued that, unlike a university professor's, a teacher's real material was not the subject that he imparted to his students. The teacher was supposed to use this subject only "as a tool for influencing the spiritual forces of the child."[103] Consequently, it was less important for teachers to advance their subject knowledge and much more relevant to study the "science of the human soul." The focus on child psychology was also meant to create corporate unity for the profession, combating the fragmentation of secondary school teachers into a heterogeneous collection of specialists in different subjects.[104]

At Russian universities, psychology was taught as part of philosophy, in a theoretical and mostly generalist way, while the empirical study of child and developmental psychology remained, in mainstream academia, a virtually uninhabited realm. This area therefore seemed like something that teachers could indeed claim as their own territory. It was argued that, since teachers had daily access to children, they were better placed than anyone else to carry out systematic observations of children's psychological development.[105] The most practical task for such "teacher-psychologists," which seemed perfectly in line with their educational role, would be to determine the concrete psychological traits and abilities of their students, first, in order to be able to make an objective evaluation of each child, and second, in order to decide how best to influence this child's mental and moral development. It was also speculated that some teachers could, depending on their abilities and inclinations, even go beyond purely "diagnostic" and "applied" tasks and use their observations to develop more general theories of child psychology.

This "psychological turn" was interpreted as a veritable paradigm shift in pedagogy,[106] with "pedagogical psychology" coming to dominate teacher-training manuals and lecture courses.[107] Every self-respecting textbook author in pedagogy, especially one that would refer to this discipline as a science, had to provide a detailed account of psychology as its backbone or else face scorn and rejection from reviewers.[108] In fact, the majority of these textbooks ended up being, in effect, courses in psychology, with pedagogical "consequences" and occasional practical advice tacked

---

[102] R[okov?], "Eshche o polozhenii" (cit. n. 24), 91–6. See also A. P. Nechaev, *Sovremennaia eksperimental'naia psikhologiia* (cit. n. 84), 1.
[103] R[okov?], "Eshche o polozhenii" (cit. n. 24), 91.
[104] Vagner, "Universitet i sredneobrazovantel'naia shkola" (cit. n. 30), 143.
[105] R[okov?], "Eshche o polozhenii" (cit. n. 24), 91.
[106] Sikorskii, "O postanovke prepodavania" (cit. n. 11).
[107] V. M. Ekzempliarskii, "Eksperiment v psikhologii i pedagogike i nauka o vospitanii," *VV*, 1917, no. 1:46–61, 46.
[108] See, e.g., A. F[eoktistov]'s review of V. A. Volkovich, *Pedagogika—nauka pered sudom ee protivnikov*, 1909, in "Retzenzii," *EEP* 3 (1910): 22–3; A. Nechaev's review of I. S. Andreevskii,

on as mere appendices at the end of each chapter. Psychology was understood as the official scientific foundation of pedagogy. While psychology featured as a "pure" science, pedagogy became its "application" in the (practical) field of education. Those keen to highlight pedagogy's scholarly credentials (as effectively an applied human science) would argue that, although pedagogy itself was not taught at universities, it was *founded upon* a legitimate university discipline—psychology.[109]

The problem of the teachers' role in psychology became a pivotal issue not just in the transformation of the teachers' own expertise but also in struggles between different groups of psychologists to define the scientific identity of psychology itself. Scholars specializing in psychology were aware of the huge potential of the "teachers' market" and were keen to promote their discipline among educators, invariably arguing that psychology ought to be a key part of the teachers' professional development.[110]

The field of education was important to psychologists because psychology was not exactly an autonomous academic discipline at this time—it had the reputation of being a mere "handmaiden to philosophy," while its scientificity was under constant attack from physiologists, neurologists, and psychiatrists, who sought to redefine it in line with their own disciplinary agendas.[111] In fact, it was only in the sphere of education and specifically in relation to "lowly" pedagogy that psychology emerged in a dominant role, as a perfectly respectable science in its own right. Of course, this eminent academic status of psychology in relation to education and pedagogy had to be constructed as such and constantly maintained, which is how teachers became *the* most important "interested" public to whom psychologists promoted the idealized visions of their discipline.

Russian psychologists were by no means united on the issue of psychology's disciplinary identity. On the one hand, there were the university professors of philosophy (with Georgii Chelpanov of Moscow University as probably their most outspoken representative)[112] who gave preeminence to general, theoretical psychology and the

---

Nauchnye osnovy pedagogiki (Kiev, 1903), in "Kritika i bibliografiia," *RS*, 1904, no. 2:1–2; I. Kostin, "Shkola i eksperimental'naia psikhologiia," *Narodnoe obrazovanie*, nos. 7–8 (1912): 169–80, 169.

[109] Zenchenko, "O podgotovke prepodavatelei" (cit. n. 28), 93–4.

[110] The idea of psychology's relevance to education was used as one argument for the strengthening of its position at the universities. See speech by Professor Matvei Troitskii at the opening of the Moscow Psychological Society in 1885 (M. M. Troitskii, "Sovremennoe uchenie o zadachakh i metodakh psikhologii," *VP*, 1995, no. 4:93–107, 106).

[111] For more on the history of Russian psychology in this period, see David Joravsky, *Russian Psychology: A Critical History* (Oxford, 1989); A. V. Petrovskii, *Psikhologiia v Rossii: XX vek* (Moscow, 2000); E. A. Budilova, *Bor'ba materializma i idealizma v russkoi psikhologicheskoi nauke: Vtoraia polovina XIX–nachalo XX v* (Moscow, 1960).

[112] Chelpanov was by no means opposed to experimental psychology itself—quite the contrary, he was one of its main promoters in Russia (conceiving it in Wundtian fashion). He eventually founded the Moscow Institute of Psychology, which sported one of the most advanced psychological laboratories at that time. However, Chelpanov always saw experimentation as only *one* subordinate part of "general psychology" and sought to develop psychology from its traditional institutional base in philosophy. He criticized the physiologists, the neurologists, the psychiatrists, and people such as Nechaev for glorifying objective, experimental methods for their own sake. For more on Chelpanov, see Alex Kozulin, "Georgy Chelpanov and the Establishment of the Moscow Institute of Psychology," *Journal of the History of the Behavioral Sciences* 21 (1985): 23–32; Joravsky, *Russian Psychology* (cit. n. 111), 107–10; S. A. Bogdanchikov, "Nauchno-organizatsionnaia deiatel'nost' G. I. Chelpanova," *VP*, 1998, no. 2:126–35; T. D. Martsinkovskaia and M. G. Iaroshevskii, "Neizvestnye stranitsy tvorchestva G. I. Chelpanova," *VP*, 1999, no. 3:99–106. See also A. A. Nikol'skaia, "Osnovnye etapy razvitiia nauchnoi deiatel'nosti psikhologicheskogo instituta," *VP*, 1994, no. 2:5–21. Another key university-based promoter of experimental psychology was N. N. Lange, who taught in Odessa.

method of "introspection" (*samonabliudenie*), confining the value of objective methods or "experimentation" to the most elementary psychic functions, especially those on the boundary with physiology (e.g., the study of sensations). They saw experimental techniques and statistical analyses of experimental data as promising but still underdeveloped, and they were therefore very cautious about "facts" produced this way, which they suspected of being potentially rooted in an epistemologically blind form of neopositivism.

On the other hand, there was the younger batch of psychologists, some with a background in philosophy (such as Nechaev), and others with education in medicine (such as Lazurskii), who, especially during their postgraduate training abroad, became steeped in the culture of laboratory experimentation that reigned at German universities. They were keen to introduce this experimental, laboratory culture to Russian psychology and to work on developing a still newer and yet more controversial form of psychological technology—mental and personality tests—which were also being pioneered in France and the United States.

Both of these groups of psychologists courted teachers, doing their utmost to get the teachers to back their opposing disciplinary paradigms, although the teachers' support seemed rather more important to the younger experimenters, such as Nechaev and Lazurskii, because they had no institutional following *other* than the trainee or novice teachers whom they tutored in teacher-training establishments. Teachers were therefore cast in the role of this group's supposed *scientific* "disciples," as opposed to just a passive audience with a general interest in psychology, or as narrowly practical consumers of an otherwise purely academic expertise, which is how the university professors tended to relate to them. Of course, even people such as Nechaev were perfectly clear that teachers could not seriously think of simultaneously becoming full-time psychologists. But the legitimation of Nechaev's experimental program, which was being developed outside mainstream academia, depended greatly on persuading teachers that their involvement in psychology at teacher-training establishments represented a direct, and perfectly legitimate, involvement in science itself.

Nechaev's recruitment of teachers as active participants in the development of scientific psychology was an open challenge to the monopoly of university professors of philosophy over defining psychology's legitimate identity. University academics, such as Chelpanov, became greatly alarmed by this strategy and rushed to defend the university's authority over psychology by questioning the teachers' qualifications to carry out any form of psychological research in schools.[113] They warned that such practice would expose psychology to dangerous dilettantism, wrecking the discipline's fragile institutional status and public image.

Nechaev and his colleagues responded with arguments that what was decisive in making particular psychological research scientific was not who carried it out but whether a properly scientific method was applied.[114] For them, what defined "scientificity" in psychology was the application of a certain kind of methodology, which they identified with "the experiment" on the model of nauka as "natural science," in a deliberately polemical counterdistinction to nauka as "institutionalized academy"

---

[113] On Chelpanov's stance, see especially his collection of articles G. I. Chelpanov, *Sbornik statei (Psikhologiia i shkola)* (Moscow, 1912). On his pedagogical ideas, see A. A. Nikol'skaia, "Psikhologo-pedagogicheskie vzgliady G. I. Chelpanova," *VP*, 1994, no. 1:36–42.

[114] See N. E. Rumiantsev, "Psikhologiia i shkola," *EEP* 5 (1912): 56–75.

(emblematized by their opponents—the university professors). They argued that, since teachers already had prime access to the required psychological "material" (schoolchildren), what psychologists had to do, in order to further psychological research in this area, was to empower teachers with experimental methodology and equip them with the required technology to carry out concrete empirical studies. They implied that such research would not only advance psychology but also help teachers to better understand their students and thereby improve their teaching.

Nechaev and his group at the St. Petersburg laboratory devised a special, inexpensive "collection of instruments" (*kollektsiia priborov*) in experimental psychology (simplified apparatuses and test cards) that they marketed partly as teaching aids for the new secondary school course in psychology (introduced into the secondary school curriculum in 1905)[115] and partly as a mini-laboratory that could be used by teachers for their self-initiation into scientific pedagogy based on psychological experimentation on schoolchildren.[116] Nechaev and his group promoted these kits in their teacher-training courses and at the various conferences, prompting a number of enthusiast teachers to purchase them and to set up "psychology cabinets" in their schools.

The thorny issue of whether, and in what sense, teachers could be expected to "do" psychology, and what exact relation consequently obtained between psychology and pedagogy (inextricably intertwined with the fight over what counted as legitimate psychology itself), became the most important bone of contention at the First and Second Conferences in Pedagogical Psychology in St. Petersburg (1906 and 1909), organized on the initiative of Nechaev and his supporters, but with the vocal presence of their opponents, especially Chelpanov.[117]

The title of these two conferences did not imply a rigorous definition of "pedagogical psychology." Its connotations captured above all the exciting new alliance of psychology and pedagogy, or rather, the promising and hugely seductive, if still hypothetical and controversial, image of the Russian educator as a "teacher-psychologist."[118] Teachers (mostly those working in secondary education) flocked from all parts of the empire with great expectations and much enthusiasm, their numbers reaching several hundred. The audience also included school directors and inspectors, doctors and psychiatrists, and a number of professors and privatdocents from philosophy departments. The papers read at these conferences dealt with a large variety of "child study" topics, but the most important ones revolved around the problem of the scientific identity and mutual relation of psychology and pedagogy, and the place of psychology in education.

---

[115] See Andy Byford, "Psychology at High School in Late Imperial Russia (1881–1917)," *History of Education Quarterly* 48, no. 2 (2008): 265–96, 288–90.

[116] N. E. Rumiantsev, "Shkol'nyi psikhologicheskii kabinet," *RS*, 1908, "Otdel eksperimental'noi pedagogiki," 4:147–73; Dzh. A. Grin, "Eksperimental'noe issledovanie pedagogicheskikh problem v Rossii," *EEP* 3 (1910): 1–9, 3–4.

[117] For background on these conferences, see M. V. Sokolov, "Voprosy psikhologicheskoi teorii na russkikh s"ezdakh po pedagogicheskoi psikhologii," *VP*, 1956, no. 2:8–22; E. A. Budilova, "Polemika o psikhologicheskom eksperimente na Vserossiiskikh s"ezdakh po pedagogicheskoi psikhologii," in Lomov, *Istoriia stanovleniia i razvitiia* (cit. n. 83), 76–86. For some participant responses see: K. T., "Chto dal nam pervyi vserossiiskii s"ezd po pedagogicheskoi psikhologii? (Vpechatleniia uchastnika)," *RS*, 1906, nos. 5–6:255–67; M. K-v, "Vtoroi s"ezd po pedagogicheskoi psikhologii v S.-Peterburge: 1–5 iunia 1909 g.," *RS*, 1909, "Pedagogicheskaia khronika," no. 10:58–73.

[118] This comes across in the plan/advertisement for the first conference. See "Proekt pervogo russkogo s"ezda po pedagogicheskoi psikhologii v g. S.-Peterburge," *VFP* 82 (1906), "Izvestiia i zametki": sec. 2, 130–3.

At the end of the first conference, in 1906, Chelpanov's status as a university academic was sufficiently authoritative to ensure that, while highlighting the importance of psychology for teachers, the conference still called for moderation when it came to the teachers' own involvement in psychological research. Nevertheless, a considerable number of teachers were seduced by the promises of the experimental technology Nechaev was offering them, and many were prompted to purchase his equipment and manuals and to start using them in their schools.[119] However, by the time of the second conference, in 1909, a number of these teachers had already made some overambitious claims about the supposedly "scientific" studies they had carried out on their students, allowing Chelpanov and others to ridicule their work, exposing blatant methodological flaws.[120] By demonstrating the teachers' lack of technical mastery and the gross unreliability of their equipment, Chelpanov continued to argue not only against a premature "democratization" of the scientific pursuit of psychology within education but also against the entire radical "experimental" agenda that Nechaev and his group were pursuing outside the university.[121]

Although in the closing speech at the 1909 conference the chairman—Bekhterev—proclaimed diplomatically that there were no winners in this debate,[122] from roughly this point on everyone became much more cautious about the whole concept of the "teacher-psychologist." Nechaev was still aggressively defending his experimental agenda and the involvement of teachers in its development. But he had to shift the debate to the role of his lab and the kind of research training it provided,[123] conceding that the most that ordinary practicing teachers could realistically produce in terms of psychological research was the work of mere enthusiast amateurs. Still, Nechaev thought their efforts commendable because contributing to the mass popularization of experimental psychology at the grassroots level was likely to help the discipline in the long run. More significantly, however, many teachers themselves now started to lose confidence in the idea that psychology per se was a discipline that they would ever be permitted legitimately to assimilate as *the* most important part of *their* expertise.

THE "NEW PEDAGOGIES": EXPERIMENTAL PEDAGOGY AND PEDOLOGY

Indeed, the teachers' involvement in psychology was always going to be only secondhand and reliant on the authority of academic specialists, who used the education profession's craving for "science" as a platform to settle their own scores.[124] The expertise that *was* directly sold to teachers was a new "science of education" that was supposed be erected "on the ruins" of old, traditional pedagogy. The two principal

---

[119] Liubomudrov, "Psikhologiia i pedagogika" (cit. n. 100), 66.
[120] G. I. Chelpanov, "Zadachi sovremennoi psikhologii," *VFP* 99 (1909): sec. 1, 285–308. Particularly revealing of the kind of experiments teachers were carrying out is A. Feoktistov, "Ob odarennosti," *EEP* 2 (1909): 1–15.
[121] Kostin, "Shkola i eksperimental'naia psikhologiia" (cit. n. 108), 169.
[122] V. M. Bekhterev, "Obshchie itogi deiatel'nosti 2-go s"ezda po pedagogicheskoi psikhologii," *VPKAG*, 1909, no. 4:225–30, 227.
[123] See A. P. Nechaev, "K voprosu o s"ezde po pedagogicheskoi psikhologii i komentariiakh G. I. Chelpanova," and G. I. Chelpanov, " 'Nuzhny li psikhologicheskie laboratorii dlia samostoiatel'nykh issledovanii pri srednikh uchebnykh zavedeniiakh?' Otvet A. P. Nechaevu," *VFP* 100 (1909), "Polemika": sec. 2, 805–14.
[124] See, e.g., A. Krasnovskii, "Eksperimental'noe napravlenie v pedagogike," *Nachal'noe obuchenie*, 1912, "Otdel neoffitsial'nyi": no. 1:6–14; no. 2:47–53; no. 3:75–82; no. 4:106–12.

labels under which this new, properly "scientific," form of pedagogy was promoted, were *experimental pedagogy* and *pedology*.[125]

Both notions were imported from the West.[126] Experimental pedagogy came from Germany and was associated primarily with the work of Wilhelm Lay and Ernst Meumann.[127] Pedology originated in the United States, and its recognized father was Granville Stanley Hall.[128] The underlying conceptions of the two disciplines were by no means identical, but in Russia they were usually promoted in unison, by the same group of people and as part of a unified campaign to turn pedagogy into something resembling science.

Experimental pedagogy was still promoted primarily as an "alliance of the teacher and the psychologist,"[129] and the emphasis was mainly on the fact that the methods used relied strictly on "experimental" (mental testing) techniques. Since it fashioned itself as a form of "pedagogy," the discipline was marketed as expertise relevant to educators. The term "experimental" (apparently acquiring a magical aura in the eyes of some teachers)[130] served as a synonym of the word "scientific" itself (implying the natural-scientific model). Crucially, though, this adjective referred not to experimentation with different teaching techniques but to the experimental study of children's psychology, which meant that "experimental pedagogy" was, ultimately, still a camouflaged expansion of experimental psychology into education.[131] It was argued that *experimental psychology* in its original, orthodox form was rigorously lab based and that it could therefore deal only with relatively simple psychological processes in an

---

[125] For more detail, see Romanov, *Opytno-eksperimental'naia pedagogika* (cit. n. 83), 10–40.

[126] For a helpful survey of the emergence of the "child study" movement in the West, see Marc Depaepe, "Experimental Research in Education, 1890–1940: Historical Processes behind the Development of a Discipline in Western Europe and the United States," *Aspects of Education* 47 (1992): 67–93. Russia was strongly influenced by these developments. Translations of works by Alfred Binet, Édouard Claparède, Gabriel Compayré, Granville Stanley Hall, Victor Henri, William James, Wilhelm Lay, Ernst Meumann, Hugo Münsterberg, Paul Natorp, Wilhelm Preyer, Wilhelm Stern, and James Sully were published regularly and often in successive editions. Although Russian academics usually had access to foreign works in the original, translations were important to the wider population of practicing teachers.

[127] See M. P., "Vozniknovenie i tseli eksperimental'noi pedagogiki," *Pedagogicheskii sbornik*, 1901, no. 6:508–40; and L. S., "Metody, zadachi i nekotorye iz itogov eksperimental'noi pedagogiki," *VV*, 1908, no. 5:46–82.

[128] The term "pedology" itself was invented by Hall's student Oscar Chrismann in 1893. When pedology was first popularized in Russia, its provenance was not always clear, and some presented it as a French invention (see L. E. Obolenskii, "Novye issledovaniia i mysli v oblasti pedagogicheskoi psikhologii," *RS*, 1904, nos. 7–8:116–31; and the preface to I. A. Sikorskii and I. I. Glivenko, *Pedagogicheskaia mysl': Izdanie kollegii Pavla Galagana*, vol. 1 [Kiev, 1904], v–vi). There was also some confusion about how best to spell the term. See L. E. Obolenskii, "Ideal'naia shkola na nachalakh novoi nauki, tak nazyvaemoi 'paidologii,'" *Vospitanie i obuchenie*, 1903, no. 3:97–117; N. Vinogradov, "Pedagogika, kak nauka i kak iskusstvo," *VFP* 113 (1912): sec. 1, 190–210, 202; Ivanovskii, "Kafedra pedagogiki v Sorbonne" (cit. n. 78).

[129] M. P., "Vozniknovenie i tseli eksperimental'noi pedagogiki"; L. S., "Metody, zadachi i nekotorye iz itogov eksperimental'noi pedagogiki." (Both cit. n. 127.)

[130] S. O. Seropolko, "'Dve psikhologii,'" *Pedagogicheskii listok*, 1910, no. 3:193–201, 198. For another denunciation of the "magical" qualities that the term "experiment" had acquired in pedagogy, see M. M. Rubinshtein, "Pedagogika ili pedagogicheskaia psikhologiia?" *VFP* 113 (1912): sec. 2, 418–36, 420 and 435.

[131] Indeed, at the First Conference in Experimental Pedagogy, many speakers kept making mistakes and spoke of experimental psychology when they wanted to say experimental pedagogy. See O. Fel'tsman, "Pervyi vserossiiskii s"ezd po eksperimental'noi pedagogike v S.-Peterburge oot 26 dek. po 31 dek. 1910 g. (Vpechatleniia)," *Psikhoterapiia*, 1911, "Korrespondentsii," no. 2:84–94.

isolated way.[132] By contrast, *experimental pedagogy* supposedly involved research into the full complexity of children's psychology in a real-life environment—the school—and primarily for the practical needs of education. This was sometimes expressed through arguments that while *psychological* experimentation was predominantly "analytical," *pedagogical* experimentation (though still dealing with children's psychology) was essentially "synthetic."[133]

As for pedology, its promotion during the the first two decades of the twentieth century did not always make it clear whether this was a new discipline in its own right, a collaboration of already established sciences and professions, or a very loose movement devoted to the study of children from a variety of methodological perspectives, with different dominant approaches often being associated with different national traditions.[134] Most often, though, pedology was also explicitly associated with the above experimental movement in pedagogy and psychology, the idea being that pedology, too, gave priority to "experimentation" as its main method of investigation.[135]

Significantly, pedology was popularized in Russia not just as a general "science of the child" but precisely as a *new pedagogy* that aimed to put the old "art of education" on positive, scientific foundations.[136] In this context, pedology was regularly associated with Ushinskii's 1860s idea of the need to study man as "an object of education."[137] The similarity in the sound of the terms "pedology" and "pedagogy" sometimes led to confusion, with people occasionally mistakenly calling training courses that were officially dubbed "pedological"—"pedagogical."[138] However, when deployed strategically, the term "pedology" was (not unlike the adjective "experimental") invariably used to symbolize *scientificity*, in important distinction to "pedagogy," pure and simple, which tended to connote educational *practice*.[139] However, for this same reason, the term "pedology" was likely to appear potentially alienating to practicing teachers,[140] which is why "experimental pedagogy" probably had greater currency in the educational realm during the 1910s, as evidenced in the

---

[132] See P. F. Kapterev, *Istoriia russkoi pedagogii* (1915; repr., St. Petersburg, 2004), 533–56.

[133] Ekzempliarskii, "Eksperiment v psikhologii i pedagogike" (cit. n. 107), 54.

[134] A. P. Nechaev, ed., *Pedagogika i pedologiia* (St. Petersburg, 1904); N. E. Rumiantsev, *Pedologiia, ee vozniknovenie, razvitie i otnoshenie k pedagogike* (St. Petersburg, 1910).

[135] Krasnovskii, "Eksperimental'noe napravlenie v pedagogike" (cit. n. 124).

[136] See, e.g., L. E. Obolenskii, "Novosti pedologii," *Vospitanie i obuchenie*, 1900, no. 8:252–65.

[137] Ushinskii's program was at times presented as an early Russian germ of the pedological movement (see Sikorskii and Glivenko, *Pedagogicheskaia mysl'*, vol. 1 [cit. n. 128], vi). However, Ushinskii himself was a popularizer of Western ideas (see I. Skvortsov, "K. Ushinskii kak psikholog," *VV*, 1894, no. 6:176–208), and what is more, although he continued to be revered as the "spiritual leader of pedagogues," his psychological theories were recognized as inevitably out of date (see P. F. Kapterev, "Ushinskii ob obshchestvennykh i antropologicheskikh osnovakh vospitaniia," *RS*, 1895, no. 12:65–73).

[138] "Proekt pervogo russkogo s"ezda po pedagogicheskoi psikhologii" (cit. n. 118), 131.

[139] Traditional pedagogy supposedly relied on the "educator's instinct" (*chut' e vospitatelia*), rather than on rigorous scientific methodology. See A. Krasnovskii, "Ob eksperimental'noi pedagogiki," *Nachal'noe obuchenie*, 1911, no. 1:12–14; Krasnovskii, "Eksperimental'noe napravlenie v pedagogike" (cit. n. 124); L. S., "Metody, zadachi i nekotorye iz itogov eksperimental'noi pedagogiki" (cit. n. 127).

[140] See Ivanovskii, "Kafedra pedagogiki v Sorbonne" (cit. n. 78). Ivanovskii acknowledged that pedology should be the scientific basis of pedagogy but also argued that the "science of the child" needed to be *applied* to teaching practice, and he had doubts about how easily this could be achieved, given the scientists' more theoretical, "cooler" attitude toward children.

renaming of the three sequels to the above two conferences in pedagogical psychology into Conferences in Experimental Pedagogy (in 1910, 1913, and 1916).[141]

Thus, in the early 1900s, both "experimental pedagogy" and "pedology" were marketed aggressively as an overturning of "old" pedagogy and a radical transformation of the teachers' "science." However, in the course of the 1910s, the gap between the scientists who were developing this new pedagogy and the teachers, whose professional expertise it was supposed to become, appeared to be constantly widening.[142]

Of course, the differences of perspective between "experts" and "practitioners" existed from the start—they were noticeable already in the late 1890s–early 1900s,[143] and, as we have seen above, they were at times deployed strategically as a way of enhancing the scientific credibility of the "new pedagogies." However, the experts' strategy at that point was *also* to make the most of *actively engaging* practicing teachers (through both rhetoric and training schemes) in the development of this new science, and to cast the teaching masses, at least provisionally, as their immediate scientific following.

By the 1910s, however, following the appearance of establishments such as the Pedagogical Academy and the Psycho-Neurological Institute, as well as several other research institutes and laboratories devoted to experimental psychology and child study more generally,[144] the experimenters were able to create a sufficiently large corps of young researchers to no longer need the *direct* involvement of the wider teaching masses. On the contrary, as we have seen above in Chelpanov's denunciation of the teachers' dilettante efforts at performing psychological experiments in schools, such involvement of nonspecialists became embarrassingly compromising.

The experimenters still relied greatly on the educational field as the main beneficiary or consumer of their expertise, but from the 1910s onward, teachers were no longer expected to play the role of their scientific disciples. Instead, the education profession was now courted as "clients" who effectively "licensed" a ready-made psychopedagogical technology, produced and owned by an increasingly specialized group of researchers in experimental pedagogy, based in labs, institutes, and sanatoriums.

Throughout the 1910s, these researchers worked predominantly on designing sets of mental and personality tests, packaged as user-friendly "methods" of psychological evaluation, usually under "brand name" labels, such as Binet, Sante-de-Sanctis,

---

[141] For background on these conferences, see M. V. Sokolov, "Kritika metoda testov na russkikh s"ezdakh po eksperimental'noi pedagogike (1910–1916)," *VP*, 1956, no. 6:16–28. The conferences were initiated by the Society for Experimental Pedagogy, which Nechaev and his followers formed soon after the 1909 conference. See N. E. Rumiantsev, "Pervyi vserossiiskii s"ezd po eksperimental'noi pedagogike (S 26 do 31 dekabria 1910 g.)," *EEP* 4 (1911): 29–77.

[142] Fel'tsman, "Pervyi vserossiiskii s"ezd po eksperimental'noi pedagogike" (cit. n. 131); "Pervyi vserossiiskii s"ezd po eksperimental'noi pedagogike," *Narodnoe obrazovanie*, 1911, no. 1:81–89; N. Georgievskii, "Pervyi vserossiiskii s"ezd po eksperimental'noi pedagogike v S.-Peterburge (Kratkii otchet i vpechatleniia)," *Pedagogicheskii vestnik Moskovskogo uchebnogo okruga*, 1912, no. 3:3–26. See also K. Tikhomirov, "Chto dal nam pervyi s"ezd po eksperimental'noi pedagogike?" *EEP* 4 (1911): 78–89, who mentions how the audience at the First Conference in Experimental Pedagogy in 1910 scarcely concealed its animosity toward "Nechaevism" (*nechaevshchina*).

[143] See, e.g., A. P. Nechaev, "K voprosu o vzaimnom otnoshenii pedagogiki i psikhologii," *RS*, 1899, no. 3:45–51.

[144] In Moscow, there was the psychological laboratory affiliated with the Moscow Pedagogical Assembly, run by A. N. Bernshtein, and also the Institute for Child Psychology, run by G. I. Rossolimo. See *Trudy psikhologicheskoi laboratorii pri Moskovskom pedagogicheskom sobranii* (Moscow, 1909); and "Institut detskoi psikhologii i nevrologii pri Pedagogicheskikh Kursakh v Moskve," *Voprosy pedagogicheskoi patologii*, 1912, no. 2:157–60.

and Rossolimo.[145] Teachers were then encouraged to use these tests in various forms of student evaluation—in entrance exams, in compiling student reports, or when deciding whether to relegate a student to a special school. What is important to note here is that teachers were being empowered not with psycho-pedagogical "expertise" (which remained in the hands of the experimental psychologists) but simply with forms of "black-boxed" psychometric *instruments*.

The consequence of this was that experimental pedagogy and pedology became increasingly dissociated from pedagogy, and the "experimenters" found it more and more difficult to persuade teachers that their expertise could *in and of itself* make pedagogy scientific.[146] Teachers started expressing ever more dissatisfaction and disappointment with these new "sciences of education," and their complaints ranged from lamentations that alien professions were monopolizing what was meant to be their domain,[147] to reproaches that the new "experts" were consistently failing to answer the teachers' most urgent practical needs.[148] This put the "experimenters" on the defensive, and fearing that they were going to alienate the teachers irretrievably, they were now at pains to stress that their science did not, after all, seek to refute and replace pedagogy itself but simply created a suitable scientific foundation on which pedagogy should be developed.

This withdrawal allowed the experimenters' academic rivals, based mostly at the universities' philosophy departments, to start arguing that "experimentation" could indeed at best create only a value-neutral "empirical base" for pedagogy and that it was *philosophy* (as a form of social ethics) that was best positioned to define the *ideals* of education. In their scheme, while experimental pedagogy and pedology could perhaps suggest particular "means" for achieving certain educational goals, only philosophy and ethics could specify what these goals actually were.[149] Moreover, by arguing that pedagogy itself was by definition "normative" (in contrast to pedology, which, as an empirical science, was purely "descriptive"), they claimed that it was not the pedologists but the *philosophers* who should serve as the teachers' ultimate guides.[150]

The philosophers' argument appealed to those teachers who thought that philosophical reflection over educational ideals (in contrast to the prohibitively technical science that experimental pedagogy was rapidly becoming) was something to which

---

[145] A. M. Shubert, "Opredelenie umstvennoi otstalosti detei po sistemam Bine i Simona: Viegandta, Norsvordsa-Goddarda, Pitstssoli, Rossolimo, Sanktisa i Tsigena," in *Defektivye deti i shkola*, ed. V. P. Kashchenko (Moscow, 1912), 12–58.

[146] On the crisis of experimental pedagogy and attempts to reestablish the autonomy of pedagogy away from it, see Ekzempliarskii, "Eksperiment v psikhologii i pedagogike" (cit. n. 107); Rubinshtein, "Pedagogika ili pedagogicheskaia psikhologiia?" (cit. n. 130); V. Volyntsevich, "S"ezd po eksperimental'noi pedagogike," *Pedagogicheskii sbornik*, 1912, no. 9:240–63.

[147] See V. Komarnitskii, "Sovremennyi pedagogicheskii konflikt," *Pedagogicheskii kruzhok studentov pri Imperatorskim Varshavskom universitete*, 1911, 32–7.

[148] Krasnovskii, "Ob eksperimental'noi pedagogiki" (cit. n. 139); Krasnovskii, "Eksperimental'noe napravlenie v pedagogike" (cit. n. 124). For a highly technical account of "applications" of psychological theories to education, see A. O. Makovel'skii, "Psikhologiia viurtsburgskoi shkoly i ee pedagogicheskie vyvody," *Voprosy pedagogiki*, 1913, no. 2:56–71.

[149] Such arguments usually referred to works by Münsterberg and Natorp, namely, G. Miunsterberg, *Psikhologiia i uchitel'* (Moscow, 1910); P. Natorp, *Filosofiia kak osnova pedagogiki* (Moscow, 1910); Natorp, *Sotsial'naia pedagogika, teoriia vospitaniia voli na osnove obshchnosti* (St. Petersburg, 1911).

[150] Vinogradov, "Pedagogika, kak nauka i kak iskusstvo" (cit. n. 128), 202; V. M. Ekzempliarskii, "K voprosu o putiakh razvitiia nauchnoi pedagogiki," *Shkola i zhizn'*, 15 Sept. 1914, no. 37:961–7.

the educators themselves could perhaps contribute more readily, alongside the philosophers. The philosophers were also aided in their counterattack by the teachers' complaints about the chaos of conflicting pedagogical agendas that characterized the Russian education scene at this time.[151] The teachers' uncertainty about which authority to turn to led to calls to reintroduce a system of dogmatic educational norms that would curb the fashion for uncontrolled, and therefore potentially dangerous, experimentation and innovation.

And yet, given the teachers' increasing frustration with *everyone* who, from higher, academic vantage points, claimed authority over their own professional territory, *both* the experimental psychologists *and* the philosophers had, in fact, to accord much more respect *to the role of the teachers themselves*, or more precisely, to the educators' pedagogical intuition, individual creativity, and knowledge of daily school realities.[152] Whereas in the early 1900s, the new pedagogies had, as we have seen, built much of their scientificity on stark denunciations of the subjective, arbitrary and grossly "unscientific" judgments of educational practitioners, their proponents now argued that they merely defined certain objective *constraints* (to do with the "laws" of child psychological development), *within the boundaries of which* teachers could then give free reign to their own pedagogical instincts.

Thus, by 1917, a certain form of jurisdictional delineation was emerging between an (empirical) "science," a (general) "philosophy," and a (practical) "art" of education.[153] Since none of the three main parties involved seemed to have the upper hand, the compromise solution appeared to be a kind of provisional peace agreement—a carving up of pedagogy between educational philosophers (whose job it was to define ultimate educational goals in the context of overall social and moral demands), experimental psychologists (who provided facts about child mental development, but devoid of value judgments[154] and quite thin on practical recommendations), and practicing teachers (whose role remained in the domain of application—namely to use their experience and pedagogical intuition to realize education on the ground, bearing in mind both the educational values stipulated by the philosophers and the demands of child development analyzed and systematized by the psychologists).

However, this division of labor remained very tense and unstable, because all parties were still in jurisdictional conflict—first, over the question of what it was that ultimately defined pedagogy as a science, and second (and no less important), over the question of which principles were the ones that ought to be used, in the last instance, to legitimize the teachers' concrete choice of educational practices or to guide

---

[151] P. Radosavlevich, "Eksperimental'nye issledovaniia psikhicheskikh protsessov v matematike, kak nauke i kak uchebnom predmete," *Obnovlenie shkoly* 5 (1911–12): 28–33; V. A. Fliaksberger, "Ocherki po istorii i teorii detskoi psikhologii," *Voprosy pedagogiki*, 1912, no. 1:113–37.

[152] For the revival of the teachers' role at the expense of the scientists, see N. Vinogradov's speech at the opening of the Moscow Shelaputin Institute on February 2, 1912 (Vinogradov, "Pedagogika, kak nauka i kak iskusstvo" [cit. n. 128], 190). For other introductions of "creativity" and "art" into pedagogy, see K. Zhitomirskii, "Pedagogicheskii modernizm," *Pedagogicheskii Sbornik*, 1909, "Chast' neoffitsial'naia," no. 2:97–112; no. 3:193–220; and the review by A. Ostrogorskii of Vl. F. Chizh, *Pedagogiia, kak iskusstvo i kak nauka* (Iur'ev, 1912), in "Kritika i bibliografiia," *Pedagogicheskii sbornik*, 1913, no. 8:142–3.

[153] See, e.g., Fliaksberger, "Ocherki po istorii i teorii detskoi psikhologii" (cit. n. 151), 116.

[154] See review of Chelpanov's and Vinogradov's papers by I. M. Solov'ev in "Kritika i bibliografiia," *VV*, 1909, no. 8:58–68.

the government's future educational reforms.[155] Both the experimenters and the philosophers were still claiming to be the ones who should ultimately be shaping pedagogy. Educators remained divided in their allegiance, with some supporting the pre-eminence of the philosophers and others highlighting the decisive role of the positive sciences.[156] Although many, of course, saw the virtues of both camps and hoped to see them forming a united front in furthering the science of education, the majority remained frustrated by the continuing uncertainty over pedagogy's legitimacy, by the lack of connection between the new pedagogies and the daily needs of education, and finally, by the question of whether educators could ever become the masters of their own professional knowledge base.

EPILOGUE

The 1917 revolution and the 1920s overhaul of the education system under the Bolsheviks could not but disrupt this precarious "peace agreement," and it did so, at first, largely in favor of the "scientists." The latter's materialist and rationalist ideology coalesced with the new regime's ambitions of radical transformation and social modernization through science, technology, and reeducation.[157] Consequently, the Bolsheviks went on to institutionalize pedology as a Soviet super science, the ultimate objective of which was to direct the social engineering of future—physically, psychologically, and ideologically "better"—generations of citizens.[158] This meant that for most of the 1920s, those interested in developing the scientific study of children and education received carte blanche from the regime to realize their ambitions, leading to a remarkable, and initially exceptionally free, institutional expansion of pedology as the umbrella discipline for virtually every approach to Soviet "child study," from the pedagogical and the psychological to the biomedical and the ethnographic.

---

[155] Indeed, in 1917, works that sought to promote the idea of *scientific* pedagogy were still reluctant to refer to "pedagogy" pure and simple, but qualified it by calling it "experimental" or "sociological" and persisted in dealing primarily with psychology rather than with teaching. See Ekzempliarskii, "Eksperiment v psikhologii i pedagogike" (cit. n. 107), 46.

[156] On the promotion of experimental pedagogy by the group around the journal *Obnovlenie shkoly*, led by A. Zachinaev and Ts. Baltalon, see A. Zachinaev, *Zlatotsvet: Kniga dlia chteniia i dlia spisyvaniia*, vol. 1 (St. Petersburg, 1911); "Ot redaktsii," *Obnovlenie shkoly*, 1911, no. 1:3–5; Ts. Baltalon, "Ob usloviiakh obnovleniia nachal'noi shkoly," *Obnovlenie shkoly*, 1911, no. 1:6–12. Pedology was also regularly perceived as an ally of the "free education" movement. See V. Byrchenko, "Sovremennaia pedagogika," *Svobodnoe vospitanie*, 1915–16, nos. 10–11:1–18; no. 12:49–66. On links between Nechaev as representative of experimental pedagogy and S. T. Shatskii as a figurehead of free education, see Romanov, *Opytno-eksperimental'naia pedagogika* (cit. n. 83). On the influence of the experimental movement in the provinces, see, e.g., *Otchet o deiatel'nosti Revel'skogo pedagogicheskogo obshchestva za 1908–9* (Revel', 1909).

[157] This, of course, was not just a Russian/Soviet phenomenon but a global one. For the international context that also includes examples from other human sciences in Russia, see, e.g., Gregg Eghigian, Andreas Killen, and Christine Leuenberger (eds.), *The Self as Project: Politics and the Human Sciences*, *Osiris* 22 (2007).

[158] On Soviet pedology, and what follows, see Nikolai Kurek, *Istoriia likvidatsii pedologii i psikhotekhniki* (St. Petersburg, 2004); E. Thomas Ewing, "Restoring Teachers to Their Rights: Soviet Education and the 1936 Denunciation of Pedology," *Hist. Educ. Quart.* 41 (2001): 471–93; Alexandre Etkind, "L'essor et l'échec du mouvement paidologique: De la psychanalyse au 'nouvel homme de masse,'" *Cahiers du Monde Russe et Soviétique* 23 (1992): 387–418; V. F. Baranov, "Pedologicheskaia sluzhba v sovetskoi shkole 20-30-kh gg.," *VP*, 1991, no. 4:100–12; Joravsky, *Russian Psychology* (cit. n. 111), 345–54; Jaan Valsiner, *Developmental Psychology in the Soviet Union* (Bloomington, Ind., 1988); Raymond Bauer, *The New Man in Soviet Psychology* (Cambridge, Mass., 1962).

The most famous product of this movement is arguably the theoretical work of the cognitive psychologist Lev Vygotskii, whose "cultural-historical" approach to developmental and educational psychology was particularly influential in Soviet pedology toward the end of the 1920s.[159]

Ironically, at precisely this time, the jurisdictional conflict between the "scientists," the "philosophers," and the "teachers" reemerged, although with increasingly ominous overtones. The philosophers in this case were *the ideologues of the regime*, who lashed out against the pedologists' increasing institutional autonomy and reliance on specialist scientific authority. In counterattack, they denounced the pedologists' uncritical openness to Western "bourgeois science," their ideologically ambiguous conclusions about Soviet educational achievements, and their damning diagnoses of the condition of the Soviet child population. Critics of pedology also specifically invoked the disassociation of the interests of the scientists from those of practicing teachers and condemned the pedologists' alleged monopoly over directing the education system.

However, this time round, the crisis failed to lead to a jurisdictional truce or negotiated settlement between different forms of expertise. Although the Soviet government made some attempts in the early 1930s to assert control over pedology and bring it in line with the ideological principles and practical demands of Stalinist socialism building (e.g., by making pedologists renounce much of their biosocial determinism and develop instead the theory of human "plasticity"), this proved unworkable in practice. Instead, in classical Stalinist fashion, pedology's previously extraordinarily bright fortunes were totally reversed by the now notorious 1936 "Party Decree on Pedological Distortions in the System of Narkompros," which condemned the discipline as an ideologically flawed and dangerous "pseudoscience." Immediately after, pedology's network of research institutes, training establishments, labs, periodical publications, and school-affiliated staff was disassembled in its entirety, and the discipline was purged from the Soviet system of sciences, never to be invoked again (until at least the 1980s–1990s), except with de rigueur "odious" connotations.

Instead, the 1930s Stalinist school reformers made much of "reinstating the teacher" as the key figure in the new education system. This, of course, implied making teachers ultimately responsible for anything that went wrong and therefore turned them into objects of greater state control, although this still meant that the teacher was finally formally cast as *the* authority in the Soviet classroom. Yet the legitimacy of this authority, and the consequent sense of intellectual and professional self-worth, no longer depended, in the final instance, on the idea of a *scientifically informed* pedagogical expertise but drew most of its strength from the formal backing of an all-powerful state. Of course, the extent to which, in the USSR, the symbolic power of science itself was distinguishable from the symbolic power of the state, is debatable, but in the case of the education profession, the teachers' sense of professionalism no longer seemed to depend on the latter's grounding in any strong notion of "science."

As we have seen in this paper, prior to this, Russia's teachers had been drawn, for almost half a century, into a tortuous, contradictory, and often seemingly vain

---

[159] The relevance of Vygotskii's ideas persists to this day. On him and his legacy, see, e.g., Loren H. Graham, *Science, Philosophy, and Human Behaviour in the Soviet Union* (New York, 1987), 168–76; René van der Veer and Jaan Valsiner, *Understanding Vygotsky: A Quest for Synthesis* (Oxford, 1991); Alex Kozulin, *Vygotsky's Psychology: A Biography of Ideas* (New York, 1990).

campaign to rise above their unflattering role of "professionals without a science." Throughout this period, at stake was the radical reshaping of pedagogy as the educators' academic knowledge base and professional specialism by rooting the legitimacy of the teachers' training institutions and working practices in the authority of science—primarily the emergent discipline of child and developmental psychology.

However, the teachers' aspirations to overcome their only semilegitimate membership of Russia's intellectual and professional elite at all times overlapped and clashed with a very different set of interests and agendas of the academic psychologists, who were themselves engaged in bitter squabbles over the legitimate identity of psychology—squabbles that, as we have seen, also involved redefinitions of the institutional model of science itself. Education proved a major arena of the psychologists' intradisciplinary conflict, and teachers were at all times drawn into this fight, not just as a passive audience but also as vitally interested protagonists.

Despite a variety of efforts in these conflicts and negotiations to bridge the gap between "science" and "education," the hierarchical relation embedded in their distinction seemed always only to be reproduced on a different level, with the new "scientists of education" persistently emerging above and beyond mere "practicing educators," and with teachers always failing to acquire proper ownership of scientific prestige in what was supposed to be the discipline that supported the dignity and independence of their profession.

In the end, in Stalinist Russia, the Gordian knot was severed, and the reproduction of this hierarchy seemed finally, if brutally, to have been halted. However, this was possible only through state violence and at the expense of the (relatively) self-sufficient legitimating power that the institution of "science" (in the broad and variable sense of nauka) exercised so pervasively in Russian professional and intellectual culture throughout the late nineteenth and early twentieth century—a role, however, which, in the context of Stalinism, seemed to have become rather redundant, at least when it came to legitimizing the teaching profession as one of the prominent segments of the new Soviet intelligentsia.

Organizational Culture and Professional Identities in the Soviet Nuclear Power Industry

By Sonja D. Schmid*

ABSTRACT

This essay describes the evolution and interaction of powerful constituencies within the Soviet nuclear energy community since the late 1950s. It elucidates the historical origins of the alleged distinction between nuclear specialists (*atomshchiki*) and power engineers (*energetiki*) and explores why there was not more disagreement with and resistance to state authorities. The emerging civilian nuclear industry faced the dual legacy of the atomic bomb project and the state electrification program, a legacy that informed attempts to regulate a complex, dual-use technology, as well as the values and routines of specialists involved with the design and operation of nuclear power reactors. The essay argues that rather than cynicism, a deep commitment to scientific rationality and an often personal conviction of the necessity of nuclear energy permeated this heterogeneous community of experts. The secrecy imposed on their work only reinforced these experts' loyalty to a state that embodied their own technocratic visions.

INTRODUCTION

The 1986 Chernobyl disaster exploded the logic of trial and error. Immediate fatalities and the specter of many more to come exceeded the capacity of explaining progress by overcoming difficulties.[1] In the postcrisis search for a scapegoat, the chief reactor designers introduced a car analogy to illustrate what they portrayed as clear proof of operator error. When a reckless driver hits a tree and ruins the car, so the story goes,

---

*Center for International Security and Cooperation, Encina Hall, Stanford University, Stanford, CA 94305-6065; sschmid@stanford.edu.

I want to thank the three editors of this special issue, as well as Peter Holquist, Asif Siddiqi, and Rebecca Slayton for helpful comments and suggestions. The Center for International Security and Cooperation and the Program in Science, Technology, and Society at Stanford University generously funded a research trip to Moscow and provided a wonderful research environment. Between 2002 and 2006, I interviewed more than two dozen senior nuclear specialists mostly in Moscow and Obninsk. To protect the anonymity of my interviewees, I provide their names only upon their explicit request; otherwise, I identify the interviews only by number, date, and location. All translations are mine unless otherwise noted.

[1] Susanne Schattenberg has characterized this logic as dialectic. Susanne Schattenberg, *Stalins Ingenieure: Lebenswelten Zwischen Technik und Terror in den 1930er Jahren* (Munich, 2002). See also Anonymous, "Gor'kii urok [Bitter lesson]," *Atomnaia Energiia* 60, no. 6 (1986): 370–1.

the car's designer cannot possibly be blamed for the accident.[2] These stories were challenged with counter stories, where the reason for the accident turns out to be the car's malfunctioning brake—an obvious design flaw. Such anecdotes did more than assign blame, however. They introduced a fundamental distinction between those engineers who designed and built nuclear reactors on the one hand, and those who operated these reactors in commercial power plants, on the other. The stories also raised the issue of who could lay claim to being a true, that is, capable, diligent, and responsible engineer.

This essay explores the evolution and interaction of different powerful constituencies within the nuclear energy community since the late 1950s and elucidates the historical origins of the alleged distinction between nuclear specialists and power engineers. The Chernobyl disaster is important for this history because in its aftermath the distinction between reactor designers and nuclear plant operators became household terms. To distinguish highly qualified, diligent, and disciplined *atomshchiki* (reactor designers) from the civilian, and therefore supposedly less diligent, somewhat less qualified, and less disciplined nonnuclear *energetiki* (reactor operators and other engineers at nuclear power plants) was an easy way to attribute blame. The technical specialists who happened to work for the civilian Ministry of Energy and Electrification (Minenergo) at the time of the accident were depicted as "mere operators" by virtue of their institutional affiliation, rather than their actual training, their experience with nuclear reactors, or their expertise in nuclear physics or engineering. Nikolai Krementsov has characterized this emphasis on institutional structures, rather than on individual biographies, as a consequence of the Second World War, which profoundly affected "the structural and functional dynamics of the Stalinist science system."[3] At the core of these changes was the emergence of interest groups within the Soviet scientific community (and, by extension, within the technical intelligentsia) who knew how to work the system and who knew how to use shifts in political authority and industrial policy to their avail. These changes were retained throughout, and reinforced by, the cold war.

The essay, then, explores how the development of professional identities was intertwined with organizational arrangements within the nascent Soviet nuclear power industry. It describes the collaboration of diverse communities of technical specialists in the civilian nuclear power industry and argues that this industry produced its own community of cadres, who shared ideas, goals, training, success, and setbacks, and who were part of a set of organizations continuously modified to meet the specific needs of their particular branch of industry. The position of nuclear power reactors at the boundary between commercial electricity production and classified military processes created specific problems of attributing expertise, sharing experience, and transferring knowledge. The cooperation among nuclear energy cadres proved rhetorically and organizationally flexible, and involved constant boundary work, most prominently between atomshchiki (designers) and energetiki (operators). I use *atomshchiki* and *energetiki* to refer to the specialists working for the secret Ministry of Medium Machine Building (usually abbreviated as Minsredmash, or simply Sredmash),

---

[2] See, e.g., Anatolii P. Aleksandrov, "Izmeniat', chto izmenit' eshche vozmozhno . . . ," *Ogonek* 35 (Aug. 1990): 6–10.
[3] Nikolai L. Krementsov, *Stalinist Science* (Princeton, N.J., 1997), 288–9. See also Ethan Pollock, *Stalin and the Soviet Science Wars* (Princeton, N.J., 2006).

and for the Ministry of Energy and Electrification (Minenergo), respectively.[4] My aim in this essay is to contribute to our understanding of the heterogeneous internal structure of Soviet nuclear energy cadres and to explore why there had not been more disagreement with and resistance to the state authorities. By examining in greater detail how the institutional history of Soviet nuclear energy was linked to the status of technical expertise in the Soviet state, I hope to shed some light on this substratum of the technical intelligentsia.[5]

The essay is divided in three parts. First, it situates the Soviet civilian nuclear industry in a political system sympathetic to technocratic visions. It describes the roots of the civilian nuclear industry in the atomic bomb project on the one hand, and in the Soviet electrification program on the other, and discusses the implications of this dual legacy for the relationship between the civilian nuclear specialists and the Soviet state. Over time, the cadres in the civilian nuclear industry developed a specific combination of skills and expertise. In contrast to the nuclear weapons program, the nuclear power industry was part of the Soviet industrialization effort and was subject to enormous economic pressure. Minenergo in general was not as shielded from state interference as the nuclear weapons community, and although the power industry enjoyed the patronage of high political authorities, it could not provide its employees with privileges available to technical specialists who worked for Sredmash.[6] Nevertheless, the construction, management, and operation of nuclear power plants (in addition to conventional power plants) increased the political clout of Minenergo and gave its minister unusual leeway in developing and pursuing his own agenda.[7]

Second, the essay defines the distribution of tasks among the civilian nuclear energy specialists. Against the backdrop of a very abbreviated institutional history, the next section identifies their tasks, responsibilities, and professional identities, as they started cooperating in the construction and operation of nuclear power plants. Building on the previous section, this part shows how values and routines historically embedded in the military nuclear program and in the task of electrifying the entire country link up with the emerging bureaucratic structures in the civilian nuclear industry.

And finally, the essay tries to answer the question of why it took a disaster of such scope to generate dissent among the technical experts working in the civilian nuclear power sector. If Chernobyl was, as some observers have claimed, a disaster waiting to happen, why did these highly qualified specialists not challenge the political leadership much earlier?[8] While I do not imply that Chernobyl was somehow the logical endpoint of a disastrous Soviet nuclear energy policy, the post-Chernobyl debate al-

---

[4] These concepts are fundamentally actors' categories that involve a variety of associations; it is therefore important to note that as analytical categories, I use them only to refer to the institutional affiliations of these specialists.

[5] See Kendall E. Bailes, *Technology and Society under Lenin and Stalin: Origins of the Soviet Technical Intelligentsia, 1917–1941* (Princeton, N.J., 1978), 6–7, 15.

[6] David Holloway, "Physics, the State, and Civil Society in the Soviet Union," *Historical Studies in the Physical and Biological Sciences* 30 (1999): 173–93.

[7] Minister Petr Neporozhnii maintained friendships with the long-term chairman of the Council of Ministers, Aleksei Kosygin, and with the State Planning Commission (Gosplan) functionary in charge of the country's energy complex, Veniamin Dymshits. Petr S. Neporozhnii, *Energetika strany glazami ministra: Dnevniki, 1935–1985 gg.* (Moscow, 2000).

[8] See, e.g., Paul R. Josephson, "The Historical Roots of the Chernobyl Disaster," *Soviet Union/Union Soviétique* 13, no. 3 (1986): 275–99; Josephson, "Atomic Energy and 'Atomic Culture' in the USSR: The Ideological Roots of Economic and Safety Problems Facing the Nuclear Power Industry after Chernobyl," in *Soviet Social Problems*, ed. T. Anthony Jones, David Powell, and Walter Connor

lows unique insights into earlier, and more general, attitudes in the nuclear energy community. The intensity that still characterizes debates about Chernobyl suggests that rather than cynicism, a deep commitment and often personal conviction of the necessity and rationality of nuclear energy permeated this heterogeneous community of experts.[9] Furthermore, this essay argues that rather than seeking privileges, bonuses, and social mobility, many of the Soviet engineers involved in the nuclear power industry perceived their skills as grounds for pride, their work as contributing to the greater good, and their role in society as part of human progress in general.[10]

NUCLEAR EXPERTS AND THE SOVIET STATE

In recent years, studies on Soviet science have substantially revised earlier assumptions on the unfavorable character of the relationship between science and the Soviet state. It has been argued that, by and large, this relationship was rather close and symbiotic and that scientists and particularly engineers were more often than not willing participants in the Communist project.[11] While the Communist Party did attempt to ensure that science would serve the state's economic and political objectives, by the late 1940s, scientists, for their part, had learned to "play the system," and emerging interest groups within the scientific community increasingly competed against each other as they pursued their own intellectual agendas.[12]

The relationship between political power and technical intelligentsia changed fundamentally when nuclear weapons appeared on the scene. As David Holloway has put it, the atomic bomb saved Soviet physics from Stalin's persecution.[13] Scientific autonomy and the claim to definitional authority over what was considered scientific truth had been a concern for the Soviet regime from the outset, as its legitimacy rested "on the assertion that it was guided by a scientific understanding of the laws of societal development."[14] In combination with the victory in the World War, the explosion of a nuclear device in 1949 allowed Soviet physicists to lay unprecedented claim to ideological authority: their awesome weapons enabled them to dodge the control of political ideologues and state functionaries over their work.[15] Nuclear physicists and engineers played a determining role in the debate over the separation of scientific facts from political authority, a debate that became even more virulent in the ensuing years

---

(Boulder, Colo., 1991), 55–77; Josephson, *Red Atom: Russia's Nuclear Power Program from Stalin to Today* (New York, 1999); David R. Marples, *Chernobyl and Nuclear Power in the USSR* (New York, 1986); Marples, *The Social Impact of the Chernobyl Disaster* (London, 1988).

[9] Recent examples include K. E. Baskin, L. P. Drach, and A. I. Glushchenko, *Eshche mozhno spasti!* (Moscow, 2006); and Anatolii S. Diatlov, *Chernobyl': Kak eto bylo* (Moscow, 2003). For a documentation of cynicism in the Soviet Youth League (Komsomol), see Steven L. Solnick, *Stealing the State: Control and Collapse in Soviet Institutions* (Cambridge, Mass., 1998).

[10] This attitude reflects Schattenberg's findings of engineers in the 1930s more than Fitzpatrick's analysis of the 1920s and early 1930s. Schattenberg, *Stalins Ingenieure* (cit. n. 1); Sheila Fitzpatrick, *Education and Social Mobility in the Soviet Union, 1921–1934* (Cambridge, UK, 1979).

[11] Krementsov, *Stalinist Science* (cit. n. 3); Holloway, "Physics, the State, and Civil Society" (cit. n. 6); Konstantin Ivanov, "Science after Stalin: Forging a New Image of Soviet Science," *Science in Context* 15 (2002): 317–38; Pollock, *Stalin and the Soviet Science Wars* (cit. n. 3), 4.

[12] Krementsov, *Stalinist Science* (cit. n. 3), 105, 287–9.

[13] David Holloway, *Stalin and the Bomb: The Soviet Union and Atomic Energy, 1939–1956* (New Haven, Conn., 1994), 206–13.

[14] Holloway, "Physics, the State, and Civil Society" (cit. n. 6), 175. See also Ivanov, "Science after Stalin" (cit. n. 11).

[15] Krementsov, *Stalinist Science* (cit. n. 3), 100.

and culminated in the late 1950s.[16] These struggles ultimately increased the Academy of Sciences' autonomy vis-à-vis Communist Party philosophers and eroded, according to Holloway, "the legitimacy of the Soviet regime by weakening its claim to rule on the basis of a scientific understanding of society."[17] But technocratic ideas were not necessarily disconnected from liberal political views.[18] In fact, technocratic visions, understood as practical, applied versions of scientific development under the direction of scientific experts, were widely shared by Soviet intellectuals in general and by the nuclear energy community in particular. The nuclear weapons umbrella sanctioned a certain degree of opposition and autonomy for these scientists, but at the same time they were part of Stalin's technocratic intelligentsia and, as such, consenting participants in the operations of the Soviet state.

Research on radioactive materials in Russia had started in the early 1910s and was institutionalized with the foundation of the Radium Institute in Petrograd in 1922. That same year, Vladimir Vernadskii, a geochemist and the main promoter of this early research, predicted "a great revolution in the life of humankind ... when man will get atomic energy in his hands."[19] In 1934, Sergei Vavilov created the Physics Institute of the Academy of Sciences (FIAN) in Moscow, four years before the discovery of fission, in 1938. In 1940, the Academy of Sciences set up a Commission on the Uranium Problem, which conducted and coordinated research in the area of isotope separation and nuclear fission, and started systematic exploration of uranium deposits. But during the war, most Soviet nuclear scientists abandoned their research and joined the war effort.[20] The sudden and uniform absence of open publications on nuclear fission in the U.S. scholarly press alerted physicist Georgii Flerov, who warned Stalin in a letter that the Americans were working to build a nuclear weapon.[21] When intelligence material on nuclear weapons research abroad confirmed Flerov's suspicion, Stalin restarted nuclear research, and in early 1943, the State Defense Committee (GKO) initiated a nuclear project with the physicist Igor Kurchatov as its scientific director. A laboratory (Laboratory No. 2) was set up for Kurchatov, nominally under the Academy of Sciences.[22]

What had started as a modest undertaking during the war gradually intensified after the United States tested an atomic bomb on July 16, 1945, at Alamogordo in New Mexico.[23] Following the bombing of Hiroshima, the Soviet State Defense Committee appointed a Special Committee on the Atomic Bomb to launch a fully funded crash program. Lavrentii Beria, the chief of Stalin's secret police, headed the committee, which consisted of specialists from science and industry, as well as representatives

---

[16] Ivanov, "Science after Stalin" (cit. n. 11); and Holloway, "Physics, the State, and Civil Society" (cit. n. 6).

[17] Holloway, "Physics, the State, and Civil Society" (cit. n. 6), 191–2; Ivanov, "Science after Stalin" (cit. n. 11).

[18] Holloway's interpretation is a matter of some dispute; for a brief overview, see, e.g., Slava Gerovitch, "'Mathematical Machines' of the Cold War: Soviet Computing, American Cybernetics, and Ideological Disputes in the Early 1950s," *Social Studies of Science* 31 (2001): 253–87.

[19] Quoted after Holloway, *Stalin and the Bomb* (cit. n. 13), 32. One of my interviewees read an extended version of this quote (in Russian) to me as an introduction to our conversation about the Soviet civilian nuclear industry (Interview #21, July 2004, Moscow).

[20] Holloway, *Stalin and the Bomb* (cit. n. 13), 75.

[21] Ibid., 78.

[22] Kurchatov's laboratory functioned quite autonomously from the academy and received its assignments directly from the Council of People's Commissars.

[23] Holloway, *Stalin and the Bomb* (cit. n. 13), 116.

from party, state, and the secret police, but no military men.²⁴ The Special Committee was assisted by other newly created organizations in the effort to turn a laboratory project into a full-fledged industry. The Soviet Union's first experimental nuclear reactor, F-1, reached criticality on December 25, 1946, in Kurchatov's laboratory in Moscow, and the first Soviet production reactor ("Annushka") was started up in June 1948 at Cheliabinsk-40. On August 29, 1949, the Soviets tested a plutonium bomb.

But even before the successful detonation, Kurchatov had turned his ambitions to civilian applications. His efforts to develop a civilian nuclear industry significantly contributed to the scope and scale of the Soviet nuclear energy program.²⁵ The first official reference to peaceful applications of nuclear energy is documented in 1946, when Kurchatov ordered testing of the possibility of using graphite-water reactors for power production.²⁶ In 1948, several design proposals were under discussion,²⁷ and on May 16, 1950, Stalin signed a government decree stipulating the construction (by 1951) of an experimental power plant with a power output of 500 kW, consisting of three reactors operating on enriched uranium.²⁸ The power plant, which was referred to as installation V-10 (*ustanovka* V-10),²⁹ was built at the Laboratory V at Obninsk (the later Institute of Physics and Power Engineering, FEI). Of the three reactors, only one was eventually realized, the graphite-water reactor (AM, *atom mirnyi*, "peaceful atom"³⁰). The plant was started up on June 27, 1954, and became known as "the World's First Nuclear Power Plant."³¹

All research related to atomic energy, military and civilian, was carried out under the aegis of the First Chief Administration (Pervoe Glavnoe Upravlenie, PGU), which was accountable personally to Beria.³² Soviet nuclear physicists used the momentum and success of the nuclear weapons project, as well as their status gained from it, to initiate the development of a civilian nuclear industry under the umbrella of the

---

²⁴ The committee's members were Georgii Malenkov, one of the Central Committee secretaries, Nikolai Voznesenskii, head of Gosplan, three industrial managers (Vannikov, Zaveniagin, and Pervukhin), the scientists Kurchatov and Kapitsa, and General Makhnev from the People's Commissariat of Internal Affairs (NKVD) (Holloway, *Stalin and the Bomb* [cit. n. 13], 135). On Beria, see Amy Knight, *Beria: Stalin's First Lieutenant* (Princeton, N.J., 1993).

²⁵ Holloway, *Stalin and the Bomb* (cit. n. 13); Raisa V. Kuznetsova, ed., *Kurchatov v zhizni: Pis'ma, dokumenty, vospominaniia (Iz lichnogo arkhiva)* (Moscow, 2002); Viktor A. Sidorenko, ed., *Istoriia atomnoi energetiki Sovetskogo Soiuza i Rossii*, vol. 1 (Moscow, 2001).

²⁶ Viktor A. Sidorenko, "Vvedenie k 1-mu vypusku," in Sidorenko, *Istoriia atomnoi energetiki Sovetskogo Soiuza i Rossii* (cit. n. 25), 1:5–15, 5; Vladimir G. Asmolov, Andrei Iu. Gagarinskii, Viktor A. Sidorenko, and Iurii F. Chernilin, *Atomnaia energetika: Otsenki proshlogo, realii nastoiashchego, ozhidaniia budushchego* (Moscow, 2004), 8.

²⁷ A graphite-moderated, helium-cooled reactor proposed by the Institute for Physical Problems (Moscow); a high-temperature vessel reactor with beryllium oxide as moderator and helium coolant (Laboratory V, Obninsk); a graphite-water reactor (Kurchatov's Laboratory No. 2), and a breeder reactor with sodium coolant (Laboratory V, Obninsk). Tellingly, the option proposed by Kurchatov's laboratory was chosen for Obninsk. Sidorenko, "Vvedenie" (cit. n. 26), 5; Asmolov et al., *Atomnaia energetika* (cit. n. 26), 8.

²⁸ Asmolov et al., *Atomnaia energetika* (cit. n. 26), 12; Sidorenko, "Vvedenie" (cit. n. 26), 5; Lev A. Kochetkov, ed., *Ot Pervoi v Mire AES k atomnoi energetike XXI veka: Sbornik tezisov, dokladov i soobshchenii* (Proceedings of the Tenth Annual Conference, 28 June–2 July 1999, Obninsk) (Obninsk, 1999).

²⁹ Kochetkov, *Ot Pervoi v Mire AES* (cit. n. 28), 15.

³⁰ Ibid., 25. According to Boris Dubovskii, this abbreviation originally referred to marine applications *(atom morskoi)*. Interview with Boris G. Dubovskii, Interview #10, April 2003, Obninsk.

³¹ Lev A. Kochetkov, "K istorii sozdaniia Obninskoi AES," in Sidorenko, *Istoriia atomnoi energetiki Sovetskogo Soiuza i Rossii* (cit. n. 25), 1:96–101. Kochetkov, *Ot Pervoi v Mire AES* (cit. n. 28).

³² Peter DeLeon, *Development and Diffusion of the Nuclear Power Reactor: A Comparative Analysis* (Cambridge, Mass., 1979); Arkadii K. Kruglov, *Shtab atomproma* (Moscow, 1998).

military. They created a powerful set of images and slogans, and as Paul Josephson has shown in his analysis of Soviet newspapers and journals from 1945–1975, they managed to establish a broad consensus on the feasibility and necessity of a civilian nuclear industry by rhetorically blending the promises of nuclear power with utopian visions of the Communist future.[33]

After Stalin's death and Beria's arrest in 1953, the PGU was elevated to the Ministry of Medium Machine Building (Sredmash) and put in charge of supervising all work connected to atomic energy.[34] Viacheslav Malyshev and Boris Vannikov, both of whom had been prominently involved in the Soviet atomic bomb project, became Sredmash's first minister and first deputy minister, respectively.[35]

In 1955, Malyshev was promoted and replaced as head of Sredmash by Avraamii Zaveniagin, once a close collaborator of Beria's, who had proven himself at the steel town of Magnitogorsk.[36] Zaveniagin died the following year and was temporarily replaced by Vannikov.[37] A few months later, in April 1957, Mikhail Pervukhin replaced Vannikov as minister. Pervukhin had significant technical and economic expertise and had been an important manager of the nuclear project, but he fell into disgrace after only a few months and was removed as head of Sredmash. The way was now clear for Efim Slavskii, another engineer molded during the atomic bomb project. Slavskii, born in 1898 into a rural Ukrainian family, had joined the party in 1918, served in the Red Army for ten years (1918–1928), and after completing his studies at the Moscow Mining Academy, had worked in the nonferrous metals industry. His political career started in 1945, when he was appointed deputy people's commissar, and later deputy minister, of nonferrous metals. In April 1946, he was assigned to the atomic project as deputy director of the First Chief Administration (PGU). From 1947 to 1949, he was director of the Kombinat No. 817, one of the secret military facilities at

---

[33] Paul R. Josephson, "Rockets, Reactors, and Soviet Culture," in *Science and the Soviet Social Order*, ed. Loren Graham (Cambridge, Mass., 1990), 168–91, 174. In fact, they were so successful that only "the Chernobyl disaster and the decline of the Soviet Union were to shake the foundations of Soviet nuclear culture" (Josephson, "Atomic-Powered Communism: Nuclear Culture in the Postwar USSR," *Slavic Review* 55 [1996]: 297–324, on 322). On nuclear power exhibitions at the Exhibition of the Achievements of the People's Economy (VDNKh) in Moscow, see Sonja D. Schmid, "Celebrating Tomorrow Today: The Peaceful Atom on Display in the Soviet Union," *Soc. Stud. Sci.* 36 (2006): 331–65. On revolutionary and utopian dreams, see Richard Stites, *Revolutionary Dreams: Utopian Vision and Experimental Life in the Russian Revolution* (New York, 1989).

[34] Viktor A. Sidorenko, "Upravlenie atomnoi energetikoi," in Sidorenko, *Istoriia atomnoi energetiki Sovetskogo Soiuza i Rossii* (cit. n. 25), 1:217–53, 218. Beria was arrested on June 26, 1953, and executed after his trial in December 1953. Kramish asserts that this arrest marked "a general lessening of tensions within the Soviet atomic energy program." Arnold Kramish, *Atomic Energy in the Soviet Union* (Stanford, Calif., 1959), 177.

[35] Vannikov's career had been truly astounding: a former people's commissioner for armament, he had repeatedly fallen out of Stalin's favor and, in fact, received his appointment as deputy chairman of the State Committee of Defense in prison. The reasons for his arrests remain obscure, but according to Grabovskii's plausible (although fictional) account, Vannikov had been arrested for publicly disagreeing with Stalin. Mikhail P. Grabovskii, *Nakanune avrala* (Moscow, 2000), 80–91; see also Vitalii P. Nasonov, *B. L. Vannikov: Memuary, vospominaniia, stat'i* (Moscow, 1997).

[36] Malyshev became chairman of Gostekhnika, the revived Committee for New Technologies under the Council of Ministers, where he exercised technical control on a higher level, but he died on February 20, 1957, of leukemia (Kramish, *Atomic Energy in the Soviet Union* [cit. n. 34], 178). On the history of Gostekhnika and its later incarnations, see Stephen Fortescue, *Science Policy in the Soviet Union* (London, 1990). For Zaveniagin's time at the Magnitogorsk Metallurgical Complex from 1933–1936, see M. Ia. Vazhnov and I. S. Aristov, eds., *A. P. Zaveniagin: Stranitsy zhizni* (Moscow, 2002); and Stephen Kotkin, *Magnetic Mountain: Stalinism as a Civilization* (Berkeley, Calif., 1995).

[37] Sidorenko, "Upravlenie atomnoi energetikoi," 218; Kramish, *Atomic Energy in the Soviet Union*, 178. (Both cit. n. 34.)

Cheliabinsk-40, before rising to deputy, and then first deputy director of the PGU.[38] Slavskii had been deputy minister of Sredmash since its inception in 1953, and following his appointment as minister in 1957, he stayed in office for almost thirty years. Under his aegis, the Ministry of Medium Machine Building retained its centralized, closed character and became a largely autonomous organization. Slavskii masterfully used the political clout his ministry had inherited from the atomic bomb project to secure scarce resources and to strengthen his ministry's standing.

Let's return to the early years of the civilian nuclear program, however. One year after the successful launch of the Obninsk reactor, Kurchatov and Anatolii Aleksandrov, then director of the Institute of Physical Problems, proposed the construction of four large nuclear power plants, two immediately, and two in the near future.[39] They argued that the design, construction, and operating experience with these four types would allow physicists and economic planners to choose the ideal directions for the extensive development of nuclear power over the next two decades.[40] Soviet nuclear physics had received international recognition at the first United Nations International Conference on Peaceful Uses of Nuclear Energy in 1955, in Geneva, and Kurchatov skillfully used references to other countries as leverage in negotiating the development of a Soviet civilian nuclear program in the highest political councils.

In 1956, Kurchatov proposed an ambitious strategy for nuclear power during the sixth five-year plan (1955–1960), but only a year later, and after Zaveniagin's death, he realized that work was not moving ahead as planned. He and Aleksandrov wrote a letter to the new minister, Pervukhin, and his first deputy, Slavskii, asking for support. But their efforts were unsuccessful, and the construction of nuclear power plants was further delayed.[41] Their optimistic plan had to compete for resources with other state priorities, and the technical feasibility, as well as the projected profitability of nuclear power, was fiercely contested. In addition, Soviet industry was not yet capable

---

[38] Sidorenko, "Upravlenie atomnoi energetikoi" (cit. n. 34), 218. On Slavskii, see also Igor' A. Beliaev and German G. Malkin, eds., *E. P. Slavskii: 100 let so dnia rozhdeniia* (Moscow, 1999); and Vitalii P. Nasonov, *E. P. Slavskii: Stranitsy zhizni* (Moscow, 1998).

[39] V. V. Goncharov, "Pervyi period razvitiia atomnoi energetiki v SSSR," in Sidorenko, *Istoriia atomnoi energetiki Sovetskogo Soiuza i Rossii* (cit. n. 25), 1:16–70. Aleksandrov had taken over the position as head of the Institute of Physical Problems from Petr Kapitsa in 1946. In 1955, he became Kurchatov's deputy at the Institute for Atomic Energy (IAE); after Kurchatov died in 1960, he became director (the institute was then renamed in honor of Kurchatov). In addition, in 1975 Aleksandrov became president of the Soviet Academy of Sciences, and he remained in these two functions until Chernobyl, when he resigned as president of the Soviet Academy of Sciences. He stayed director of the IAE until 1988.

[40] "The construction of large atomic power plants and their operation will also make it possible to determine which one of the installations are most harmless and safest for the surrounding population." (Igor' V. Kurchatov, *Nekotorye voprosy razvitiia atomnoi energetiki v SSSR/Some Aspects of Atomic Power Development in the USSR* (Russian & English) [Moscow, 1956], 17.) This is also the first reference to civilian nuclear safety that I am aware of. See also Goncharov, "Pervyi period" (cit. n. 39), 20.

[41] Kurchatov gave a passionate speech about the peaceful future of nuclear energy at the Twentieth Convention of the Communist Party of the Soviet Union (CPSU) in 1956 (Igor' V. Kurchatov, "Rech' tovarishcha I. V. Kurchatova [Akademiia nauk SSSR]," *Pravda* 53, 22 Feb. 1956, 7); on June 21, 1958, he wrote a letter to Leonid Brezhnev about delays on the Novo-Voronezh nuclear power plant construction site (f. 5, op. 40, d. 107, rolik 7213, l. 64–66, Russian State Archive of Contemporary History [hereafter cited as RGANI], Moscow; reproduced in Goncharov, "Pervyi period" [cit. n. 39], 65–6). In the spring of 1959, he wrote to the secretaries of the party's Central Committee, F. R. Kozlov (April 16) and A. I. Kirichenko (April 6), as well as to the Council of Ministers (Aleksei Kosygin, April 24), again promoting the Novo-Voronezh nuclear power plant (Goncharov, "Pervyi period" [cit. n. 39], 67–70).

of keeping up with the ambitious plans of nuclear energy promoters. Kurchatov died before the nuclear power program was restarted in 1962.[42] The first industrial-scale reactor blocks were completed at the Beloiarsk and the Novo-Voronezh sites only in 1964, four years after his death.[43] These reactors were nevertheless experimental stations, and after more than a decade of experimenting with a variety of different reactor designs, the Soviet program still faced the risk of being abandoned. The civilian nuclear power industry would attain a significant place in the country's energy strategy only by the early 1970s. This was accomplished not only by successfully integrating visions of a nuclear future into the "iconography of Soviet culture"[44] but by incorporating these nuclear visions into the country's short- and long-term plans, and into the Soviet state's bureaucratic structures.

In addition to its military prehistory, the emerging nuclear power industry built on the traditions of GOELRO, the state electrification plan launched in 1920. This plan tied into overall industrialization efforts and relied on ideas of scientific management that had reached the Russian electrical engineering community even before 1917.[45] The October Revolution had brought "a regime to power that believed in the promise of machines to liberate."[46] Lenin wholeheartedly supported the plan, and with his famous slogan "Communism equals Soviet power plus the electrification of the whole country," he consciously linked technological progress with social transformation.[47] The engineers behind the GOELRO plan were "bourgeois specialists,"[48] but they "saw the new government removing the restraints of tsarism and supporting their modernizing mission."[49] In its actual implementation, GOELRO relied heavily on prison labor. As a result, by 1932 the total output of Soviet power plants had more than tripled, and it continued to grow until 1956, when the prison labor system was abolished.[50] The involvement of the Secret Police (People's Commissariat of Internal Affairs, NKVD) in the construction of power plants is reflected in the confusing reorganization of the People's Commissariat of Power Plants and Electrical Industry that was first created in 1939, as an offspring of the People's Commissariat of Machine Building.[51] After more than two decades of continuously regrouping responsi-

---

[42] Goncharov, "Pervyi period" (cit. n. 39), 9, 49–51.

[43] Kurchatov was still alive when the first dual-use reactor, the "Siberian nuclear power plant," came online in 1958. For a history of this reactor, see Mikhail P. Grabovskii, *Vtoroi Ivan: Sovershenno sekretno* (Moscow, 1998); and L. A. Alekhin and G. V. Kiselev, "Istoriia sozdaniia pervogo v SSSR i v mire dvukhtselevogo uran-grafitovogo reaktora EI-2 dlia odnovremennogo proizvodstva oruzheinogo plutoniia i elektroenergii," *Istoriia nauki i tekhniki* 12 (2003): 2–35.

[44] Josephson, "Atomic-Powered Communism" (cit. n. 33).

[45] Jonathan Coopersmith, *The Electrification of Russia, 1880–1926* (Ithaca, N.Y., 1992), 169–70. In fact, some critics argue that the revolution actually hampered an already ongoing process of electrification and that the GOELRO plan was too often fetishized. Iurii I. Koriakin, *Okrestnosti iadernoi energetiki Rossii: Novye vyzovy* (Moscow, 2002).

[46] Coopersmith, *Electrification of Russia* (cit. n. 45), 189.

[47] Holloway, *Stalin and the Bomb* (cit. n. 13), 10.

[48] "Burzhuaznye spetsy," in Koriakin, *Okrestnosti*, 29; Coopersmith, *Electrification of Russia*, 175. (Both cit. n. 45.)

[49] Coopersmith, *Electrification of Russia* (cit. n. 45), 190.

[50] Koriakin, *Okrestnosti* (cit. n. 45), 31.

[51] The Narodnyi Komissariat Elektrostantsii i Elektropromyshlennosti was created on January 24, 1939 (f. 7964, op. 15, t. 1, l. 5, Russian State Archive of the Economy [hereafter cited as RGAE], Moscow). See also "Elektroenergetika: nekotorye vazhneishie sobytiia," prepared by Vanguard Ltd. as part of the project Elektromysl', in March and April of 2001, which contains a summary of the history of this industry from the nineteenth century to the early 1990s (http://www.sapov.ru/consul/reports/electro/el-sense_06.htm; last accessed 7 Feb. 2008).

bilities for the country's power plants and electricity grid, in 1962 all organizations relating to power plants, their construction, and their connecting power lines were united in the Ministry of Energy and Electrification, in short Minenergo.[52]

In 1956, the Chief Administration for the Construction of Atomic Power Plants, Glavatomenergo, was created under the Ministry of the Construction of Power Plants (one of Minenergo's predecessors), and Konstantin Lavrenenko, an engineer who was prominently involved in the management of the postwar electric power industry, was appointed its director.[53] In 1957, this newly created organization was put in charge of supervising construction of the first industrial-scale nuclear power plants.[54] Lavrenenko immediately launched several personnel initiatives, since according to the original schedule the start-up of the Beloiarsk and the Novo-Voronezh nuclear power plants was imminent. His team examined "management schemes of nuclear power plants, the structure of workshops and departments, [and] the distribution of personnel within workshops, which is very different from common power plants."[55] Following a decree issued by the Council of Ministers on April 4, 1957, Glavatomenergo started training specialists for the design, assembly, and operation of nuclear power plants in courses taught at the Moscow Power Engineering Institute (Moskovskii Energeticheskii Institut, MEI).[56] Apparently, the students of these special courses at MEI were experienced industry specialists: all of them had higher education degrees; most of them were mechanical or electrical engineers, some of them physicists and chemists.[57] Over the course of one or two years, they received a broad introduction to nuclear technologies, with the goal of conducting independent work in the future. For this purpose, teachers were recruited from leading scientific-technical institutes and specialized laboratories. In addition, a number of specialists would undergo a more detailed practical training at the World's First Nuclear Power Plant at Obninsk. The future chief engineer of a nuclear power plant, in particular, was supposed to serve in any and all positions before reaching his designated position.[58]

In 1962, Petr Neporozhnii, an experienced hydropower engineer and loyal party member, was appointed head of Minenergo.[59] Neporozhnii, who had been first deputy

---

[52] In its post-1962 incarnation, it came to life through the Order by the Central Committee and the Council of Ministers No. 985, 21 Sept. 1962, "Ob organizatsii soiuzno-respublikanskogo Ministerstva energetiki i elektrifikatsii SSSR" (mentioned in f. 4372, f. 66, d. 231, RGAE).

[53] Lavrenenko, born 1908 in Ukraine, was trained at the Kiev Polytechnic Institute before working at the Berezniki thermal plant. He began working in the Energy Industry Administration in 1938 and became deputy minister after the war. From 1956 to 1958, he was head of Glavatomenergo, before moving on to the State Planning Commission, and the State Committee on Science and Technology. See also Schattenberg, *Stalins Ingenieure* (cit. n. 1); Schattenberg was able to locate an unedited manuscript of Lavrenenko's memoirs (f. 9592, op. 1, d. 404, RGAE). The Glavatomenergo collection is f. 9599, RGAE.

[54] Original plans had been for a conventional plant at the Beloiarsk site, according to Glavatomenergo's annual capital investment report for 1957. F. 7964, op. 3, d. 1881, l. 2, RGAE.

[55] F. 7964, op. 3, d. 1881, l. 11, RGAE.

[56] Ibid., l. 12.

[57] As of January 1, 1958, there were ninety-four students enrolled. Ibid.

[58] Ibid., l. 13. This corresponds to what other nuclear specialist have described: they went through several positions, getting to know the nuts and bolts of a given reactor, and gradually worked their way up the professional hierarchy. Grabovskii, *Vtoroi Ivan* (cit. n. 43); Viktor A. Sidorenko, *Ob atomnoi energetike, atomnykh stantsiiakh, uchiteliakh, kollegakh i o sebe* (Moscow, 2003); Interview #1, February 2003, and Interviews #2 and #7, March 2003, Moscow.

[59] In the early 1960s, as a consequence of Khrushchev's reforms, the ministry was called the State Production Committee for Power and Electrification (Gosudarstvennyi proizvodstvennyi komitet po

minister since 1959, was born in 1910 near Kiev, in a rural family. In 1933, he graduated from the Leningrad Institute of Navigation Engineering (Leningradskii Institut Inzhenerov Vodnogo Transporta) as a hydropower engineer. After having served in the Soviet Navy from 1933 to 1935, he became involved in managing the construction of large hydropower plants. In 1954, he started his political career, first in Ukraine, and a few years later on the all-union level. By 1962, his ministry was overseeing and coordinating every type of conventional power plant designed, built, and operated in the territory of the Soviet Union, the entire Soviet electricity grid, as well as the centralized heat supply for industry and households. In addition, Minenergo assumed responsibility for the construction of nuclear power plants by incorporating the aforementioned Glavatomenergo.[60] Initially indifferent to nuclear energy, Neporozhnii became a dedicated advocate for nuclear power plants.[61] During his long time in office (he retired in 1985, at age seventy-five), he pursued an aggressive, and remarkably successful strategy of negotiating with the State Planning Commission (Gosplan) for more money, especially for nuclear power plants.[62]

In the late 1950s, the Soviet technical leadership had relatively small stakes in the success or failure of civilian nuclear power: neither the nuclear weapons complex nor the power plant industry had to justify its respective authority through nuclear power. Slavskii's ministry got its legitimacy from its military successes and considered civilian applications a by-product of its activities; Neporozhnii's ministry was in charge of the entire country's electrification, a task of monumental scale and a national priority since the GOELRO initiative.[63] Both Slavskii and Neporozhnii had worked their way up the social ladder through engineering careers and service in party organizations.[64] Their spheres of action were protected by their status as cornerstones of the Soviet economy and foreign policy, respectively, and they remained relatively untouched by ongoing economic reforms.[65] And yet, starting in the early 1960s, both ministers chose to promote nuclear power over the skepticism of planners and took on civilian nuclear energy as part of their organizational culture and professional identity.[66]

---

energetike i elektrifikatsii SSSR), with the same responsibilities, and with Neporozhnii as its chairman. F. 4372, op. 65, d. 765, l. 79–88, RGAE.

[60] F. 7964, op. 15, t. 1, l. 8, RGAE.

[61] In 1964, the Novo-Voronezh and Beloiarsk plants started up, but they were still perceived as experimental stations, not as powerful industrial power plants. Electricity produced by nuclear power plants remained negligible, even in the plans for 1966–1970, when nuclear power was forecast to contribute less than 1.5 percent of the overall electricity production (f. 4372, op. 65, d. 769, l. 60–3, RGAE). In late March of 1965, neither the Novo-Voronezh nor the Beloiarsk nuclear power plants were seriously considered as industrial energy production facilities (f. 4372, op. 66, d. 233, l. 245–50, RGAE). Sidorenko, "Upravlenie atomnoi energetikoi" (cit. n. 34), 217; Neporozhnii, *Dnevniki* (cit. n. 7).

[62] In 1964, for example, he requested authorization to use flexible schemes for reimbursing contractors and autonomy in amending not just monthly plans but three-month plans as well. F. 4372, op. 65, d. 771, RGAE.

[63] According to Bailes, Lenin referred to the state electrification plan as "our second party program." Bailes, *Technology and Society under Lenin and Stalin* (cit. n. 5), 416.

[64] In Balzer's words, they were "leaders with special traits—political acumen, blue-collar credentials, and superb networks." Harley Balzer, "Engineers: The Rise and Decline of a Social Myth," in Graham, *Science and the Soviet Social Order* (cit. n. 33), 141–67, 153.

[65] Alec Nove, *The Soviet Economy: An Introduction*, 2nd rev. ed. (New York, 1969).

[66] Sonja D. Schmid, "Envisioning a Technological State: Reactor Design Choices and Political Legitimacy in the Soviet Union and Russia" (PhD diss., Cornell Univ., 2005).

Not unlike their counterparts of the late 1920s and early 1930s, whom Sheila Fitzpatrick has described, the technical specialists who promoted a civilian nuclear industry demonstrated a pragmatic attitude that did not prioritize intellectual freedom—quite in contrast to the intelligentsia of the nineteenth century. Their area of expertise provided them with social prestige and unusual leeway in pursuing their interests; in addition, their technocratic views of societal development converged with much of the Soviet state's industrial and foreign policies.[67] According to Loren Graham, engineers in particular stood by the implicit agreement struck with the party in the late 1920s, that they "would not raise basic political questions, but instead carry out the Party's orders" in exchange for promotions in industry, agriculture, and the military forces.[68] Even specialists who criticized certain aspects of the Soviet energy program (I will return to these below) were and remained fundamentally committed to nuclear energy.[69]

The debate about "old" or "bourgeois" specialists that Kendall Bailes has chronicled for the first decades of Soviet rule had mostly subsided by the time nuclear engineering was established as a college program in the 1950s.[70] Many of the founding fathers of the nuclear industry were loyal engineers who had served in party functions, and although the deep and institutionalized mistrust of the previous decades no doubt persisted, these specialists were perceived as less of a threat to the political establishment than their colleagues in the 1920s had been.[71] During and immediately following the war, scientists and engineers joined state agencies in increasing numbers, reversing the dominant trend of the 1930s, "when high-level party-state officials had become members of the scientific establishment."[72]

The nascent civilian nuclear industry initially had to rely on technical cadres with little specialized knowledge and recruited young specialists from general technical colleges to complement the cadre of experienced specialists and industrial managers. It is important to keep in mind that the entire first generation of cadres for the Soviet nuclear program was not formally trained in nuclear engineering.[73] A prominent example is Nikolai Dollezhal, a mechanical engineer from a very modest background, who advanced from working in the chemical industry to designing the

---

[67] "In political terms, the intelligentsia leadership [of the late 1920s and early 1930s] came from the Academy of Sciences and the high-salaried specialists and consultants associated with the government commissariats; and for these men the issue of intellectual freedom was secondary to the issue of political influence and specialist input in government policy-making." Fitzpatrick, *Education and Social Mobility* (cit. n. 10), 84–5.

[68] Loren R. Graham, *Science in Russia and the Soviet Union: A Short History* (Cambridge, UK, 1993), 164. See also Valerie Bunce, *Subversive Institutions: The Design and the Destruction of Socialism and the State* (Cambridge, UK, 1999), 33.

[69] See, e.g., the famous article Nikolai A. Dollezhal' and Iurii I. Koriakin, "Iadernaia elektroenergetika: Dostizheniia i problemy [Nuclear energy industry: Achievements and problems]," *Kommunist* 14 (1979): 19–28. Intended as a proposal to further increase the economic efficiency of the nuclear industry by suggesting remotely located "nuclear energy complexes," it was perceived as a criticism of current practice. Interview #20, December 2003, Moscow.

[70] Bailes, *Technology and Society under Lenin and Stalin* (cit. n. 5); Jonathan Coopersmith, "The Dog That Did Not Bark during the Night: The 'Normalcy' of Russian, Soviet, and Post-Soviet Science and Technology Studies," *Technology and Culture* 47 (July 2006): 623–37.

[71] Bailes, *Technology and Society under Lenin and Stalin* (cit. n. 5), 158. See also Holloway, "Physics, the State, and Civil Society" (cit. n. 6), 176.

[72] Krementsov, *Stalinist Science* (cit. n. 3), 97.

[73] Arkadii K. Kruglov, *Kak sozdavalas' atomnaia promyshlennost' v SSSR* (Moscow, 1995); Kruglov, *Shtab atomproma* (cit. n. 32).

first Soviet production reactor and who subsequently directed one of the country's most important, and most secret, research and development institutes for nuclear technologies.[74]

Nuclear engineering as a discipline became institutionalized at a time when an engineering career was seen as the preferred way of social advancement.[75] A career as an engineer allowed ambitious individuals from working-class backgrounds to advance quickly into the intelligentsia and into the political elites.[76] Scholars have compared the role of engineering education in the Soviet Union of the 1920s, in particular the practice of *vydvizhenie*, the promotion of individuals from working-class background, with the already existing recruitment patterns for the Communist Party.[77] Many of the engineering graduates were considered politically loyal precisely because they had benefited most from the Soviet system in terms of upward social mobility.[78]

Nuclear engineers were trained in specialized programs within existing engineering departments, and entire curricula for nuclear power engineers were set up simultaneously with the launch of the atomic bomb project.[79] Theoretical instruction at the country's leading technical colleges was complemented by practical training at research and design institutes, at the World's First Nuclear Power Plant in Obninsk, and at secret nuclear facilities (*ob"ekty*).[80] For the purpose of extended collaboration among several ministries, which became characteristic for large-scale projects throughout the Soviet industry, a new type of technical intelligentsia increasingly gained importance: economic strategists, planners, and managers with enough technical background to be able to evaluate the proposals from scientists and engineers and assess them as realistic or not. As late as 1968, there were still not enough engineers who also had expertise in industrial planning. The first graduates from such training programs were not expected until 1972.[81]

When the technical intelligentsia managed to improve its standing vis-à-vis the Communist Party in the 1950s, nuclear engineers, civilian or otherwise, reaped the benefits in terms of training, privileges, and relative intellectual freedom. A training in nuclear engineering was prestigious, partly because it involved sensitive knowledge that was potentially of military relevance, but the ideological component in

---

[74] Dollezhal was drawn into the atomic bomb project by Kurchatov in 1946 (Holloway, *Stalin and the Bomb* [cit. n. 13], 183). In 1952, Dollezhal's construction bureau was expanded to form a new institute, NII-8 (later NIKIET), with him as the director.

[75] Harley Balzer, ed., *Russia's Missing Middle Class: The Professions in Russian History* (Armonk, N.Y., 1996); Fitzpatrick, *Education and Social Mobility* (cit. n. 10).

[76] Balzer, "Engineers" (cit. n. 64), 141.

[77] Fitzpatrick, *Education and Social Mobility* (cit. n. 10), 183–4; Michael David-Fox, "What Is Cultural Revolution?" *Russian Review* 58 (April 1999): 181–201.

[78] Graham, *Short History* (cit. n. 68), 164; see also Sidorenko, *Ob atomnoi energetike* (cit. n. 58).

[79] Sidorenko, *Ob atomnoi energetike* (cit. n. 58), especially 232–56. These programs were initially highly restricted, and even lecture notes were subjected to classification. Interview #3, March 2003, Obninsk.

[80] Among those who received training at the Obninsk plant were military nuclear submarine crews. When the framework of this training was first being discussed late in 1954, the director of the Obninsk plant, Nikolai A. Nikolaev, suggested a one-year placement for mastering the operation of a nuclear reactor. The military officers settled for a three-month training program, after which the officers had to pass an exam that would certify them to operate a nuclear reactor on their own. Boris A. Fain, *Aktivanaia zona: Povest' ob atomnom institute* (Moscow, 1998), 39–40.

[81] These specialists were referred to as *inzhenery-organizatory proizvodstva*. F. 7964, op. 15, d. 64, l. 89, RGAE.

choosing an education is often overstated.[82] Rather, a university program in nuclear engineering, just as any other technical education, appealed to individuals inclined to do applied work, and it served as a prime vehicle for upward social mobility, especially for aspiring engineers from the provinces.[83] Few of them anticipated what awaited them in the emerging nuclear industry.

INSTITUTIONAL HISTORY, ORGANIZATIONAL CULTURES, AND PROFESSIONAL IDENTITIES

As the previous section has shown, both the nuclear weapons program's culture of secrecy and the demands of industrial production embodied in the state electrification plan (GOELRO) shaped professional identities and organizational culture in the civilian nuclear industry. This dual legacy also informed the division of labor in the civilian nuclear industry, first and foremost between the two ministries introduced above.[84] The distinction between atomshchiki and energetiki that was so adamantly advocated after Chernobyl built on the emerging division of labor and the corresponding organizational delineation of responsibilities.

The planners of the civilian nuclear program originally envisioned a close collaboration between a number of ministries and enterprises, in an effort to integrate nuclear power plants into the country's ambitious electrification plans.[85] Lavrenenko's efforts at training nuclear energy cadres for these plants in the 1950s is evidence that nuclear power plants were taken seriously as an addition to the country's energy generating efforts, and that the ministry in charge of the Soviet Union's electrification program was eager to take on this new task. However, the advocates of nuclear energy in the Ministry of Medium Machine Building (Sredmash) insisted that specific processes in the nuclear fuel cycle, such as uranium enrichment and plutonium reprocessing, were better handled by specialists familiar with radioactive materials, that is, by specialists trained in the nuclear weapons program.[86] Whether this was in the interest of nuclear security, or a mundane issue of competition and power play, the result was an uneasy cooperation between two ministries, one associated with nuclear weapons (Sredmash), the other with conventional power plants and the electricity grid (the later Minenergo).[87]

---

[82] This corresponds to Nikolai Krementsov's argument that ideology had played a subordinate role in concrete science-policy decision making even in the 1930s and the 1940s. Krementsov, *Stalinist Science* (cit. n. 3), 288–9.

[83] The fact that relatively unrestricted research and experimenting could be justified with reference to its usefulness for the state was not lost on promoters of the civilian nuclear industry. Schmid, *Envisioning a Technological State* (cit. n. 66); Holloway, *Stalin and the Bomb* (cit. n. 13); Krementsov, *Stalinist Science* (cit. n. 3), 287–8.

[84] There were more ministries and organizations involved, but as for cadres, the two I am focusing on here were the most relevant.

[85] Overall, the cooperation to develop a nuclear power industry involved more than 400 engineering and scientific research institutes, construction bureaus, and other organizations—for example the ministries of petrochemical engineering (Minneftemash), of tool industry (Minpriborprom), of ferrous and nonferrous metals (Minchermet and Mintsvetmet), and of chemical industry (Minkhimprom). (See "Letter to Comrade Baibakov, N. K." by A. Pavlenko, K. Vinogradov, and A. Nekrasov, No. 22–9, 10 Jan. 1969, f. 4372, op. 66, d. 3215, l.1–3, RGAE.) The State Planning Committee, in particular its Department of Power Engineering and Electrification, served as the coordinating agency.

[86] Sidorenko, "Upravlenie atomnoi energetikoi" (cit. n. 34), 223.

[87] The initial collaboration among these entities did not clearly define or assign responsibilities—a feature of the Soviet system that Jeremy Azrael has identified as Stalinist in origin. Jeremy R. Azrael,

Distinct professional identities within the cadre of specialists working in the civilian nuclear industry started forming when nuclear power plants were transferred from the Ministry of Medium Machine Building (Sredmash) to the Ministry of Energy and Electrification (Minenergo) in August 1966. Responsibilities for nuclear power plants, both operating and under construction, were now assigned to Minenergo. The revitalized Chief Administration of Nuclear Power Plants within Minenergo, Glavatomenergo, was put in charge of the construction and commercial operation of nuclear power plants, while the Ministry of Medium Machine Building, Sredmash, continued supervising a number of key tasks.[88] Sredmash kept control over the design and manufacturing of nuclear reactors and over the production and handling of nuclear fuel.[89] Sredmash also kept the authority to decide whether a nuclear power plant or a research institute would be transferred to Minenergo; in addition, the country's leading nuclear research and reactor design institutes would remain under the aegis of Sredmash.[90]

The decision to entrust the Ministry of Energy and Electrification with operating nuclear power plants at that time was predicated on the assumption that Soviet nuclear specialists had learned to contain the risks of controlled nuclear fission, that the available designs and the already working reactors proved this accomplishment, that appropriate organizational structures were in place, and that reliable technical cadres would guarantee the safe operation of nuclear power plants. The transfer of nuclear power plants to Minenergo thus signified the mastery and consequently the "normalization" of nuclear technology and indicated that nuclear power reactors no longer needed the special attention of a closed, militarized organization but were fit to be operated by civilian engineers whose work ethics reflected a different institutional culture, that of industry and production.

The transfer came more than a decade after the first Soviet nuclear power plant had been launched in July 1954 at Obninsk, and two years after two larger power reactors (at the Beloiarsk site in the Urals, and at the Novo-Voronezh site in southern Russia) had started operation in 1964.[91] By the end of the 1960s, both ministers, Slavskii and Neporozhnii, headed a "state within the state." They not only oversaw production facilities but also controlled research and development institutes, design and construction bureaus, and educational institutes. They maintained excellent networks among high functionaries in the Council of Ministers, the Central Committee, and the State Planning Commission, and although they shared technocratic visions of progress, they remained deeply suspicious of specialists trained by the other, as

---

Managerial Power and Soviet Politics (Cambridge, Mass., 1966), 117; see also Kotkin, *Magnetic Mountain* (cit. n. 36).

[88] Sidorenko, "Upravlenie atomnoi energetikoi" (cit. n. 34), 219. I was not able to see the actual transfer document.

[89] Ibid., 220.

[90] The power plants that were not transferred to Minenergo were all prototypes of new designs: the first RBMK-1000 at Leningrad (launched in 1973), the first RBMK-1500 at Ignalina (Lithuania, launched in 1983), and the first fast breeder reactor at Shevchenko (Kazakhstan, launched in 1973).

[91] Obninsk had a power output of 5 MW; by contrast, the graphite-water reactor at Beloiarsk-1 had 100 MW, and the pressurized water reactor at Novo-Voronezh 210 MW. Although the Chief Administration for the Construction of Atomic Power Plants under Lavrenenko had been in charge of the construction at the two latter sites, the severe budget cuts for nuclear power plants and the continuous reorganization of the ministry itself diffused responsibilities, and the archival documents pertaining to Lavrenenko's Chief Administration end in 1959.

well as of the other's work.⁹² For example, Slavskii's employees frequently complained about the inadequate construction practices in Minenergo; Neporozhnii's engineers, in turn, expressed frustration about delays and last-minute changes to the construction plans imposed by Sredmash's design and engineering institutes. The redistribution of responsibilities following this transfer entailed a reliance on different sets of engineering cadres for different tasks and facilitated an association either with a secret, quasi-military organization (atomshchiki) or with the competitive, production-oriented profession of power engineers (energetiki).

Gradually, a division of labor took shape that aimed at standardizing the design, construction, and operation of nuclear power plants, consistent with the Taylorist division of labor between those who plan and those who execute the work. As will be explained in detail below, this process entailed a tripartite organization of labor and a distinction between scientific director (*nauchnyi rukovoditel'*), chief design engineer (*glavnyi konstruktor*), and chief project manager (*glavnyi proektirovshchik*).⁹³ Within this tripartite structure, Sredmash maintained control over the scientific and design engineering tasks. Thus, even after the outsourcing of nuclear power plants, Sredmash continued to control the actual reactor design and manufacturing, as well as the design, production, and handling of nuclear fuel.

The scientific director for most Soviet nuclear power plants was the Institute for Atomic Energy. After Kurchatov's death in 1960, Anatolii Aleksandrov succeeded him as the institute's director, and even after being elected president of the Academy of Sciences in 1975, he headed the institute until 1988.⁹⁴ Due to his influential positions, he was a key figure in the development of nuclear energy.⁹⁵ Aleksandrov closely collaborated with Minister Slavskii, and any suggestion, any proposal, in the area of nuclear energy had to pass the review by this powerful duo.⁹⁶ It is noteworthy that another research institute had performed well as scientific director in nuclear power engineering: Laboratory V at Obninsk was scientific director for the nuclear reactors at the Beloiarsk, Bilibino, and Shevchenko nuclear power plants. Following Kurchatov's death, Laboratory V was renamed the Institute of Physics and Power Engineering (FEI) in September 1960. It gradually lost its leading role and was relegated to a specialized institute for breeder reactor technology.

In close cooperation with the scientific director, the organization appointed as chief design engineer was responsible for core computations, for coordinating the production of all parts pertaining to the nuclear part of the plant (most important, the fuel

---

⁹² This mutual distrust between ministries or people's commissariats has a long tradition; see, e.g., Balzer, "Engineers" (cit. n. 64), 152.

⁹³ This was reportedly standard operating procedure in other spheres of the Soviet industry as well.

⁹⁴ Aleksandrov was scientific director of nuclear reactors, that is, not just nuclear power plant reactors but also reactors for submarines, icebreakers, space vehicles, and so on (Sidorenko, *Ob atomnoi energetike* [cit. n. 58]). He stepped down as president of the Academy of Sciences after Chernobyl. See also Sidorenko's essay on the nature of scientific leadership in the Soviet nuclear industry (Viktor A. Sidorenko, "Nauchnoe rukovodstvo v atomnoi energetike," in *Istoriia atomnoi energetiki Sovetskogo Soiuza i Rossii*, vol. 2, ed. V. A. Sidorenko [Moscow, 2002], 5–28). On Aleksandrov's role in the development of Soviet power reactor designs, see Kramish, *Atomic Energy in the Soviet Union* (cit. n. 34), 182; V. K. Ulasevich, ed., *Sozdano pod rukovodstvom N. A. Dollezhalia: O iadernykh reaktorakh i ikh tvortsakh (K 100-letiiu N. A. Dollezhalia)* (Moscow, 2002), 42.

⁹⁵ Paul Josephson refers to him as "Mr. Atom"; Aleksandrov's contemporaries referred to him as "AP," short for Anatolii Petrovich. Josephson, *Red Atom* (cit. n. 8).

⁹⁶ N. S. Khlopkin, ed., *A. P. Aleksandrov: Dokumenty i vospominaniia; K 100-letiiu so dnia rozhdeniia* (Moscow, 2003).

elements), for subjecting all parts to rigorous quality control, and for supervising the implementation of their technologies (both material and processes) on site. The latter task was also referred to as *avtorskii nadzor*, a kind of supervision and ongoing quality control: inspectors from the chief design engineering institute would spend months on the construction site of a new nuclear power plant, supervising every step of the process of construction and assembly, and signing off certificates required to proceed to the next step in construction work.[97] The engineering and construction bureaus that performed the role of the chief design engineer in the civilian nuclear industry were subordinate to the Ministry of Medium Machine Building. One of the leading institutes was the Research and Development Institute of Power Engineering, NIKIET. Under its director, Nikolai Dollezhal, NIKIET's staff had designed the reactor for the World's First Nuclear Power Plant at Obninsk, several military reactors, submarine reactors, and starting in the late 1960s, the graphite-water reactor that operated at the Chernobyl plant, among others.[98] The Specialized Construction Bureau Gidropress in Podolsk became chief design engineer for pressurized water reactors, the other major design for the Soviet civilian nuclear industry.[99]

Finally, the chief project manager's task was to plan, supervise, and administer the construction and assembly of the nonnuclear parts of a nuclear power plant. The organization taking on the chief project management was usually subordinate to the Ministry of Energy and Electrification, and although it outsourced many of its tasks, it acted as the overall coordinator. In particular, the project manager was in charge of the logistics regarding construction materials for a plant and of designing a "sanitary safety zone" (*sanitarnaia zona*) around the plant's territory. The project manager also coordinated human resources (including recruitment and wages) and oversaw the construction of the plant's entire infrastructure (roads connecting the plant with other industrial sites and nearby cities, electrical wiring, a sewage system, specialized cleaning and decontamination services, and so on). And finally, the project manager was responsible for building the plant's satellite town for operators and their families, including child care facilities, schools, shops, canteens, basic medical infrastructure, clubs and movie theaters, and public transportation.

This assignment of tasks shows that even after the decision to transfer nuclear power plants, Sredmash stayed in charge of the "hearts" of these plants—the nuclear reactors themselves. Research and engineering institutes of Sredmash designed the reactors, manufactured the fuel elements, and provided general technical supervision. Minenergo was responsible for the construction of all nonnuclear installations and, once construction was completed, for the plants' routine operation. This division of responsibilities was no doubt an attempt at ensuring that the unique requirements of nuclear power plants, which were considered beyond the knowledge and capabilities

---

[97] When the inspectors did not sign these certificates, the process had to be stopped and a meeting with the scientific director called. Apparently, this happened at least once at the Kursk nuclear power plant: NIKIET's inspectors blocked work from proceeding, and their objections were found valid by the scientific director. Fain, *Aktivanaia zona* (cit. n. 80), 114–7.

[98] On NIKIET and Dollezhal', see Nikolai A. Dollezhal', *U istokov rukotvornogo mira: Zapiski konstruktora*, 3rd ed. (Moscow, 2002); V. K. Ulasevich, ed., *O iadernykh reaktorakh i ikh tvortsakh: Prodolzhenie traditsii (K 50-letiiu NIKIET im. N. A. Dollezhalia)* (Moscow, 2002); Ulasevich, *Sozdano pod rukovodstvom* (cit. n. 94); Fain, *Aktivanaia zona* (cit. n. 80).

[99] V. P. Denisov and Iu. G. Dragunov, *Reaktornye ustanovki VVER dlia atomnykh elektrostantsii* (Moscow, 2002).

of Minenergo, were being met. However, this particular constellation also resulted in asymmetries in the access to and the distribution of information.

Consider the civilian engineers, whom planners had envisioned performing the task of operating nuclear power plants on a day-to-day basis, that is, a cadre of young technical specialists with or without specialized training in *nuclear* engineering. The civilian nuclear industry profoundly relied on the sound judgment of these engineer-operators. But their judgment was built into a sociotechnical system of rules, technologies, and practices. Among the devices used to ensure reliable performance of nuclear power plant operators were voluminous instruction manuals, and strict adherence to the written rules and procedures was expected.[100] In addition, annual exams were intended to ensure operators' proficiency. A detailed system of rewards and reprimands was tied to test scores and performance on the job and made up a considerable part of an operator's salary. In case of a mishap, or a violation of work discipline, these rewards could be pulled and formal reprimands issued.[101] Yet practices such as simulator training, especially accident simulations, were deemed unnecessary for the operation of civilian power reactors. Redundancy of technical systems was considered superfluous, a Western obsession. The idea of a "foolproof" design was considered all but an insult for Soviet specialists. "Training on the job," with older, more experienced operators supervising their younger peers for lengthy periods of time, seems to have been common practice during the early years of the nuclear industry and a persistent part of operator training in later years.[102]

The operators for Soviet nuclear power plants, by and large employees of the Ministry of Energy and Electrification, were thus being trained for an almost schizophrenic task: as highly qualified engineers, they should be competent to make correct decisions under difficult conditions, but at the same time, they were expected to limit their decisions to what designers had put in writing. The work of nuclear power plant operators is often dull and routine—not exactly the challenging environment a highly qualified specialist would be looking for. But as Constance Perin has pointed out, operators are not merely "'monitoring,' 'surveilling,' and 'regulating' a technology that will operate as designed and licensed." Rather, they have to use their experience and best judgment to handle "ongoing design changes and spontaneous, puzzling system interactions, not to mention accidents and near-accidents."[103] Instead of seeing operators as passively adjusting to new devices, Perin has shown that operators are highly skilled decision makers who actively and efficiently develop strategies to maneuver

---

[100] Anthropologist Constance Perin has pointed out for the U.S. context that there is a persistent ambivalence between rule following and situated expert judgment: "[S]afety may depend on the exercise of individual and/or collective judgment. Nevertheless, the dominant risk-handling strategy remains strict adherence to rules and procedures. The industry frames this dilemma as one of a choice between the proceduralization of global, algorithmic knowledge and the localism of professional experience and knowledge, or between routinization and latitude for 'disciplined improvisation.'" Constance Perin, "Operating as Experimenting: Synthesizing Engineering and Scientific Values in Nuclear Power Production," *Science, Technology, and Human Values* 23 (1998): 98–128, on 105. See also Perin, *Shouldering Risks: The Culture of Control in the Nuclear Power Industry* (Princeton, N.J., 2005).

[101] In the archival documents, I found a maximum of 29 percent of the salary consisting of rewards, and according to one interviewee, at times this percentage rose to 60 percent. Administrators criticized a 29 percent bonus (*premiia*) as too low. F. 7964, op. 15, d. 7267, ll. 73–4. RGAE; Interview #15, October 2003, Visaginas, and personal communication.

[102] See, e.g., Grabovskii, *Vtoroi Ivan* (cit. n. 43).

[103] Perin, "Operating as Experimenting" (cit. n. 100).

around design deficiencies.[104] The tacit skills acquired by technical cadres in the nuclear industry, both in the operation and in the management of nuclear power plants, reached far beyond what had been codified in operation manuals, but the significance of these skills was long underestimated.[105]

By contrast, those nuclear specialists who did not move into production and who stayed in research and development became, in one way or another, employees of the Ministry of Medium Machine Building.[106] They developed new reactor designs, experimented with research reactors, and pursued new paths in reactor engineering. More often than not, they were familiar with both military and civilian applications. The older generation of designers had helped create the industrial base necessary for detonating the first Soviet atomic bomb in 1949. They had been highly decorated, generously rewarded, and granted liberties unprecedented in Soviet society.[107] Some of them moved up and into important political or managerial functions in the Soviet polity; others used their clout to influence decision making in the country's highest political circles.[108] Several taught at top engineering colleges in Moscow, training the very specialists who would eventually join research and design institutes or would work as leading specialists at nuclear power plants.

Characteristically, research and design institutes tended to be concentrated in Moscow, with a few other research centers in the country.[109] They worked together closely with the engineering bureaus and the general construction authorities at a given plant during the development process, but once a reactor was assembled and brought to the designed power output level (a process that could take years), the designers retreated from the process of operation, unless the operating personnel reported problems that required the designers' attention.

The most striking difference between the two organizational cultures, Sredmash

---

[104] Discussions about the "human factor" were not new in Soviet industry. Fitzpatrick has identified the idea of operators being "cogs in a machine" as part of a longstanding institutional conflict between the People's Commissariat of Enlightenment (Narkompros) and the Supreme Council of the People's Economy (Vesenkha), that culminated in 1928, when Narkompros argued that "workers were simply conditioned *(trenirovany)* to become efficient cogs in the industrial machine" and concluded: "Certainly the Soviet Union must adopt modern industrial methods, including the conveyer belt and the assembly line. But this did not mean that workers could be treated as automata." Fitzpatrick, *Education and Social Mobility* (cit. n. 10), 127.

[105] The international nuclear community has finally realized that the training of nuclear cadres is a major issue and that an aging workforce poses serious problems in terms of knowledge transfer. See, e.g., C. R. Clark, A. Kazennov, A. Kossilov et al., "Achieving Excellence in Human Performance in the Nuclear Industry through Leadership, Education, and Training," in *Fifty Years of Nuclear Power— The Next Fifty Years* (proceedings of an international IAEA conference held in Moscow and Obninsk, 27 June–2 July 2004 [IAEA-CN-114/F-8]) (Vienna, 2004).

[106] I am hedging here because some institutes, especially the secret *iashckiki* (literally, "boxes," institutes referred to only by numbers), for the purpose of this essay could be considered as part of Sredmash, although they were technically subordinate to other ministries or entities.

[107] Holloway, *Stalin and the Bomb* (cit. n. 13).

[108] Several recent Russian publications celebrate this older generation of designers as heroic spirits and ingenious creators, who altruistically devoted their lives and talents to their fatherland. See, e.g., Kuznetsova, *Kurchatov v zhizni* (cit. n. 25); Dollezhal', *U istokov rukotvornogo mira* (cit. n. 98); Fain, *Aktivanaia zona* (cit. n. 80); Ulasevich, *Sozdano pod rukovodstvom* (cit. n. 94); Ulasevich, *O iadernykh reaktorakh* (cit. n. 98); Nasonov, *E. P. Slavskii* (cit. n. 38); Beliaev and Malkin, *E. P. Slavskii* (cit. n. 38); Anatolii P. Aleksandrov, ed., *Vospominaniia ob Igore Vasil' eviche Kurchatove* (Moscow, 1988); Petr A. Aleksandrov, *Akademik Anatolii Petrovich Aleksandrov: Priamaia rech'*, 2nd ed. (Moscow, 2002); Khlopkin, ed., *A. P. Aleksandrov* (cit. n. 96).

[109] For example, so-called science cities, including Obninsk, Dubna, and Melekess (Dimitrovgrad), to mention but a few. Among the most prominent engineering bureaus were the aforementioned NIKIET and the Specialized Construction Bureau (OKB), Gidropress, in Podolsk.

on the one hand, Minenergo on the other, appears to have been an awareness of the risks involved with nuclear energy. The military operators I interviewed professed without exception that they were conscious of the consequences a severe accident at a nuclear power plant might have; several of them had witnessed, managed, and mitigated nuclear accidents at military installations prior to the accident at Chernobyl in 1986. In fact, Sredmash had experienced some serious accidents before, at military as well as civilian installations. To name but two: there was an explosion at a radioactive waste storage facility near Kyshtym in 1957, and a partial core meltdown at the Leningrad nuclear power plant in 1975.[110] By contrast, engineers who had worked for electricity production authorities all their careers claimed that they had been led to believe Soviet nuclear power plants were completely safe. These engineers knew they were dealing with a high-risk technology, but they did not have access to classified information on past accidents. In fact, serious research on the possibility of severe accidents at civilian nuclear facilities started only in the late 1970s, especially after the accident at Three Mile Island in 1979.[111] It is unclear how much information on potentially severe accidents was passed on to the Ministry of Energy and Electrification. Neporozhnii clearly understood that nuclear power plants were "special"—if nothing else, they required the construction of safety zones.[112] Several employees of Minenergo I interviewed claimed that Sredmash had concealed from them the potentially catastrophic scale of accidents in nuclear power plants.[113]

An episode from one of the first industrial-scale nuclear power plants may illustrate how knowledge about problems and failures was channeled, sanitized, and ultimately buried in modified operating regulations. In February 1968, the operators at the "Beloiarka," as they fondly referred to the Beloiarsk nuclear power plant, were confronted with an unanticipated problem: after repair work one channel was not cooled properly and melted when power was rising. As a reaction, Glavatomenergo set up a commission to inspect the plant and its organization of labor. In March 1968, Ivan Emelianov, deputy director of the chief design engineering institute NIKIET, signed a document that admitted critical mistakes in the computation of the power distribution to individual channels. It was decided that in the future, two experts would conduct independent computations. In addition, a computer was supposed to repeat these computations twice; only if all results were in agreement could they be used for further computations. In September 1968, the chief design engineer also decided to introduce new safety rods every year, to balance out any potential computing mistakes.[114] No attempts were made to spell out this experience-based, case-by-case

---

[110] Sidorenko, "Upravlenie atomnoi energetikoi" (cit. n. 34), 224–5; Judith Perera, *The Nuclear Industry in the Former Soviet Union: Transition from Crisis to Opportunity*, vols. 1 and 2 (London, 1997).

[111] Sidorenko reports that nuclear physicists at the Sector of Nuclear Reactors at the Kurchatov Institute of Atomic Energy started analyzing the possibility of severe accidents at Soviet nuclear power plants in the late 1970s. Viktor A. Sidorenko, "Vvodnye zamechaniia k urokam Chernobyl'skoi avarii," in *Istoriia atomnoi energetiki Sovetskogo Soiuza i Rossii,* vol. 4 (Moscow, 2002), 4–16. See also J. Samuel Walker, *Three Mile Island: A Nuclear Crisis in Historical Perspective* (Berkeley, Calif., 2004).

[112] Neporozhnii is reported to have started his day by calling every nuclear power plant; but he is also remembered as having compared a nuclear reactor to conventional fuel boilers. Personal communication with his son, December 2003, Moscow; Commemorative Seminar at the World's First Nuclear Power Plant, 16 May 2003, Obninsk.

[113] E.g., Interview #5, March 2003, and Interview #9, April 2003, Moscow.

[114] F. 7964, op. 15, d. 802, RGAE.

learning. Instead, the design engineering institutes implemented new guidelines for conducting these computations, and the operating regulations were updated. This episode is fairly characteristic of the technological optimism of both designers and operators: their behavior reflected a deep confidence that mistakes were lessons to be learned on the way to perfection.[115]

When nuclear power plants were transferred, the more flexible research-oriented work ethics of Sredmash had to compete against economic imperatives and the relentless requirements of production informing the work of Minenergo. Sredmash attempted to preserve its expertise and professional culture within the changed organizational framework by appointing individual specialists to key positions in the civilian nuclear industry, and thus to maintain the connection between nuclear power plants and their front- and back-end infrastructure (uranium enrichment facilities, fuel manufacturing, and spent fuel and waste management)—a connection that had been weakened by the transfer.[116] The architects of the Soviet civilian nuclear industry thus acknowledged the need for continuity, but the numerous reorganizations of responsibilities within the civilian nuclear industry that followed the initial transfer (which I have described elsewhere in greater detail) did not provide the foundation for a stable set of institutional routines and practices to develop.[117]

Quite to the contrary, an ability to improvise was pivotal in such a volatile institutional environment. Following the initial transfer, bureaucratic structures were created specifically to manage nuclear power plants. They were supposed to render improvisation skills less crucial, and eventually superfluous, but because these organizational structures never reached stabilization, improvisation remained imperative. The initial rationale for adopting bureaucratic structures was to create efficient, reliable processes by establishing a clear, hierarchical division of tasks. Simultaneously, however, individual nuclear experts were regarded as a balancing factor. Those cadres that had been molded in the Soviet atomic bomb project especially were expected to improvise where bureaucratic structures proved inadequate. So *in practice*, the idea that bureaucratization would guarantee a rational form of organization went hand in hand with the use of individuals capable of improvising. Such individuals, usually experienced engineers and managers, were appointed because they had access to relevant networks; they were expected to circumvent the very bureaucratic structures whose efficiency they were supposed to guarantee.

The division of labor outlined above encouraged professional identities to develop in two very different directions. Concepts of safety and reliability, as well as the organizational structures, kept changing. Overall, the atomshchiki (that is, those associated with Sredmash) tended to see the "human factor" as the main risk to reactor safety, while the energetiki perceived the capricious technology created by the designers as the main risk and as an obstacle to meeting Minenergo's production goals. Although the division of labor with regard to the design, construction, and operation

---

[115] See, e.g., Schattenberg, *Stalins Ingenieure* (cit. n. 1).

[116] Sredmash employees who were transferred to Minenergo included the deputy director of NIKIET, Boris B. Baturov (Sidorenko, "Upravlenie atomnoi energetikoi" [cit. n. 34], 222). The former director of the Leningrad and the Ignalina nuclear plants, Nikolai F. Lukonin, was appointed minister of atomic power engineering (MAE) in 1986 (ibid.); and Evgenii O. Adamov, former director of NIKIET, was appointed minister of atomic energy in 1998. For a rather skeptical view of shifting personnel, see Balzer, "Engineers" (cit. n. 64), 198–201.

[117] Schmid, *Envisioning a Technological State* (cit. n. 66), especially chap. 3.

of nuclear power plants was not conflict free, it seems to have worked in general: Sredmash kept control over the manufacturing of the key components of nuclear plants and a watchful eye on difficult transitory regimes, such as start-up, repair, and shutdown processes. Minenergo operated the reactors on a day-to-day basis, knowing that this technology was risky, but dependent on the reassurance of Sredmash's designers that these reactors were safe to handle. Ultimately, it has been argued, the Soviet nuclear power industry became a hostage of its own organizational design.[118] But did this really escape the technical intelligentsia's attention? Were there no critics at all among the country's nuclear energy cadre?

MISSING DISSENTERS: OR WHY WERE THERE NOT MORE SAKHAROVS?

Technical training in the U.S.S.R. "led more often than not to political leadership rather than to political dissidence," and nuclear specialists were no exception.[119] They decided to cooperate with the Soviet state, disappointing U.S. hopes that any intellectual would eventually turn against the Communist Party.[120] But if technical specialists in the nuclear industry struck a deal with the political leadership, were they really part of the intelligentsia? Doesn't a true *intelligent* have to be oppositional?[121] It appears that technocratic visions of progress and relative intellectual freedom within their own scientific community were more appealing to the nascent nuclear energy cadres than ideas of a liberal polity. David Holloway has pointed out that "Russian intellectuals before and after the October Revolution regarded science as a force for rationality and democracy. They believed that it had a cultural value in and of itself, above and beyond the knowledge that was accumulated."[122] The scientists and engineers who organized the Soviet nuclear industry were largely part of Stalin's new intelligentsia of "white-collar technocrats." The fact that they cooperated with the Soviet regime, and the Stalinist regime for that matter, raises the question of whether this association corrupted and discredited science, especially in the context of the Chernobyl disaster.[123]

The post-Chernobyl debate over whom to blame, operators or designers, occurred during a period of unprecedented openness, Mikhail Gorbachev's glasnost. The debate became the first public controversy over nuclear energy in the Soviet Union. The loosening and eventual elimination of censorship prompted several nuclear specialists to use new avenues for voicing their concerns, which in turn allows the analyst to consider published counter narratives to the official accident explanations.[124] It is safe to assume that these published accounts were only the tip of the iceberg and

---

[118] Bunce, *Subversive Institutions* (cit. n. 68).

[119] "The scientific rather than the technical intelligentsia has been the more vocal force in dissident activity." Balzer, "Engineers" (cit. n. 64), 167.

[120] Albert Parry, "Science and Technology Versus Communism," *Russ. Rev.* 25 (July 1966): 227–41.

[121] See Martin Malia, "What Is the Intelligentsia?" in *The Russian Intelligentsia*, ed. Richard Pipes (New York, 1961), 1–18; James H. Billington, "The Renaissance of the Russian intelligentsia," *Foreign Affairs* 35 (1957): 525–30.

[122] Holloway, *Stalin and the Bomb* (cit. n. 13), 5.

[123] Ibid.

[124] See, e.g., Glenys A. Babcock, "The Role of Public Interest Groups in Democratization: Soviet Environmental Groups and Energy Policy-Making, 1985–1991" (PhD diss., RAND Graduate School, 1997); Sonja D. Schmid, "Transformation Discourse: Nuclear Risk as a Strategic Tool in Late Soviet Politics of Expertise," *Sci. Tech. Hum. Val.* 29 (2004): 353–76. For official accounts, see, e.g., Sidorenko, *Istoriia atomnoi energetiki Sovetskogo Soiuza i Rossii*, vol. 4 (cit. n. 111).

that most reports on the Chernobyl disaster were submitted and handled internally.[125] Many accounts were published after attempts to use official channels first had failed, which in part explains their emotional, often accusatory tone.[126]

In the early 1970s, for instance, Vladimir Volkov, an employee of the Institute of Atomic Energy, had warned the chief reactor designers in an internal report about the insufficient experimental data for the graphite-water design that had been chosen for the Leningrad nuclear power plant and cautioned against its operation.[127] His concerns were mostly ignored, but they proved justified in 1975, when the first reactor at the Leningrad plant experienced the problems Volkov had predicted.[128] After the Chernobyl accident, Volkov turned his earlier report over to the Prosecutor's Office. Almost instantaneously, by May 1, 1986, he was banned from entering his workplace, the Institute of Atomic Energy, whose director was one of the inventors and the scientific director of this reactor type.[129] Volkov then wrote a letter directly to Gorbachev, and on February 27, 1990, the State Nuclear Regulatory Agency (Gosatomenergonadzor) created a commission to reinvestigate the Chernobyl accident in the light of Volkov's report.[130] This commission, under Nikolai Shteinberg, presented its report in January 1991, concluding that this reactor design had serious flaws that significantly contributed to the disaster.[131]

Other early critics of the Chernobyl type reactor, who decided to go public only after the disintegration of the Soviet Union, included Boris Dubovskii and Viktor Sidorenko. Dubovskii, an experienced nuclear physicist who had worked with Kurchatov at starting up the first Soviet experimental reactor in Moscow, was pushed into retirement when he stated publicly that he considered flaws in the reactor design responsible for the Chernobyl accident and that two predecessors of the Chernobyl accident should have alerted the designers.[132] Dubovskii had served as chief safety

---

[125] Evidence are references to these internal reports in the open press, e.g., Diatlov, *Chernobyl'* (cit. n. 9).

[126] Diatlov, *Chernobyl'* (cit. n. 9); Alla Yaroshinskaya, *Chernobyl: The Forbidden Truth* (Lincoln, Neb., 1995); and many popular publications.

[127] Gennadii A. Shasharin, "Chernobyl'skaia tragediia," in *Chernobyl' desiat' let spustia: Neizbezhnost' ili sluchainost'?* ed. Aleksandr N. Semenov (Moscow, 1995), 75–132, 123–4. See also Diatlov, *Chernobyl'* (cit. n. 9), 11. The Russian original of I. F. Polivanov's report is at http://treeofknowledge.narod.ru/chernob.htm (last accessed 7 Feb. 2008).

[128] The former deputy chief engineer of the Chernobyl nuclear power plant, Anatolii Diatlov, who was sentenced to ten years in prison, argues that these skeptical voices were not heard at all (Diatlov, *Chernobyl'* [cit. n. 9], 118–9). Ulasevich (*Sozdano pod rukovodstvom* [cit. n. 94], 41–2), however, lists some delays in implementing the RBMK due to a series of concerns and suggestions that had to be considered.

[129] Diatlov, *Chernobyl'* (cit. n. 9), 80–1, 184.

[130] This commission also considered a report written and circulated by A. A. Iadrikhinskii, a nuclear specialist from the Kursk nuclear power plant, which I have not been able to read in full (Prikaz predsedatelia Gosatomenergonadzora No. 11, 27 Feb. 1990, cited in Diatlov, *Chernobyl'* [cit. n. 9], 178). The delay between Volkov's letter to Gorbachev and the creation of the commission can be explained by the ongoing fight between party officials who wanted to save face vis-à-vis the international community and technical specialists who argued that Western specialists had already figured out for themselves what had happened at Chernobyl and might as well be told the plain truth. Sidorenko, *Ob atomnoi energetike* (cit. n. 58), 249.

[131] Gosudarstvennyi Komitet SSSR po nadzoru za bezopasnym vedeniem rabot v promyshlennosti i atomnoi energetike, "O prichinakh i obstoiatel'stvakh avarii na 4 bloke Chernobyl'skoi AES 26 aprelia 1986 g. Doklad Komissii Gospromatomnadzora SSSR," 1991. Sidorenko, *Istoriia atomnoi energetiki Sovetskogo Soiuza i Rossii* (cit. n. 111), 4:333–409.

[132] Vladimir Anisinov, "AES: Stepen' riska (Beseda s B. G. Dubovskim)," *Smena* 10 (1994): 55–61, 56. See also Sidorenko, *Istoriia atomnoi energetiki Sovetskogo Soiuza i Rossii* (cit. n. 111), 4:44; Nadezhda Nadezhdina, "Zalozhniki reaktora," *Trud,* 3 April 1996.

inspector in the Ministry of Medium Machine Building (Sredmash) from 1958 to 1973 and subsequently directed the Sector for Nuclear Safety at the Institute of Physics and Power Engineering (FEI) in Obninsk. Although he had not been granted access to the documentation on the 1975 accident at the Leningrad nuclear power plant, Dubovskii conducted his own analysis and concluded that the reactor lacked a mechanism for handling excess steam and a fast emergency shutdown system. Reportedly, Aleksandrov agreed with Dubovskii's assessment but did not follow through with having Dubovskii's remedies implemented until after Chernobyl.[133] Apart from technical design issues, Dubovskii was most critical of the management system in the civilian nuclear industry. He contrasted it to management in the military project, when everyone had been personally accountable for their work.

Viktor Sidorenko had joined the Institute of Atomic Energy in 1952, right after graduating from the Moscow Power Engineering Institute (MEI). He was among the chief designers of the Soviet pressurized water reactor and later became one of the leading nuclear safety experts in the Soviet Union. In 1983, Sidorenko's colleagues in the Kurchatov Institute's Sector of Nuclear Reactors reached a disconcerting conclusion: their computations showed that the graphite-water reactor that was being implemented at nuclear plants all over the country had a real, if improbable, potential to suffer a severe accident due to a design flaw in the control rods.[134] Their alarming report was met with conspicuous indifference. The reactor's design engineers at NIKIET declared that this problem was well known but that it was highly unlikely to trigger an accident, therefore, mitigation was postponed until all control rods would eventually undergo refurbishment.[135] The ministry's authorities, who had received the report as well, did not react either, and no design changes were made. When Chernobyl happened, one of Sidorenko's colleagues reportedly told him, "that very thing happened" (*sluchilos' to samoe*).[136] Sidorenko, who by 1986 had become deputy director of the State Nuclear Regulatory Agency, Gosatomenergonadzor, recounts not without a grim sense of humor what this evidence meant in the context of the post-Chernobyl, early glasnost Soviet Union. Evgenii Kulov, Sidorenko's superior, amicably informed him that the existence of this earlier report would not exonerate him but more likely would be held against him. And as a matter of fact, the Communist Party's committee in charge of investigating the accident charged Sidorenko with unscrupulousness (*besprintsipnost'*): he knew and yet did not act on it.[137]

As these three examples illustrate, prior to glasnost, dissent and opposition among nuclear specialists remained limited to very narrow technical circles, a practice sustained by the nuclear specialists themselves, not just imposed on them through

---

[133] Anisinov, "AES: Stepen' riska," (cit. n. 132), 57.

[134] In particular, the physicists noted an initially *positive* surge of reactivity in the reactor's emergency shutdown system and a positive void coefficient of reactivity.

[135] Sidorenko, "Vvodnye zamechaniia," (cit. n. 111), 8.

[136] Sidorenko, *Ob atomnoi energetike* (cit. n. 58), 249.

[137] Ibid. The fourth volume Sidorenko edited on the history of nuclear energy in the Soviet Union contains Sidorenko's July 1986 "explanatory note" (*ob"iasnitel'naia zapiska*) to the CPSU's Accident Investigation Committee. In this short essay, he admits to "moral guilt," to shortcomings in the regulatory organization's recruitment policies, and he agrees with the government commission's conclusion that he had often been too accommodating when cooperating with ministries and other agencies, which undermined overall discipline. Viktor A. Sidorenko, "Ob"iasnitel'naia zapiska Sidorenko V. A. v komitet partiinogo kontrola pri TsK KPSS," in Sidorenko, *Istoriia atomnoi energetiki Sovetskogo Soiuza i Rossii* (cit. n. 111), 4:44–7.

pressure from above.[138] Nuclear specialists participated in the Communist project in a very pragmatic manner. There was a clear overlap in technocratic goals, they shared a confidence in technical feasibility, and they relied on a similar rhetoric of progress.[139] The oppositional stance associated with the cultural intelligentsia was generally at odds with the technical intelligentsia's sense of duty, honor, and responsibility for defending the homeland, especially among the specialists who had been involved in the Soviet atomic bomb project. Their goal was to fit in seamlessly, to function elegantly as part of a large machine.[140] The fact that Soviet nuclear power plants were built without the redundant safety features required in most Western plants had to do with the way the human operators were imagined: a nuclear power plant in the Soviet Union was understood as a sociotechnical system that could rely on technical specialists and their efficient performance *as* redundancy features. Reliable human operators were meant to relieve the designers of the need to implement technical redundancy systems. And yet at the same time, a rigid and contradictory system of accountability restricted these operators in their actions and created systematic distrust.

Resistance on the part of the nuclear specialists, of course, was lacking not only because of unconditional support for the Communist project. David Holloway has captured the atmosphere of feverish haste, tight secrecy, and fear of failure and repression during the Soviet atomic bomb project.[141] Beria's system had ensured conformity quite brutally, and these traditions informed practices of sharing and controlling information among different organizations in the emerging civilian industry as well. For example, secrecy established as a norm in the military nuclear program governed even internal technical communication in the civilian nuclear industry, often in areas of knowledge that many technical specialists considered unproblematic. When in the early 1980s the newly established State Nuclear Regulatory Agency, Gosatomenergonadzor, attempted to create a database of incidents and accidents at nuclear power plants to help operators learn from mistakes and avoid them in the future, the initiative was categorically rejected by one of the leading design engineering institutes, NIKIET. Its representatives argued that information about accidents with the graphite-water reactors was confidential and was not to be shared outside of the organizations of the Ministry of Medium Machine Building.[142] This was despite the

---

[138] The Interdepartmental Technical Council (MVTS) is a good example: the most powerful decision-making body for the Soviet Union's nuclear energy policy, it was intended to serve as a forum for open expert discussion. (On the creation of this council in 1971, and its demolition after the Chernobyl disaster, see Sidorenko, "Upravlenie atomnoi energetikoi" [cit. n. 34], 237.) Another example are publication patterns, in which instead of publishing controversial replies to a provocative argument, follow-up articles were published that provided the "correct interpretation" of the original thesis (see, e.g., Dollezhal' and Koriakin, "Iadernaia elektroenergetika," [cit. n. 69]). On censorship and self-control in the Soviet mass media, see, e.g., Gayle D. Hollander, *Soviet Political Indoctrination: Developments in Mass Media and Propaganda since Stalin* (New York, 1972); and Ellen Mickiewicz, *Split Signals: Television and Politics in the Soviet Union* (New York, 1988).

[139] Ivanov, "Science after Stalin" (cit. n. 11); Bailes, *Technology and Society under Lenin and Stalin* (cit. n. 5); Schattenberg, *Stalins Ingenieure* (cit. n. 1).

[140] When the system failed, when responsibility was to be assigned, the first charge was usually "wrecking," or "sabotage," and when this proved unfounded, "personnel error." Bailes, *Technology and Society under Lenin and Stalin* (cit. n. 5).

[141] Holloway, "Physics, the State, and Civil Society" (cit. n. 6). See also Mikhail P. Grabovskii, *Atomnyi avral* (Moscow, 2001).

[142] Sidorenko, "Upravlenie atomnoi energetikoi" (cit. n. 34), 226. See also Viktor A. Sidorenko, "Nuclear Power in the Soviet Union and in Russia," *Nuclear Engineering and Design* 173 (1997): 3–20, 7; and Sidorenko, "Vvedenie" (cit. n. 26), 10–1.

fact that the Ministry of Energy and Electrification had started operating these very reactors.

But what about cynicism and the "pervasive apathy of the Brezhnev era"?[143] In his analysis of the Soviet Youth League, Steven Solnick found rampant cynicism among active members of the Komsomol and other leading Soviet institutions.[144] However, he also found that this cynicism did not threaten the stability of the regime.[145] Komsomol activists, according to Solnick, simply refrained from promoting the regime's political rhetoric, and instead nurtured an attitude of indifference to it.[146] Based on his evaluation of the pervasiveness of cynicism, Solnick dismissed ideological decay as a reason for the collapse of Soviet institutions in the late 1980s. By contrast, published memoirs, as well as my own interviews, suggest that at least up until the Chernobyl disaster in 1986, cynicism was rare among nuclear specialists.

Nuclear specialists considered themselves privileged participants in the construction of a rational society and as able contributors to the industrialization of their country.[147] Confident in their technical expertise and technocratic legitimacy, most of them saw more opportunities in contributing constructively to state policies than in opposing them. Evidence that this commitment was more than justificatory rhetoric is the fact that the Chernobyl debate is still going on with so much heat and heart. Both groups, atomshchiki as well as energetiki, acknowledged the significance of nuclear power for the country's, and ultimately the world's, energy supply and saw their task as helping to realize the vision of a large-scale nuclear power industry.[148]

CONCLUSION

The post-Chernobyl debate about whom to blame for the worst accident ever recorded at a civilian nuclear facility breached a tradition of secrecy deeply engrained in Soviet culture. Starting in 1986 and 1987, the mass media, and as a consequence ordinary Soviet citizens, were discussing the role and legitimacy of a ministry that officially had never existed. Emergent environmentalist groups publicly scolded the military-industrial complex in general and the Ministry of Medium Machine Building, Sredmash, in particular and accused them of having prepared the grounds for the accident with their incommensurate secrecy. They demanded not only transparency

---

[143] Balzer, "Engineers" (cit. n. 64), 159. See also Moshe Lewin, *The Gorbachev Phenomenon: A Historical Interpretation* (Berkeley, Calif., 1988).

[144] Solnick, *Stealing the State* (cit. n. 9), 54–5. "Komsomol secretaries who were supposedly in charge of indoctrination were actively squelching any members' expressions of zeal that might require special attention on their part. VUZ [college] and labor ministry officials were criticizing university graduates for shunning job assignments in their chosen field even as these same officials inflated labor demand figures and wasted those specialists who did show up. Military officers in charge of conscription similarly castigated youth who sought to evade service even as they dispatched ill or lame young men to work alongside hardened criminals in brutal conditions." Ibid., 218–9.

[145] Ibid., 58.

[146] He concludes that "actors who served in official organizations seem to have been no less cynical—and were perhaps more so—than their counterparts in the general population." Solnick, *Stealing the State* (cit. n. 9), 59.

[147] Loren Graham has referred to this attitude as the "[d]evelopment of productive forces school of thought" that first became popular during the New Economic Policy (NEP) period. He showed that the centrally planned economy that took shape in the 1920s was extraordinarily compatible with the technocratic tendencies of the time. Graham, *Short History* (cit. n. 68), 159–60.

[148] One of my interviewees told me that he had "fallen in love" with nuclear power early on in his career. Interview #9, April 2003, Moscow.

but also public participation on all levels of decision making.[149] Nuclear specialists were overwhelmed by the sudden outburst of public opposition to nuclear power, but in addition to this public debate, they found themselves entangled in machinations carefully shielded from the public. In July 1986, the Communist Party's Central Committee concluded an extensive investigation into the causes of the accident and issued a decree that put blame equally on designers, operators, and regulators. And although selected portions were published in *Pravda*, the decree itself, which entailed the dismissal of a number of high-ranking officials, including Minister Slavskii, was promptly classified, and the report to the International Atomic Energy Commission in August 1986 was manipulated.[150] Even a year later, the party was still blocking disclosure of all the facts.[151]

The post-Chernobyl debate about whom to blame for the accident, although outside the current volume's overall time frame, is crucial for understanding the earlier history of Soviet nuclear cadres. Precisely because this event was so unique and unusual, it allows us to draw conclusions about the "normal" conduct of technical intellectuals in the nuclear industry. It revealed not only who these nuclear specialists were, and how diverse this community was, but also how dissent was supposed to be handled by illustrating clearly which kinds of disagreement were considered unacceptable.

The distinction between *atomshchiki* and *energetiki* served as powerful rhetorical resource in the post-Chernobyl nuclear crisis: it was invoked by both sides to accuse the respective other, as the car analogies at the beginning of this essay illustrated. Some nuclear specialists insisted on their professional authority as atomshchiki, to distinguish themselves from the civilian operators, the energetiki. They reminisced about a sense of personal accountability and the possibility of exchanging information even under conditions of intense secrecy, which they had experienced in the military nuclear program under Beria's rule.[152] They perceived themselves as a handpicked elite and as highly qualified specialists who were deeply loyal to their country. The energetiki, for their part, insisted on their capability and desire to function as reliable parts in the Soviet economy. They stressed that they were unassuming participants and compliant executors of rational orders but that they had been mollified and ill-informed about the tasks at hand. The distinction, however, allowed easy black-and-white painting, and this was not lost on other constituencies. For example, antinuclear groups concluded that if the designers were right, and the accident happened due to human error, all operating nuclear power plants should be shut down immediately, since human error would never be eliminated completely.

In practice, this distinction between atomshchiki and energetiki and their respective skills is hard to maintain, as Sidorenko points out: "[This judgment] can be easily challenged by recollecting events that happened during the development of the military atomic project (suffice it to remember the Kyshtym accident, where a nuclear waste storage facility exploded), as well as by recalling the accidents at the Leningrad nuclear power plant, which was under the control of [Sredmash] (not with the

---

[149] See Babcock, *Role of Public Interest Groups*; and Schmid, "Transformation Discourse" for references. (Both cit. n. 124.)

[150] The *Pravda* article appeared on July 20, 1986, under the heading "V Politbiuro TsK KPSS."

[151] Sidorenko, *Ob atomnoi energetike* (cit. n. 58), 249; reports in Sidorenko, *Istoriia atomnoi energetiki Sovetskogo Soiuza i Rossii*, vol. 4 (cit. n. 111). The decree was declassified during the Communist Party trial in the early 1990s and is now available at RGANI.

[152] Sidorenko, "Upravlenie atomnoi energetikoi" (cit. n. 34), 225.

same catastrophic consequences as at Chernobyl unit-4, but clearly a 'harbinger' of the Chernobyl accident that did not receive due attention)."[153] Maybe Sredmash's personnel were better trained, reacted faster, and made better decisions; but maybe they were just lucky or able to cover up mishaps.

As I have outlined in more detail above, the attempt at creating a unified professional workforce responsible for the civilian nuclear energy industry remained a half-hearted attempt to regulate a complex, dual-use technology. A division of labor was introduced that provided the backdrop for information sharing and, alternatively, justified the refusal thereof. There was no understanding that the individual parts of the nuclear industry were interdependent, or "hostages of each other," a realization that had hit home in the United States after the accident at Three Mile Island.[154] It is tricky to draw the boundary between limiting sensitive technical know-how to a small selection of specialists, on the one hand, and educating as many specialists as possible, on the other; between "black-boxing" technological processes and general transparency in the interest of safety. Soviet nuclear experts revisited this boundary on a regular basis, and the constant administrative changes in the nuclear industry testify to their attempts at improving and refining the existing division of labor and the resulting principles of information control. Ultimately, they held on to a conservative model, giving a "need to know" approach preference over a "need to tell" attitude, which privileged those with better access to information.

The imbalances in access to information on accidents played a major role for the different perceptions of risk among employees of Sredmash and Minenergo. It was only after the collapse of the Soviet system that the tradition of secrecy inherited from the nuclear weapons program was identified as a major stumbling block for the creation of a professional culture in the civilian nuclear industry that included awareness of risk and of the fallibility of both humans and machines.[155] But was secrecy justified in some cases and not in others? Was the restrictive information control pursued by the Ministry of Medium Machine Building vis-à-vis the Ministry of Energy and Electrification based on a legitimate incommensurability of technical expertise or simply on bureaucratic protectionism?

As the example of the nuclear energy experts demonstrates, secrecy was both: it was the glue that bonded a heterogeneous group of technical specialists and, at the same time, a divisive element that introduced doubt and suspicion. The nuclear specialists understood the secrecy attached to any and all Sredmash projects, even civilian ones, in the context of the cold war, in the context of not only a political showdown but also a race for technical supremacy, and economic development under competing doctrines. Secrecy seemed necessary, and disagreement only justified when in the interest of this overall framework. For example, Minister Neporozhnii reportedly resisted the implementation of the Chernobyl-type reactor in nuclear plants under his ministry, but he had based his opposition on economic, not technical, grounds.[156]

[153] Ibid., 224–5.
[154] Joseph V. Rees, *Hostages of Each Other: The Transformation of Nuclear Safety since Three Mile Island* (Chicago, 1994).
[155] Sidorenko, "Nuclear Power," (cit. n. 142); Kruglov, *Shtab atomproma* (cit. n. 32); Kruglov, *Kak sozdavalas'* (cit. n. 73). But this assessment is still problematic because it does not take into account that nuclear power plants went from total secrecy to quite comprehensive civilian control and back, in tight interconnection with organizational responsibilities, overall economic reforms, and the adoption of specific technical choices. See Schmid, *Envisioning a Technological State* (cit. n. 66).
[156] Sidorenko, *Ob atomnoi energetike* (cit. n. 58), 217–8.

The technical specialists of the late 1950s and 1960s, who were involved in the construction of a civilian nuclear industry, worked in a state that had taken on their very own vision of progress and rational development, a state that had adopted a technocratic attitude. Based on scientific ideas and on tangible achievements, this attitude endowed the state with authority, which in turn provided legitimacy to the technocratic worldview. In the process, individual technical specialists were at once empowered (after all, the state had embraced their worldview and represented them more than it did anyone else in Soviet society) *and* subjected to the relentless rules of a technocratic doctrine. There was no room for other, humanizing, voices in this model; it represented the technocratic worldview so well that it becomes clear why dissidents more often than not came from nontechnical backgrounds.

Resistance and opposition from nuclear energy specialists seem to have focused on decisions that they perceived as "political," as opposed to rational. The three cases introduced above illustrate that resistance was linked to knowledge: all three of them were nuclear scientists who had discovered mistakes that they tried to correct, and all three trusted the official channels first. Only after those official channels had clearly failed did these men pursue other ways to publicize "the facts" and to castigate those who covered them up for "political" reasons. For them, "political" was synonymous with irrational and often associated with personal ambition. Rational, by contrast, was a level-headed deliberation about the nuts and bolts of substantiating scientific progress, technical modernization, and societal development. According to this culturally specific definition of "political," concepts such as "progress," "modernization," and "development" were considered not political but universal.[157] The specialists tended to discard the opposition they met from the party-state apparatus as "political" and trusted that conflicts would ultimately be resolved through logic and reason.

These nuclear specialists had embodied diligence, responsibility, and a deep faith in scientific rationality, values that made them less vulnerable to apathy and cynicism than other groups and institutions. Ironically, these values may have been reinforced by the regime of secrecy imposed on their work. This secrecy implied that their work, although ultimately aimed at benefiting the public good, was something to be shared only with the select few. In other words, secrecy meant protection in more ways than one: it protected specialists and their work *from* public scrutiny, but it also protected it *for* the common good.[158] Functioning as part of a state that embodied their professional ethos, they lived to work. Had they turned cynical on the state, they would have turned cynical on their own work, their own values.[159] By suddenly disrupting and challenging these values, the Chernobyl tragedy demonstrated that these ideals were

---

[157] This attitude prevails even today. Several of my interviewees, for example, dismissed the decision to shut down Lithuania's nuclear reactors in exchange for membership in the European Union as "political" and therefore unsound and irrational.

[158] See Holloway's argument about the nuclear physicists' freedom amid the Stalinist terror (Holloway, "Physics, the State, and Civil Society" [cit. n. 6]). See also Asif A. Siddiqi, "Within the First Circle: Science and Engineering in the Gulag" (paper presented at the Annual Convention of the American Association for the Advancement of Slavic Studies [AAASS], Washington, D.C., 16–19 Nov. 2006); Michael A. Dennis, " Secrecy and Science Revisited: From Politics to Historical Practice and Back," in *The Historiography of Contemporary Science, Technology, and Medicine: Writing Recent Science*, ed. Ronald E. Doel and Thomas Söderqvist (London, 2006), 172–84.

[159] Very few managed to separate the "political" from the "rational" as Andrei Sakharov did. His boundary drawing allowed him to criticize the Soviet "political" regime (the lack of intellectual freedom) while remaining faithful to its "rational" values: modernization, progress, scientific rationality.

intact. Regardless of their institutional affiliation, nuclear specialists experienced the fierce public opposition to nuclear power that followed the accident, as a personal attack. Even those who had previously criticized certain aspects of the Soviet civilian nuclear program or the policies surrounding it remained fundamentally committed to nuclear energy. They saw nuclear energy as an inexhaustible source of energy and thus as a solution to future energy scarcity. They kept faith in science and technology as driving technical, as well as societal, progress and ultimately as something clearly distinct from politics.

INTELLIGENTSIA AS POLITICAL AGENT

The Phenomenon of Soviet Science

By Alexei Kojevnikov*

ABSTRACT

The grand "Soviet experiment" constituted an attempt to greatly accelerate and even shortcut the gradual course of historical development on the assumption of presumed knowledge of the general laws of history. This paper discusses the parts of that experiment that directly concerned scientific research and, in fact, anticipated or helped define impórtant global changes in the functioning of science as a profession and an institution during the twentieth century. The phenomenon of Soviet, or socialist, science is analyzed here from the comparative international perspective, with attention to similarities and reciprocal influences, rather than to the contrasts and dichotomies that have traditionally interested cold war–type historiography. The problem is considered at several levels: philosophical (Soviet thought on the relationship between science and society and the social construction of scientific knowledge); institutional (the state recognition of research as a separate profession, the rise of big science and scientific research institutes); demographic (science becoming a mass profession, with ethnic and gender diversity among scientists); and political (Soviet-inspired influences on the practice of science in Europe and the United States through the social relations of science movement of the 1930s and the *Sputnik* shock of the 1950s).

SCIENCE AND SOVIET VALUES

The fact that the Soviet Communist regime placed extraordinarily high value and expectations upon science is, of course, rather well known. So much so, perhaps, that it has usually not been seen as a historical problem but has been taken for granted as something natural that does not ask for further discussion or inquiry. Behind the cover of obviousness, however, one can find a complex combination of historical choices and heterogeneous reasons—some ideological, some pragmatic, some accidental—that together may offer an explanation of why, among all the various political regimes and movements of the twentieth century, Communism, especially in its initial Soviet incarnation, happened to be the one most favorably predisposed toward science, believing most utterly, up to the point of irrationality, in science's power and value.

To begin with, the Soviets mounted their belief in science on top of a preexisting and rather high foundation. The cult of science flourished across Europe at the

---

*Department of History, University of British Columbia, Vancouver, Canada; anikov@interchange.ubc.ca.

I am thankful to John Krige, Jessica Wang, participants at the seminars at the University of Michigan, Aarhus University, University of Vienna, and at the conferences "Intelligentsia: Russian and Soviet Science on the World Stage, 1860–1960" (Athens, Georgia, 2004) and "Writing the History of the Physical Sciences after 1945: State of the Art, Questions, and Perspectives" (Strasbourg, 2007) for useful criticism and comments.

© 2008 by The History of Science Society. All rights reserved. 0369-7827/08/2008-0005$10.00

beginning of the twentieth century. It happened to be particularly prominent in the Russian empire, which had only recently embarked upon industrialization and modernization. Almost all parts of the political spectrum bought into it, although for different reasons. For Russian liberals, science was synonymous with economic and social progress; for the radical intelligentsia, including the yet utterly insignificant and marginal Bolsheviks on the very left, it was the closest ally of the revolution. Many among the monarchists, too, placed high hopes on modern science as a remedy for the country's relative economic backwardness vis-à-vis Germany, France, and Britain (other European countries rarely figured in the comparison). After the Great Reforms of the 1860s, they helped institutionalize science and promote the research imperative at Russian universities, hoping that at the very least it could distract unruly students from pursuing dangerous political temptations.[1]

World War I challenged, and in several important ways undermined, scientism in most of Europe. During the Great War, all major belligerent countries de facto recognized science as a matter of national concern and state policy by establishing special governmental offices to coordinate research with military needs. Yet in the wake of the European catastrophe—which is how the war began to be perceived ever more widely after its end, not only among the losers but often among the victors as well—many intellectual circles started doubting the common assumption of the preceding half century that material and social progress naturally followed from the development of science and technology. The thesis that knowledge results in vastly increased power reconfirmed itself strongly in the course of the military conflict, but this time in the form of much deadlier weapons. Unlike in the prewar decades, however, the great advances in positive knowledge and technical ingenuity produced not an amelioration of the human condition and social life but destruction, killing, and suffering on a previously unimaginable scale. These sacrifices appeared increasingly meaningless to the disillusioned postwar public, especially in the countries that had lost the war. The disaster and its absurdity made many feel that something was wrong with the previously optimistic worldview, with its uncritical belief in progress, reason, positive knowledge, civilization, and representative democracy.

Postwar economic troubles, revolutions, and counterrevolutions (and subsequently the Great Depression) further reinforced the public perception of a genuine crisis of modern society, which affected, at least temporarily during the interwar period, almost all countries with advanced science and technology. Various lines of thought disagreed with each other about which elements of the preceding worldview to retain and which to reject, to restrict in validity, or to mix in varying combinations with elements of an equally loose neoromantic worldview that could include the sense of tragedy, disruption, irrationality or mysticism, will, violence, and direct action. The spectrum of attitudes toward science reflected those disagreements. Some authors held science ultimately responsible for the European debacle, either as the social Darwinian inspiration for the war or as an utterly materialist, soulless Frankenstein leading humanity toward self-destruction. Others identified the main problem in the

---

[1] On Russian universities in the late imperial period, see G. I. Shchetinina, *Universitety v Rossii i ustav 1884 g.* (Moscow, 1976); and specifically on the institutionalization of research imperative, A. M. Korzukhina, *Ot prosveshcheniia k nauke: Fizika v Moskovskom i S.-Peterburgskom universitetakh vo vtoroi polovine XIX–nachale XX v.* (Dubna, 2006).

gap between humanity's advanced powers to manage the external world and its lagging capacity for moral and social restraint, and they called for a temporary moratorium on scientific research to give society a chance to catch up. However different and opposing these views were, they all implied some kind of malfunction in the relationship linking science and society.

Military defeat, collapse of imperial ambitions and empires themselves, and postwar economic meltdown made the feeling of crisis especially strong in Germany and parts of the former Austria-Hungary. German-speaking academics suffered a painful blow to their social prestige and economic prosperity. Formerly enjoying a dominant position in many branches of world science, they became international outcasts. Formerly envied by foreign colleagues for their privileged social status as civil servants and for the superior financial support they received from the German states, they saw both status and support severely undermined by postwar economic instability and hyperinflation. Alienated from the new democratic order, German professors nostalgically recalled the golden age of the prewar empire. While still adhering to the ideal of the "pure" and "apolitical" scholar, most tended to sympathize with reactionary opposition to the Weimar Republic. They accepted the conservative ideological argument that the mechanistic and materialistic science was at least partly to blame for the deep crisis in society. Talk about the "crisis in science"—meaning not only the undermined status of the academic profession but also serious doubts about fundamental methods and foundations of existing knowledge—became commonplace.[2]

Although Germany had been and continued to be, during the Weimar period, the source of much cultural inspiration in Russia, in their attitudes toward science, the Russian intelligentsia chose not to follow their neighbor's example, generally drawing instead almost the opposite conclusions from the traumatic war experience. It tended to see the source of the country's poor war performance and other social woes in the lack of, rather than in the excessive development of, science and technology and in insufficient support for them under the monarchical regime. The geologist and geochemist Vladimir Vernadsky, who during the war emerged as one of the main spokesmen for the academic community, was both reflecting upon and helping to define this general understanding when he argued in 1915 that

> regardless of the outcome of the war, both winners and losers will have to direct their thought toward further development of scientific applications to the military and navy affairs . . . One of the consequences—and also one of the causes—of Russia's economic dependence on Germany is the extraordinary insufficiency of our knowledge about the natural productive forces which Nature and History have granted to Russia. After the war of 1914–1915 we will have to make known and accountable the natural productive forces of our country, i.e., first of all to find means for broad scientific investigations of Russia's nature and for the establishment of a network of well equipped research laboratories, museums and institutions. . . . This is no less necessary than the need for an improvement in the conditions of our civil and political life, which is so acutely perceived by the entire country.[3]

---

[2] Paul Forman, "The Financial Support and Political Alignment of Physicists in Weimar Germany," *Minerva* 12 (1974): 39–66; Forman, "The Environment and Practice of Atomic Physics in Weimar Germany: A Study in the History of Science" (PhD diss., Univ. of California, Berkeley, 1968); Daniel J. Kevles, "'Into Hostile Political Camps': The Reorganization of International Science in World War I," *Isis* 62 (1971): 47–60.

[3] V. I. Vernadsky, *Ocherki i rechi*, 2 vols. (Petrograd, 1922) 1:5, 131–2, 140.

The end of the war, the revolution, and the civil war that followed did not invalidate his conclusions but greatly reinforced them not only in Vernadsky's own eyes, but also in the eyes of his political enemies, the Bolsheviks. Belief in science remained one of the few things intelligentsia on the opposite sides of revolutionary barricades still agreed about. In the course of the civil war the new Bolshevik government embraced and endorsed Vernadsky's vision of science and in 1922 reprinted as extremely important the collection of his addresses and proposals regarding science policy, including the ones quoted above.

No longer marginal and unimportant, but successfully clinging to their claim to rule the country and guide the revolution, the Bolsheviks drew similar lessons from the war and strengthened them with their own peculiar combination of ideological, cultural, and pragmatic reasons. From their intellectual roots—the Enlightenment, classical Marxism, and the cultural tradition of the intelligentsia—the Bolsheviks inherited a perception of science as an important ideological ally and a major force of not merely economic but also social and political progress. They expected the rational scientific worldview to unseat the power of religion and superstition over the minds of the people. At least in the Communists' own minds, the belief in the power of science as a panacea for many social problems substituted for the religious belief per se or in any case rose to similar levels of intensity and devotion. Similar perceptions and high expectations were also extended to social sciences and the humanities, in part because of the wider meaning of the words for "science" in Russian (*nauka*) and German (*Wissenschaft*), which embraced all fields of scholarship. Marxism, too, belonged to science in this wider sense, as it modeled itself upon natural science and applied the naturalistic method of explanation to the study of human society and history. Its conclusions were therefore, to Russian Marxists, as certain as scientific truths about nature. This linguistically reinforced linkage between science and Marxism helped further support their perceived alliance and mutual prestige in the Bolshevik worldview.[4]

In contrast to Germany, in Russia the prestige of science did not suffer but rose significantly during and after the war and the revolution. The Bolsheviks, in particular, placed the blame in the science-society equation squarely on society—more precisely, capitalist society. The European crisis, in their interpretation, was that of the bourgeois social order and capitalist economy, which misused knowledge toward inhumane goals. Science participated in the crisis only insofar as it was bourgeois, that is, adversely affected by irrational and mystical intellectual currents, militarism, and nationalism. Hence the motto "crisis of science" became in the Soviet translation the "crisis of bourgeois science." Under socialist principles, the Bolsheviks claimed, "all the marvels of technology, all the accomplishments of culture will belong to everyone, and never again will the human mind and ingenuity be turned into instruments of violence and the means of exploitation."[5] After attaining political power, Bolsheviks in Soviet Russia not only maintained but further promoted the intelligentsia's virtually unrestricted cult of science and technology as the key to achieving their primary economic objectives—industrialization and modernization of the country.

---

[4] P. V. Alekseev, *Revoliutsiia i nauchnaia intelligentsia* (Moscow, 1987); Helena Sheehan, *Marxism and the Philosophy of Science: A Critical History; The First Hundred Years* (Amherst, N.Y., 1993).

[5] III Vserossiiskii s'ezd sovetov rabochikh, soldatskikh i krest'yanskikh deputatov, 10–18 (23–31) yanvaria 1918 (citation from K. M. Bogoliubov, ed., *Lenin i KPSS o razvitii nauki* [Moscow, 1981], 96).

In "The Immediate Tasks of the Soviet Power," written in March 1918, a few months after the revolution, Lenin contemplated the challenges arising before the Communist Party as the ruling party. He acknowledged the lack of scientific expertise and technical cadres among the Bolsheviks as the main obstacle to the difficult task they faced of running the country and its economy and suggested a policy of cooperation and compromise with predominantly noncommunist and nonsympathizing "bourgeois intelligentsia"—educated experts of all kinds, including scientists, engineers, medical doctors, educators, and military professionals. He even included some "former industrial chiefs, former bosses and exploiters," who, according to Lenin's plan, "must take on the role of technical experts, managers, consultants and advisers."[6] He went so far, he admitted, as to compromise Communist principles by proposing to pay "bourgeois experts" excessive salaries because their collaboration was an absolutely necessary condition for the Soviet regime's economic and political survival and a key to building socialism.

As a result of this policy, throughout the 1920s, many nonparty scientists and engineers were invited to join (and did so) various offices of the Soviet state to work in responsible positions alongside Communist officials. A monarchist general and chemistry professor, Vladimir Ipatieff, directed the nationalized chemical industry and military-oriented research. A liberal, Vladimir Vernadsky, watched the Bolshevik government adopt and implement his vision of ambitious geologic exploration, in particular of the country's remote northeastern territories. A Communist and electrical engineer, Gleb Krzhizhanovsky, drafted the plan for national electrification—GOELRO—an ambitious technocratic project that was enthusiastically adopted and promoted by the Bolshevik government and became a political symbol of the entire revolutionary experiment. Lenin's related slogan, "Communism is the Soviet power plus the electrification of the entire land!" metaphorically recognized the dual nature of the early Communist regime as grounded upon a political pact between the party and "bourgeois intelligentsia."[7]

The strengthened cult of science and technology emerged as a kind of national consensus, stretching across the Red-White divide and uniting the intelligentsia of practically all political stripes amid violent political disagreements of the revolutionary era. The revolution and the civil war inflicted upon Russian scientists enormous material hardships, but at the same time, they saw their social prestige and political importance skyrocket to previously unthinkable levels. Scientists and engineers in postrevolutionary Russia achieved an unprecedented direct access to political power, perhaps the closest approximation in real historical time of the utopian project of "philosopher-guardians" in Plato's *Republic*. Over the next ten years, although noncommunist, they became de facto an active part of the governing elite and influenced many large and small policies of the early Soviet regime and day-to-day management of the nation's economy. Together with the Bolsheviks, "bourgeois experts" invented and built up the new Soviet state as a nationwide technocratic, modernizing project.

---

[6] V. I. Lenin, "Pervonachal'nyi variant stat'i "Ocherednye zadachi sovetskoi vlasti," in *Polnoe sobranie sochinenii*, vol. 36 (Moscow 1969), 137–42.

[7] Jonathan Coopersmith, *The Electrification of Russia, 1880–1925* (Ithaca, N.Y., 1992); V. N. Ipatieff, *The Life of a Chemist* (Stanford, Calif., 1946); A. V. Kol'tsov, *Sozdanie i deiatel' nost' Komissii po Izucheniiu Estestvennykh Proizvoditel' nykh Sil Rossii, 1915–1930 gg.* (St. Petersburg, 1999); S. A. Fediukin, *Velikii oktiabr' i intelligentsiia: Iz istorii vovlecheniia staroi intelligentsii v stroitel'stvo sotsializma* (Moscow, 1972).

In the process of which, they also embarked upon creating a new type of infrastructure for scientific research and development.

SCIENCE AS PROFESSION

On a rainy day in October 1912, a royal cavalcade descended on a quiet Berlin suburb and brought Kaiser Wilhelm II to the inauguration of two chemical institutes, the first ones of the newly organized research society, the Kaiser-Wilhelm-Gesellschaft. The society, funded jointly by the state and private industry, planned to establish several such institutes, each one for an exceptional scientist of "genius" rank (according to the so-called Harnack Principle), who would be freed from teaching as university professors and granted the privilege of engaging in full-time research in the direction of his choice with the help of the most up-to-date equipment and research assistants. Proud of their leadership in many fields of science, German professors convinced the state to grant science one extra level of support and protection.[8]

Even before that development, the rest of the world of science envied Germany's generously funded laboratories and university institutes. Rumors about the latest plan reached Russia at a particularly critical time, when Moscow University, the oldest and largest university in the country, half destroyed itself in turmoil. As a result of a student political meeting and protest, and government intrusion into the prerogatives of the university council, about a quarter of all professors and privatdocents resigned in protest against the state violation of the principle of academic autonomy. In typically inflated intelligentsia rhetoric, biology professor K. A. Timiryazev used and exaggerated the news from Germany to contrast the obscurantism of the Russian government with the enlightened recognition of the value of science in the "entire civilized world." Describing as "backward" the fact that in Russia "all science [is] concentrated in universities," he called on Russian society to liberate science from both teaching obligations and the state by creating privately funded institutes for research.[9]

The idea attracted an enthusiastic following among the Russian academic intelligentsia, and by 1917, the Moscow society founded by the resigned professors had raised funds and established two institutes, one for biology and one for physics. The revolution and the civil war, however, channeled development in a somewhat different direction. On the one hand, "liberation from the state" was arrested as the revolutionary Bolshevik government nationalized private foundations and their research institutes. On the other hand, the socialist state proved much more willing to satisfy scientists in the second half of their agenda, the "liberation from teaching." The Bolsheviks supported the scientists' project of separate research institutes not only because it helped them win scientists as collaborators in their larger social experiment but also because they had much less respect and tolerance for the same academics in the role of university professors. As quickly became apparent, university teachers and administrators were subject to stricter political demands and controls than were employees at the new research institutes. This discrepancy, and the availability of new,

---

[8] Jeffrey Johnson, *The Kaiser's Chemists: Science and Modernization in Imperial Germany* (Chapel Hill, N.C., 1990).
[9] K. A. Timiriazev, "Novye potrebnosti nauki XX veka i ikh udovletvorenie na zapade i u nas" (1911), in *Nauka i demokratiia: Sbornik statei 1904–1919 gg.* (Moscow, 1963), 56–66.

attractive jobs, soon led to a sizable portion of the Russian scientists being employed primarily, if not exclusively, in research rather than in teaching positions.[10]

The profile of the new institutions reflected the experience of the Great War as well as specific Marxist ideas about science, which at the time sounded dangerously radical and provocative to much of the public. In particular, Bolshevik authors insisted that scientific thought, even at its most abstract, had originated from and stayed linked to the practical, especially economic, activity of the people. They thus categorically rejected the ideology of German and British (as well as the majority of Russian) academics that privileged and separated "pure science" from "applied" research and technology. To Marxists, the concept of pure science bordered on nonsense, as even the most fundamental science was worthy of its name only if it had potentially useful applications, at least in the long-term perspective. While offering scientists increased public support for research, the Bolsheviks required in return that investigations be targeted toward social and economic goals of society and the state.

With this purpose in mind, Communists rejected the principle of the autonomy of the academic profession as a closed, self-governing corporation. Instead, they promoted the ideal of science as a public profession or branch of civil service, supported by public funds and consciously serving social needs. Orientation toward producing useful knowledge required that the main directions of research and the distribution of resources for science would be rationally "planned" via institutions of the state. Such ideas were not unique to Russia or to the Bolsheviks: during the war, scientists in other belligerent countries formulated similar, if less radical and demanding, proposals in response to the need to mobilize science for wartime use. As the military conflict subsided and science returned to peacetime existence, such ideas were largely abandoned or, as in Germany, restricted by conditions of military defeat. In Soviet Russia, however, they did not dissipate after the war but were taken up and carried much further by the revolution, declared essentially socialist, and promoted by the revolutionary government as the program for the ensuing period of peace.

The Russian academic intelligentsia shared part of the above ideas. Like their wartime peers in other countries, Russian professors tried to make their knowledge useful during World War I but became painfully aware of the absence of working links between Russian academic science and industry. They developed proposals for radical reform in the country's scientific infrastructure, including calls to turn science toward practical tasks facing the nation. As the first step toward fulfillment of this goal, Vernadsky established in 1915 the Commission for the Study of Natural Productive Forces at the Petrograd Academy of Sciences, thus reorienting the academy away from its century-long preference for pure science. After the revolution, the commission and its practical economic orientation provided the basis for the first collaboration agreement between the Academy and the Bolshevik government.[11]

The new regime particularly welcomed the turn of science toward economically important research. Several dozen research institutes were established during the civil war in fields such as optics, roentgenology and radiology, aero- and hydrodynamics,

---

[10] Alexei Kojevnikov, "The Great War, the Russian Civil War, and the Invention of Big Science," *Science in Context* 15 (2002): 239–75.

[11] V. I. Vernadsky, "Ob izuchenii estestvennykh proizvoditel'nykh sil Rossii (Dolozheno v zasedanii Fiziko-Matematicheskogo otdeleniia 8 aprelia 1915 g.)," in *Ocherki i rechi* (cit. n. 3), 1:5–25; Kol'tsov, *Sozdanie i deiatel'nost' Komissii po izucheniiu estestvennykh proizvoditel'nykh sil Rossii* (cit. n. 7).

plant breeding, rare metals, and radio. Russia's economy suffered much more severely as a result of the civil war than the economies of Austria and Germany did from postwar inflation, but the rise in the social value of science compensated somewhat for the material losses. Thanks to this preferential treatment, science actually managed to advance despite poverty and starvation. During the civil war, scientists received rations essential for their physical survival amid hunger and devastation, while many of the newly established research institutes became administratively independent from the universities. The Bolsheviks were particularly enamored with the fields that combined revolutionary utopia with utilitarianism—the promise of achieving some grand practical, even if only remotely possible, goal on the basis of some modernist, revolutionary novelty, such as radioactivity, X-rays, aviation, or genetics. The corresponding state-funded institutes were often multidisciplinary and combined basic investigations in advanced science with the design and production of sophisticated technology, a "symbiosis, . . . a fusion of 'pure' science, technology, and engineering," that several decades later began to be called "big science."[12]

More important than a significant increase in size, a new definition of the research worker distinguished the new institutes from the preceding generation of university laboratories. Up until then, the usual position for a scientist was that of a university professor or assistant involved in undergraduate teaching. Employment at the new institutes created a large group of salaried scientists and research engineers who were professionally occupied full time with scientific research, training of advanced students, and the development of new technologies, with no or only incidental obligations to teach undergraduates. Effectively, the Soviet government lifted scientific research up to the status of a new mass profession and a branch of civil service, recognized as socially important and supported in its own right, rather than indirectly via higher education, by generous public funds.

Under the Bolshevik rule, scientists lost much of their autonomy and independence but acquired more social prestige and de facto influence on politically important decision making. The Soviet regime valued science more highly and allocated it a proportionally larger share of the national income than did contemporary governments in economically better developed and more prosperous countries. It strongly opposed the ideology of pure science, promoting instead the ideal of science as potentially usable—even if not always immediately applicable—knowledge about the world. The Soviets particularly excelled in their efforts to popularize scientific knowledge among the widest possible segments of the population and make science education available to broad masses.[13] They strongly pressured the existing scientific cadres into doing research related to economic and societal needs and, with this purpose in mind, emphasized the need for rational organization and planning of scientific enterprise. At least some of the above features of the Soviet model appealed to outside observers, especially on the political left, and influenced reforms in other countries. As early as the 1920s, the Nationalist (Guomindang) government in China established scientific institutions modeled upon the Soviet example. During the 1930s, scientist

[12] Dominique Pestre and John Krige, "Some Thoughts on the Early History of CERN," in *Big Science: The Growth of Large-Scale Research*, ed. Peter Galison and Bruce Hevly (Stanford, Calif., 1992), 93.
[13] James T. Andrew, *Science for the Masses: The Bolshevik State, Public Science, and the Popular Imagination in Soviet Russia, 1917–1934* (College Station, Tex., 2003); J. G. Crowther, *Soviet Science* (London, 1936).

activists in western Europe started demanding a similar public recognition for science from their own governments.

In Europe, the ideal of pure, disinterested research remained largely unchallenged in the scientific community until the 1930s. The Great Depression subjected to critical scrutiny not only the failures of capitalism but also the role of science in it, as overproduction of goods and unemployment were blamed, at least in some circles, on technological advances achieved through science. But the very scene of a society in deep crisis and of human misery alone made many scientists, regardless of their political sympathies, think more about social problems and the possible role of science in solving them. Even the British Association for the Advancement of Science, whose long-standing principle was to avoid becoming involved in issues of politics and government, agreed in 1933, after two years of hesitation and deliberation, that scientists should no longer isolate themselves from general affairs and added to its traditional fields of concern the study of the social impact of science.[14]

A crucial encounter occurred, however, in July 1931 at a rather unlikely and otherwise insignificant event, the Second International Congress of the History of Science and Technology in London. Without much advance warning, a Soviet delegation that included a high-ranking politician and several leading scientists and Marxist philosophers arrived at the congress. At a specially arranged session, Nikolai Bukharin, Boris Hessen, and others presented an unprepared audience with a different discourse on science, which to some of the attendees sounded like a Martian language and to others like a revelation. According to the Soviet view, science, past and present, was not just the intellectual pursuit of a few selected great minds but an intelligent answer to social and economic problems and a method of solving them. It did not have to be individualistic and uncoordinated but could be planned, consciously directed at socially useful goals, and pursued in collective work. Instead of isolating themselves from society, scientists could and should consciously engage in it. Immediate converts to this view formed a closely knit group of left-wing scientists, which included, among others, J. D. Bernal, Hyman Levy, J. B. S. Haldane, and Joseph Needham, who subsequently became principal activists in the social relations of science movement in Great Britain.[15]

Later in 1931, Bernal, by then a member of the Communist Party, traveled to the Soviet Union to become acquainted firsthand with the Soviet organization of science. "Is it better to be intellectually free but socially totally ineffective or to become a component part of a system where knowledge and action are joined for one social purpose?" he asked his British colleagues rhetorically, convinced by then that the time of haphazard, individualistic, and small-scale science had already passed and that research itself had entered an industrialized, massive stage. Beginning in 1932, Bernal and other left-wing scientists became the dominant voice in the Association of Scientific Workers and popularized Marxist views in general and Marxist vision of science in particular among British academics and the public. Besides raising scientists' social consciousness, the association pushed for an increase in

---

[14] William McGucken, *Scientists, Society, and State: The Social Relations of Science Movement in Great Britain, 1931–1947* (Columbus, Ohio, 1984); J. G. Crowther, *The Social Relations of Science* (London, 1941).

[15] N I. Bukharin et al., *Science at the Cross Roads; Papers Presented to the International Congress of the History of Science and Technology Held in London from June 29 to July 31, 1931, by the Delegates of the USSR*, 2nd ed. (London, 1971).

state allocations for research and for a tighter relationship between scientists and the government.[16]

The Soviet example was also important in France, where, beginning in 1930, Jean Perrin led a similar campaign to recognize scientific research as a separate profession in civil service, while the Communist Paul Langevin supplied the movement with information about Soviet science. Socialist electoral victories in 1932, and the united Popular Front government in 1936, brought scientists' political allies into power. The Caisse Nationale de la Recherche Scientifique (CNRS), established in 1935 in a merger of two modest predecessors, assumed the leading role in distributing funds for research. Irène Curie, and after her Jean Perrin, accepted an appointment in the Popular Front government to the newly created position of deputy minister for scientific research and managed to increase significantly state allocations for research. By 1939 CNRS was supporting, fully or partly, approximately 600 salaried researchers, about half of the academic scientists in France.[17]

By Bernal's estimate, the United Kingdom was spending about 0.1 percent of its national income on scientific research in the mid-1930s, compared with 0.8 percent in the USSR, for which much better statistical data was also collected and published at the time.[18] In 1937, the British government turned down Bernal's memorandum, written on behalf of the Parliamentary Science Committee, proposing reform in the national funding of research. As the war drew closer, however, even the Royal Society and the liberal opponents of Marxism agreed with the demand for government coordination of science, while Bernal and his supporters dropped their previous disdain for military research.[19] With the start of the war, the Communist Bernal entered government service and played a key role in numerous military planning and research committees, seeing in the wartime mobilization of science a realization of his Marxist proposals of the 1930s. When British scientists working on the development of radar technology decided to meet regularly for informal discussions with representatives from different branches of the military, they even called their gathering the "Sunday Soviet."[20]

The political awakening of scientists in the United States came mainly through the British example but with considerable delay. In 1937, Robert K. Merton still observed that "attempts for concerted action by English scientists contrast sharply with the apathy of scientists in this country." The mood changed over the following two years, with the establishment of two activist organizations: the liberal and antifascist American Committee for Democracy and Intellectual Freedom, and the more radically leftist adaptation of its British prototype, American Association of Scientific Workers. Ultimately, in the United States, as in Britain, the war emergency forged an otherwise improbable alliance between normally anticommunist military leaders,

---

[16] J. D. Bernal, "Science and Society," *Spectator*, 11 July 1931, 43–4, reprinted in Bernal, *The Freedom of Necessity* (London, 1949), 334–9; Gary Werskey, *The Visible College: The Collective Biography of British Scientific Socialists of the 1930s* (New York, 1978).

[17] Spencer Weart, *Scientists in Power* (Cambridge, Mass., 1979); Mary Joe Nye, "Science and Socialism: The Case of Jean Perrin in the Third Republic," *French Historical Studies* 9 (1975), 141–69.

[18] Chris Freeman, "The Social Function of Science," in *J. D. Bernal: A Life in Science and Politics*, ed. Brenda Swann and Francis Aprahamian (London, 1999), 101–31.

[19] Maurice Goldsmith, *Sage: A Life of J. D. Bernal* (London, 1980); Neal Wood, *Communism and British Intellectuals* (New York, 1959).

[20] Guy Hartcup, *The Challenge of War: Britain's Scientific and Engineering Contributions to World War Two* (New York, 1970), 24, 52.

such as General Leslie Groves, and left-wing scientists with Communist connections, such as J. Robert Oppenheimer, whose respective visions regarding the proper relationship between science and the government became compatible and overlapping to a significant degree. They worked jointly on large-scale military research projects, collaborated in the establishment of the institutions of big science, and brought about federal support for big science and the recognition of it by society and the state.[21]

SOCIAL CONSTRUCTIVISM

Marx's classical critique of bourgeois ideology argued that political, economic, philosophical, and legal doctrines reflected and represented social and class interests. At the same time, Marx generally stopped short of applying a similar analysis to the doctrines of the natural sciences. This concession to the dominant positivistic mood of the era somewhat contradicted the general Marxist assertion that science, including the most abstract concepts of mathematics, was as much a human activity as other forms of knowledge and had its origin in human life and needs.[22] Profiting from the early twentieth-century intellectual turmoil and epistemological critiques of the foundations of science by Ernst Mach and Henri Poincaré, Aleksandr Bogdanov carried the Marxist thought one logical step further in his 1918 brochure *Socialism of Science*, by introducing a perspective on science that has since become known as "social constructivism."[23]

According to Bogdanov, science, understood as an "organized social and economic experience" of the laboring humanity, reflected the needs and concerns of society, not merely the natural world. Thus it was not accidental, for example, that modern European science appeared during the sixteenth and seventeenth centuries in parallel with the rise of new capitalist societies and new economic demands of manufacturing, navigation, trade, and warfare. New scientific knowledge, such as Newtonian mechanics, in turn helped advance capitalist technology and production further, into the industrial age. The nature of scientific knowledge and activity, therefore, should be understood as closely linked with technology and industry, not separated from them. In the course of the nineteenth century, as the bourgeoisie, according to Marx, was losing its revolutionary and historically progressive role, so were the bourgeois features of contemporary science. This resulted in science's increasing abstraction, the rise of the ideology of "pure" science, science's inaccessibility to the masses and alienation from the concerns and practices of working classes, and its subordination to market relations. Bogdanov envisioned that the rise of a new, socialist society would bring along a serious reform in the existing body of scientific knowledge, a change in the viewpoint that he compared to the shift in the point of view required by the Copernican revolution in astronomy. Novel social and labor practices would

---

[21] Peter J. Kuznick, *Beyond the Laboratory: Scientists as Political Activists in 1930s America* (Chicago, 1987); Alexei Kojevnikov, "The Making of the Soviet Bomb and the Shaping of Cold War Science," in *Reappraising Oppenheimer: Centennial Studies and Reflections*, ed. Cathryn Carson and David A. Hollinger (Berkeley, Calif., 2005), 129–45. As Kuznick describes, however, American engineers and especially physicians preceded scientists in developing interest toward the Soviet experience starting in the early 1930s, particularly attracted to the system of socialized medicine.

[22] For a summary of Marx's sociology of science, see Michael Mulkay, *Science and the Sociology of Knowledge* (London, 1979), chap. 1

[23] A. A. Bogdanov, "Sotsializm nauki" (1918), reprinted in Bogdanov, *Voprosy sotsializma: Raboty raznykh let* (Moscow, 1990), 360–76.

also enrich science with this important new knowledge, "proletarian science," in Bodganov's terminology.[24]

Trotsky and Lenin rejected talk about "proletarian science" as premature and dangerously radical, denouncing it as damaging the party's pact with bourgeois specialists. For them, it was more important to recruit the already existing science and expertise into the immediate service of socialist construction.[25] Bolsheviks continued to criticize Bogdanov and reject the term "proletarian science," even though, upon closer examination, they were in agreement on most other basic values, such as demands for science to deliver practical results and be accessible to the masses. The subsequent mainstream Communist vision of science adhered unwaveringly to the principle of scientific materialism—a belief that science delivers truth, or at least relatively objective truth, about nature. At the same time, their discourse combined the former principle with important concessions toward "social constructivism"— that *nauki* (sciences in the wide sense) in general and the natural sciences in particular were subject to social and ideological influences, all the way up to the very content of scientific knowledge—but avoided explicit contradictions between the two approaches. According to Soviet Marxist views, it was quite appropriate that their time—the period of great social crises and revolutions—resulted in a crisis of "bourgeois" science and inspired revolutionary developments in the basic foundations of knowledge. The real-life difficulty was that the two aspects were so closely intertwined within the contradictory developments in contemporary science that it was hard to distinguish precisely which scholarly claims could be dismissed as reflecting the dead-end crisis of bourgeois thought and which ones needed to be supported as important and promising revolutionary breakthroughs.

Take, for example, Einstein's theory of relativity, which met an enthusiastic early welcome in revolutionary Russia, with the majority of scientists, avant-garde artists, and the public welcoming it as an overturning of the existing conceptual order, an accomplishment in science just as radical and revolutionary as the Russian Revolution's overturning of the existing order in society. The 1922 discovery by the Soviet mathematician Aleksandr Friedmann that the universe governed by Einstein's equations is not stable but can explode, collapse, and be born again, in a process subsequently called the big bang, also appealed, in a metaphorical way, to the revolutionary mentality and the existential experience of those who survived a social explosion of cosmological magnitude. But the same relativity theory had been inspired by and laden with the philosophical ideas of neopositivism, such as those by Ernst Mach, which appeared to contradict dialectical materialism, the Marxist philosophy of nature, and scientific materialism in general.[26] The contradiction could be resolved by following the example of Lenin's 1909 book *Materialism and Empirio-Criticism*, which attempted a demarcation between, on the one hand, novel and positive scientific developments and, on the other hand, often "wrong" and "idealistic" philosophical conclusions attached to them or derived from them by bourgeois ideologues and scientists. Adopting this approach, future president of the Soviet Academy of Sciences physi-

---

[24] On Bogdanov as philosopher, see: Arran Gare, "Aleksandr Bogdanov's History, Sociology, and Philosophy of Science," *Studies in History and Philosophy of Science* 31 (2000): 231–48; Zenovia A. Sochor, *Revolution and Culture: The Bogdanov-Lenin Controversy* (Ithaca, N.Y., 1988).

[25] Leon Trotsky, "What Is Proletarian Culture and Is It Possible?" (1923), http://www.marxists.org/archive/trotsky/works.

[26] Alexander Vucinich, *Einstein and Soviet Ideology* (Stanford, Calif., 2001).

cist Sergei Vavilov could, within the same line of discourse, praise Einstein's relativity as the great revolutionary achievement in science, criticize Einstein's philosophical "mistakes," and proclaim that for better methodological and heuristic guidance in their search for true knowledge, natural scientists should cling to the philosophy of dialectical materialism.[27]

Modern (Mendelian) genetics and eugenics provided yet another controversial case of novel scientific developments mixed with corrupting ideological influences that were difficult to separate from one another. Both fields, related by their common foundation in the concept of "hard" heredity and overlapping communities of scholars, initially received an enthusiastic welcome in the revolutionary Russia of the 1920s. As information about eugenics' close association with racism in the United States and the related practices of forced sterilization gradually became known in the USSR, eugenics increasingly came to be regarded as so thoroughly corrupted by bourgeois ideological influences that it abandoned the objective path of true science. The Soviets were first to come to a completely negative verdict about eugenics—as early as 1931—and to abandon all research conducted under this label. The close association of geneticists with the eugenics movement also partially undermined their reputation in the eyes of Soviet authorities. Elsewhere, eugenics continued to grow, inspired in particular by the example of legislation adopted in 1933 by Nazi Germany and partially imitated in a number of other countries in Europe and North America. It did not become discredited internationally until 1945, after shocking revelations about Nazi medical crimes; it quickly faded away, at least in name. Having effectively banned eugenics more than a decade earlier, the Soviet Union overreacted to the new revelations. In 1948, the agronomist Trofim Lysenko won political support for his claim that Mendelian genetics, too, was ideologically and scientifically untenable and racist and had to be abandoned in favor of his own, idiosyncratic, genetics based on the concept of "soft," or environmentally flexible, heredity. It would take the Soviet establishment sixteen years to officially recognize the latter decision as a serious mistake.[28]

Even before 1945, Nazi Germany provided another striking example of political influences corrupting science that inspired strong international reactions. The 1933 law banning non-Aryans and political opponents of the Nazi regime from the civil service resulted in massive dismissals of Jewish and socialist academics. In the first wave of firings, more than 1,000 university teachers, including some 300 professors, lost their jobs. The refugee crisis, the rising international tide of Fascism, and the looming threat of a new war moved the positions of liberal and leftist scientists in Europe closer to each other. Many on the liberal side who formerly subscribed to the ideal of the apolitical scholar saw science threatened by the Nazi assault on its core values and consequently became more active politically. Previously reluctant government officials recognized the threat of a possible war and moved increasingly toward a more active science policy that would allocate additional funds and coordinate

---

[27] S. I. Vavilov, "Novaia fizika i dialekticheskii materialism," *Pod Znamenem Marksizma*, 1939, no. 12:27–33.
[28] Diane B. Paul, *Controlling Human Heredity, 1865 to the Present* (Amherst, N.Y., 1995); Robert Proctor, *Racial Hygiene: Medicine under the Nazis* (Cambridge, Mass., 1988); Mark Adams, ed. *The Wellborn Science: Eugenics in Germany, France, Brazil, and Russia* (New York and Oxford, 1990). On Soviet genetics and the Lysenko case, see the article by Nils Roll-Hansen in this volume and the extensive literature cited there.

research with anticipated military needs. And many scientists, such as Bernal and his leftist allies, who had once strongly denounced the application of science to warfare, started seeing their most pressing social responsibility in turning their research to the service of defense. The emerging collaboration of the liberal and the Left in common opposition to Fascism came to be known, in politics, as the Popular Front. In France, united antifascist scientists helped create the alliance named Comité de Vigilance des Intellectuels Antifascistes, which unified parties left of center in joint defense of democracy against both foreign and homegrown Fascism. In discourse about science, the trend corresponding to what in politics was represented by the Popular Front developed under the slogan "science and democracy."

Scientists all across the Popular Front spectrum, from liberal to Communist, started pushing forward the idea that science and political democracy were closely linked and that both had to be defended against Fascism. From New Zealand, the Austrian refugee Karl Popper argued that scientific "progress depends very largely on political . . . democracy," as both are essentially grounded in freedom of critical discourse. In England, Bernal saw science as the crucial ally in the looming global struggle between "democratic and Fascist states," adding that "in its endeavour, science is communism" because in it "men have learned consciously to subordinate themselves to a common purpose, without losing the individuality of their achievements." Merton in the United States, hoping to prove that "democratic order is integrated with the ethos of science," defined the latter as consisting of four "institutional imperatives—universalism, communism, disinterestedness, organized skepticism." And in the USSR, Sergei Vavilov of the Soviet Academy of Sciences predicted that the allied victory over Fascism would result in "strengthening the role of science and democracy in the life of peoples," adding that "science serves progress only when combined with democracy." "Democracy" in all these pronouncements was understood in its antifascist meaning, rather than in the later cold war interpretation of it as anticommunist.[29]

The unrecognized father of Marxist social constructivism, Bogdanov, died in Moscow in 1928 in a scientific experiment on blood transfusion he conducted upon himself.[30] Soviet obituaries praised him as the organizer of the national system of blood transfusion but continued to criticize him ideologically. They were loath to acknowledge that Bogdanov's social constructivist argument continued in Soviet discourse about science, without, however, any reference to the author and any mention of his discredited notion of proletarian science. In 1931 social constructivism made its way to the West through the famous (or infamous, depending on the interpretation) paper by Boris Hessen, "The Socio-Economic Roots of Newton's Principia," delivered, along with other Soviet talks, at the London Congress of the History of Science.[31] It was subsequently picked up by the Marxist social relations of science movement of the 1930s, was strongly opposed as "externalism" by the anti-Marxist school of "internal logic of the development of science" during the cold war, and eventually

---

[29] Karl R. Popper, *The Open Society and Its Enemies* (Princeton, N.J., 1950), 404; Robert K. Merton, "A Note on Science and Democracy," *Journal of Legal and Political Sociology* 1 (1942), 115–26; J. D. Bernal, *The Social Function of Science* (London, 1939), 331; S. I. Vavilov, *Sovetskaia nauka na novom etape* (Moscow, 1946); David A. Hollinger, *Science, Jews, and Secular Culture* (Princeton, N.J., 1998).

[30] V. N. Yagodinsky, *Aleksandr Aleksandrovich Bogdanov (Malinovsky), 1873–1928* (Moscow, 2006).

[31] Loren R. Graham, "The Sociopolitical Roots of Boris Cherenkov Hessen: Soviet Marxism and the History of Science," *Social Studies of Science* 15 (1985): 705–22.

mutated into the currently more familiar version of social constructivism of the New Left around 1970. The latter version has since evolved quite far from the original Marxist one, both by greatly widening the "social" away from the narrow primacy of the "economic" and, more importantly, by dropping the scientific materialism part of the discourse, which for the old Left was and remains more important than the "constructivist" part. Some modern-day authors may even be loath to acknowledge the original Marxist roots of the "social constructivist" concept, hoping to keep them entirely invisible.

AFFIRMATIVE ACTION

The career of Pavel Cherenkov can serve as a guide through the Soviet policies of reverse privileges in education that encouraged representatives of formerly discriminated groups to enter the ranks of the scientific profession. Born in 1904 to a peasant family in southern Russia, with eight siblings from his father's two marriages, Cherenkov grew up in poverty. At the age of thirteen, he started working as a manual laborer, after completing just two years of elementary schooling. A Soviet secondary school opened in the village in 1920, after the revolution and the civil war, which allowed Cherenkov to resume his education while continuing to earn a living by occasional work at a grocery store. The radical reform of the entire educational system enacted by the revolutionary Bolshevik government in 1918 opened the door to further schooling.[32]

In their attempts to democratize access to higher education, the Bolsheviks did not remain satisfied with merely annulling the formal discriminatory barriers of gender, ethnicity, and religion. They tried to compensate actively for economic disadvantages and earlier discrimination by adopting a system of preferential treatment for potential students from underrepresented groups—workers, peasants, women, and ethnic minorities—broader and more radical than the analogous one currently known in the United States as "affirmative action." The measures included free tuition, quotas, and stipends for low-income students. The number of colleges and universities greatly expanded, and special crash courses, called *rabfaks* (workers' faculties), opened within all universities to provide a quick functional education substitute for students who had not had an opportunity to complete secondary schooling. These faculties and their students helped Communists and Communist sympathizers to seize administrative and political control over universities.[33]

Last but not least, the Commissariat of Enlightenment either abolished or softened the formal requirements for school certificates, diplomas, and degrees necessary for advancing from one educational level to another. It became possible, in principle, to enroll in a university without formal graduation from a high school, start a graduate program without fully completing an undergraduate education, and be hired as a professor without a PhD degree or equivalent. Many future prominent Soviet scientists

---

[32] On early Soviet education see: M. N. Pokrovsky, "Narodnoe prosveshchenie i vysshaya shkola," in *Izbrannye proizvedemiya*, vol. 4 (Moscow, 1967), 457–551; William G. Rosenberg, ed., *Bolshevik Visions: First Phase of the Cultural Revolution in Soviet Russia* (Ann Arbor, Mich., 1984).

[33] Sheila Fitzpatrick, *The Commissariat of Enlightenment: Soviet Organization of Education and the Arts under Lunacharky, October 1917–1921* (Cambridge, UK, 1971); Terry Martin, *The Affirmative Action Empire: Nations and Nationalism in the Soviet Union, 1923–1939* (Ithaca, N.Y., 2001); V. V. Mavrodin, ed., *Na shturm nauki: Vospominaniya byvshikh studentov fakul'teta obshchestvennykh nauk Leningradskogo universiteta* (Leningrad, 1971).

of the Cherenkov generation skipped one or another of these formal steps (but not all of them altogether) while embarking on their academic careers. Cherenkov, too, took advantage of promotional opportunities available for lower-class students. In 1924, apparently without earning a high-school diploma or completing secondary education, he enrolled in the Pedagogical Department of Voronezh State University.[34]

Graduation from a provincial university in 1928 enabled Cherenkov to become a teacher of physics and mathematics at an evening school for workers in the small town of Kozlov. The following year, however, a new and more radical stage of the cultural revolution broke out in the Soviet Union. For the first twelve years following the revolution, the Communist reforms were gradually enacted in higher education, but scientific institutes continued to remain the province of "bourgeois experts." Yet more militant factions among the Communists had worried all along about the power of experts and the possibility that they would take advantage of scientifically uneducated Soviet officials. In their worst case scenarios, instead of being politically guided by Soviet commissars, the bourgeois intelligentsia would exert its own political influence over Soviet decision making under the guise of scientific advice. Tensions exploded in 1929, with the start of the all-out industrialization campaign and the crash collectivization in agriculture.

The party annulled its pact with bourgeois experts and set out to replace them with a new generation of "red experts," who would ideally combine proper professional qualifications with sincere devotion to socialist values. The massive industrialization made the already acute shortage of technical cadres even more critical, yet despite this, many senior scientists and engineers educated prior to the revolution were demoted or purged on accusations of disloyalty and sabotage. Many more, however, declared themselves reformed into earnest supporters of socialism. And in the meantime, an even larger cohort of fresh graduates with loyal backgrounds was being hastily educated in engineering colleges and universities, whose numbers quadrupled.[35] This time, the cultural revolution spread wider, affecting not only colleges but also research institutions and their staffs. Renewed and intensified attempts to attract lower-class students, women, and minorities into the ranks of scientific researchers extended affirmative action to the greatly expanded *aspirantura*, or graduate studies programs.

A thorough statistical study of the nation's scientific cadres in 1930 represented the existing situation as follows. There were approximately 50,000 positions whose holders engaged in scientific activities of some sort, about 18,000 of them primarily research positions. In the meantime, the study acknowledged, the goal of "planned development of scientific thought in the USSR require[d] in the shortest time possible, to attract new and massive human contingents." Between 1917 and 1929, the proportion of women among scientific cadres had risen from 11.7 percent to 18.8 percent (to 23.3 percent among graduate students). The study mentioned the problem of the glass ceiling and the unequal representation of women in various fields of research, with the highest percentages achieved in philosophy, pedagogy, and medicine (38.9, 36.8, and 35.7 percent, respectively). Half of all scientists were concentrated in Moscow and

---

[34] A. N. Gorbunov and E. P. Cherenkova, eds., *Pavel Alekseevich Cherenkov: Chelovek i otkrytie* (Moscow, 1999).

[35] Sheila Fitzpatrick, *Education and Social Mobility in the Soviet Union, 1921–1934* (Cambridge, UK, 2002); Kendall E. Bailes, *Technology and Society under Lenin and Stalin: Origins of the Soviet Technical Intelligentsia* (Princeton, N.J., 1978).

Leningrad, but conscious efforts were under way to expand both geographic and ethnic diversity, by educating and promoting members of underrepresented minorities. In the Ukrainian Soviet Socialist Republic, for example, the proportion of Ukrainians among scientists (46.8 percent) was still significantly lower than that among the republic's population as a whole (80.0 percent), whereas the corresponding numbers for Russians and Jews in Ukraine were 27.6 versus 9.2 and 20.8 versus 5.4 percent, respectively.[36]

Cherenkov's biography once again reflected the major trends and events of the period. His peasant father was exiled as a kulak during village collectivization in 1930, while his father-in-law, a professor of Russian literature at Voronezh, was dismissed as a bourgeois professor and sent to work in a labor camp. His relatives' political troubles and possible suspicions they could have thrown on him did not prevent Pavel from being accepted, the very same year, to one of the country's most prestigious graduate programs at the Physicomathematical Institute of the USSR Academy of Sciences in Leningrad. Most likely, his peasant background worked as a strong argument in favor of admission, outweighing his relatives' problems. Soviet educational policies at the time still did not recognize academic degrees such as the PhD. Instead of writing a thesis, the job of an aspirant was to learn the trade of scientific research while working as a junior apprentice alongside established scientists.

Cherenkov's adviser, Sergei Vavilov, had the background, education, and looks of a typical bourgeois professor, but he always maintained a very loyally Soviet posture in his public pronouncements. Under Vavilov's direction, Cherenkov started working diligently on luminescence induced by radioactive gamma rays and in 1933 discovered a faint background glow, which turned out to be a heretofore unknown kind of radiation. Now called the Cherenkov radiation and used for detecting high-speed particles in accelerators and cosmic rays, it brought Cherenkov worldwide recognition, a 1946 Stalin Prize (jointly with Vavilov), and a 1958 Nobel Prize in Physics. (Vavilov had died earlier and therefore could not be nominated.)[37]

The Soviet ideologically shaped solution to the extreme shortage of technical specialists in the 1930s included a dramatic expansion of public funding to education at all levels, the abolishment of tuition along with all other fees for higher education, a radical increase in the number of specialized technical schools, and an unprecedented broadening of the demographic base of science and engineering students, with massive promotion of representatives of lower classes, women, and ethnic minorities. Ten years later, in the course of World War II, that system of training engineering cadres stood its ultimate test by proving its ability to design advanced technological weapons and outproduce Nazi Germany in tanks and aircraft. After the end of the war, the Soviet educational model exerted strong international influence, first of all on the countries of Eastern Europe, and not only on the initiative of local Communists. In 1945, Hungarian biochemist and Nobel laureate Albert Szent-Györgyi traveled to the Soviet Union to familiarize himself with the Soviet organization of science. After returning from this trip, he embarked on the transformation of Hungarian science along similar lines, with research institutes independent from the universities

---

[36] O. Yu. Shmidt and B. Ya. Smulevich, eds., *Nauchnye kadry i nauchno-issledovatel'skie uchrezhdeniya SSSR* (Moscow, 1930), 4, 17, 19, 49–52, 59.

[37] I. M. Frank, "A Conceptual History of the Vavilov-Cherenkov Radiation," *Soviet Physics Uspekhi* 27 (1984): 385–95.

and with massive specialized training of scientists and engineers. Other countries in Eastern Europe enacted similar reforms.[38]

The largest reform of this kind—some argued the largest program of technology transfer in history—developed between the Soviet Union and the People's Republic of China. In 1949, Mao Zedong ordered a reorientation of Chinese science and education exclusively toward the Soviet example. During the subsequent ten years, more than 10,000 Chinese students and scientists went to the USSR for training, and approximately the same number of Soviet experts visited and worked in China as teachers and consultants at hundreds of industrial and academic projects. Existing academic institutions in China underwent reform, and new ones came to be established based on Soviet blueprints. An entire generation of Chinese scientists was trained and a new scientific infrastructure built before 1960, when the relations between the two Communist parties worsened and, in response to Chinese accusations of "revisionism"(that is, preferring peaceful coexistence with American imperialists to world revolution), Nikita Khrushchev ordered the complete withdrawal of Soviet experts from China.[39]

In the cold war United States, the National Manpower Council and the Commission on Human Resources and Advanced Training had been issuing warnings since the early 1950s that "Russia is already far advanced on a program of mass education, and by selecting the most competent pupils at each school level is educating greater and greater numbers of students through technical institutes, college and the university."[40] The issue caught national attention, however, only after the spectacular launch of *Sputnik* in October 1957. American politicians and the public entered the debate about how the country managed to lose its scientific and technological supremacy: in 1945 they had felt comfortable in seeing the Soviets as technologically far inferior; fifteen years later, the USSR had caught up with the United States by testing atomic weapons, had come even in developing thermonuclear ones, and had actually surpassed America in missile design.

In the ensuing soul searching, some lessons and measures were easier to accept and execute than others. A further increase in government funding for scientific research was no longer politically problematic: that familiar and accepted lesson had been assimilated from the wartime experience and the Manhattan Project. This time the financial flow jumped another level, not only through military but also through civilian, or ostensibly civilian, channels such as the National Science Foundation and NASA, the newly established, Soviet-inspired government agency. The offices distributing this flow abandoned the once functional categorization of science into pure and applied research, just as the Soviets had done thirty years earlier, at a time

---

[38] Gábor Palló, "Accommodation to a New Center: Albert Szent-Györgyi's Trip to the Soviet Union," in *Travels of Learning: A Geography of Science in Europe*, ed. Ana Simões, Ana Carneiro, and Maria Paula Diogo (Boston Studies in the Philosophy of Science, vol. 233) (Dordrecht, Netherlands, 2003), 329–42.

[39] Chu-yuan Cheng, *Scientific and Engineering Manpower in Communist China, 1949–1963* (Washington, D.C., 1965); Baichun Zhang, Jiuchun Zhang, and Fang Yao, "Technology Transfer from the Soviet Union to the People's Republic of China, 1949–1966," *Comparative Technology Transfer and Society* 4 (2006): 105–71.

[40] Commission on Human Resources and Advanced Training, *America's Resources of Specialized Talent: A Current Appraisal and a Look Ahead; The Report of the Commission on Human Resources and Advanced Training* (New York, 1954); National Manpower Council, *A Policy for Scientific and Professional Manpower: A Statement by the National Manpower Council* (New York, 1953).

of equally dramatic increases in state support for science and technology. The mentality of scientists on the receiving end followed along, and the label "pure science" gradually died out in the vocabulary of American scientists. The old ethos requiring scientists to be ashamed of taking out patents started to look obsolete, and the road toward the later commercialization of science opened up.

The main bottleneck and real difficulty in connection with the *Sputnik* crisis concerned scientific manpower, for the increase in funding could not be quite matched by a corresponding increase in the training of qualified personnel, and making higher education free proved to be outside the limits of political possibility in the United States. Still, *Sputnik* helped invigorate previously stalled proposals for reform in American education through federal support. The National Defense Education Act, passed in 1958, reflected political compromise: ideological and legal conditions restricted the use of federal funds at the undergraduate level and below, effectively directing it to the graduate level and higher, with serious implications for the resulting "brain drain."

Importing qualified manpower from overseas ultimately proved to be a politically and ideologically easier solution than fixing remaining deficiencies in education at home. The memory of the special role played by foreign émigrés in the atomic bomb project was still very vivid. Changes in U.S. immigration policies—turning away from earlier racial quotas toward preferential acceptance of trained specialists—started in the 1950s and continued under the Immigration and Nationality Act of 1965. The proportion of professional and technical workers among new immigrants had increased to 17.9 percent by 1960 and to 29.4 percent by 1970, more than twice the share of such workers in the total U.S. population. By 1966, the share of foreign-born PhDs in American science constituted roughly 12 percent and continued to grow.[41]

In the early post-*Sputnik* years, about half of scientific immigration came to the United States from Europe, especially from Britain and Germany. At one point, the British government pronounced the very existence of the nation's missile program to be in danger because so many of its participants had taken jobs with NASA. A 1962 Royal Society report cried out about the loss of British scientists to U.S. immigration and coined the term "brain drain." But Britain was actually in the middle of the food chain: although it lost scientists to the United States, it also acquired them from other countries, especially from its former colonies. The dynamics of the brain market soon shifted toward bringing a proportionally larger share of immigrant scientists to the United States from India, Taiwan, South Korea, and later other regions, more arriving for graduate study in the United States than with PhDs in hand. The new arrivals, and the affirmative action policies demanded by the civil rights movement, which were similar to but more limited than earlier Soviet efforts, challenged the still widespread racist and sexist stereotypes about scientists and ultimately changed the demographics of the scientific profession in the United States. Today, a large portion of the presently retiring generation of scientists in the West and those who are replacing them in academic positions owe the existence of their jobs to the launch of the little aluminum sphere in 1957. The fact that science now is not only much larger

---

[41] Barbara Barksdale Clowse, *Brainpower for the Cold War: The Sputnik Crisis and National Defense Education Act of 1958* (Westport, Conn., 1981); Laura Fermi, *Illustrious Immigrants: The Intellectual Migration from Europe, 1930–41* (Chicago, 1968).

but also much more multiracial and multicultural than fifty years ago is probably the most important, if often overlooked, consequence of the *Sputnik*-enticed public awareness about the Soviet Union's massive training and employment of scientists with diverse ethnic, gender, and class backgrounds.[42]

CONCLUSIONS

In developing the understanding of Soviet science presented above, I tried to overcome a trio of mutually connected obstacles that have stood in the way of that understanding. The first one may be called methodological; it included an attempt to understand historical specificity without recourse to divisions into polar categories and oppositions. Instead, a style of analysis was adopted that uses as its tools interactions, connectivity, and mutual influences. The second was political, the prevalence of cold war stereotypes in academic discourse about all matters Soviet, which did not go away with the end of the cold war and the Soviet Union itself. This discourse had encouraged on both sides of the iron curtain, and to a large extent continues to encourage even now, exactly the opposite: a dualistic kind of analysis with a set of polar categories—East-West, socialist-capitalist, totalitarian-democratic, and such—postulated from the start and then providing the main reference frame for sorting out the particulars and for understanding the complexities and paradoxes of historical developments. The third may be called either linguistic or the problem of vision. The dualistic mentality instinctively tries to hide away analogous trends and dependencies, to make invisible mutual influences and borrowing, even to the point of deliberately attaching different linguistic labels to identical phenomena (for example in the official names of the professions of the "cosmonaut" and the "astronaut") lest the naïve audience be confused by their similarity. Parting with such instincts, ingrained in one's mind by more than a half century of public use, may not be an easy matter; in the above discussion I tried to battle, perhaps not entirely successfully, with them in my own thinking and to develop at least a partially alternative vision.

The alternative presents the phenomenon of Soviet science with its peculiarities and distinctiveness, but without making it the alienated "other." Instead, the Soviet case is interpreted as a part of general international developments in twentieth-century science, influenced by some, transforming and influencing in return some others. This essay specifically concentrated on those aspects of the story that made the Soviet example important elsewhere and, in a wider sense, also for "us" today, as a part, however invisible, of contemporary scientific practices. One such important twentieth-century trend is the political recognition of vast state involvement in supporting and directing scientific research. The trend appeared inconclusively in many countries during the course of WWI, demanded particularly strongly by Russian scientific intelligentsia prior to the revolution, subsequently adopted by the Bolsheviks and implemented on a massive scale within the Soviet experiment in the 1920s, then taken up by leftist scientists in Europe in the 1930s, and generally, perhaps permanently, accepted, even by anticommunists, as a lesson learned from the Second World

---

[42] The Royal Society, *The Emigration of Scientists from the United Kingdom to the United States: A Report of the Committee Appointed by the Royal Society* (London, 1963); Herbert G. Grubel and Anthony Scott, *The Brain Drain: Determinants, Measurement, and Welfare Effects* (Waterloo, Ontario, 1977); Bradley W. Parlin, *Immigrant Professionals in the United States: Discrimination in the Scientific Labor Market* (New York, 1976).

War. A related and similarly important trend involved public recognition of scientific research as a profession, which took different forms in different countries but brought about an establishment of a new infrastructure—separately funded and governed institutions for research—and a new social group, that is, salaried scientists and engineers concerned primarily with research rather than teaching.

The increased social and political importance of science inspired, in a way, a reconsideration of the general theoretical understanding of the relationship between science and society. The Soviet Marxist discourse on the matter produced an analysis that has subsequently evolved into what is currently known under the term "social constructivism." At the same time, the enormous expansion of the scientific profession in the course of the twentieth century depended upon radical changes in its demographics, namely its opening up to representatives of previously underrepresented and discriminated groups, by gender, class origin, ethnicity, or race. Such policies, pioneered in the interwar USSR, were an essential part of the Soviet effort to educate within a short period a massive contingent of professional scientists and engineers. The message of success in outproducing other scientifically advanced countries in scientific education was brought out, powerfully, by the launch of *Sputnik* in 1957. This event caused, in turn, a similarly radical expansion of the scientific profession internationally, first of all in the United States, with corresponding demographic changes in the racial and gender representation in science.

One result of this study, potentially troubling to some, is a gradual uncovering of the degree to which some features that were once thought to belong specifically to Soviet science or ideologically promoted by Communism, have contributed to, evolved into, or become part of today's generally accepted scientific practices and ostensibly anticommunist worldview. Some of them have achieved this status by the way of conscious or unconscious borrowing, others by osmosis, rivalry, negation, transformation, or simply renaming. This interconnectedness was part of a general process that changed science, and the world, in the course of the twentieth century. The realization of such a worst case nightmare would have certainly upset Senator Joseph McCarthy and others like him, who smelled and feared "communist infiltration, communist indoctrination, communist subversion and the international communist conspiracy." To the degree that we no longer share the senator's mindset and paranoia, and can treat Communism as a historical event, there is no need to remain in a state of denial.

The Conquest of Science:
Women and Science in Russia, 1860–1940

By Olga Valkova*

ABSTRACT

This essay describes the growing number of women in science in Russia from 1860 to 1940, analyzing the development of a significant community in terms of three generations. These generations are defined by the removal of various obstacles to women's participation in the sciences. The decisive transitions took place with the creation of higher education for women in the 1870s and the establishment of formal gender equality by the Bolshevik regime after 1917. To develop a composite picture, many women's careers are examined.

INTRODUCTION

This article concerns women who entered the scientific profession, as well as women who participated in science on a regular basis, in the Russian empire and later in the Soviet Union. Although both in Europe and the United States there is a considerable national literature on this topic, in Russian historiography one cannot find a single monograph considering this question. There are a few books about the very first Russian women scientists working in different disciplines (physicians, geologists, meteorologists), but the majority of them are more reminiscences than historical explorations.[1] And aside from a few brief articles, there has been nothing at all about the "women in science" question since 1917; because there was no "woman's question" in the USSR, there was nothing to talk about.[2] There are, of course, more than a few biographies of Russian women scientists, but many more women scientists have been completely forgotten. Before the early 1990s, there were no biographical handbooks about women scientists in Russia, and even today there are very few.[3]

It appears that until the 1990s, Western historiography was more interested in this topic than Russian historiography was. There exist a number of books about Russian

---

*Institute of the History of Natural Sciences and Technology, Staropansky per. 1/5, Moscow, Russian Federation; o.valkova@ihst.ru.

I should like to thank Mike Gordin, Karl Hall, and Alexei Kojevnikov for their help with editing and with the English language, and my colleagues Oleg Belozerov, N. A. Grigor'ian, S. S. Ilizarov, I. N. Iurkin, E. S. Levina, G. I. Liubina, E. N. Mirzoian, V. M. Orel, Kirill Rossiianov, O. V. Sevast'ianova, and Z. K. Sokolovskaia for interesting ideas, discussions, and moral support.

[1] A. A. Shibkov, *Pervye zhenshchiny-mediki Rossii* (Leningrad, 1961); D. V. Nalivkin, *Nashi pervye zhenshchiny-geologi* (Leningrad, 1979); and E. S. Selezneva, *Pervye zhenshchiny geofiziki i meteorologi* (Leningrad, 1989).

[2] See, e.g., V. A. Ribasov, "Russkie zhenshchiny v nauke," *Nauka i zhizn'*, 1949, no. 3:27–9.

[3] For example, there are two books published in Nizhi Novgorod: *Lichnost' v nauke: Zhenshchiny-uchenie Nizhnego Novgoroda; Sb. ocherkov i vospominanii*, 2 vols. (Nizhni Novgorod, 1997–99).

©2008 by The History of Science Society. All rights reserved. 0369-7827/08/2008-0006$10.00

women in intellectual professions in different historical periods, but there are more books about women writers than about women scientists.[4] This limited historiography is not uninteresting. For example, Ann Hibner Koblitz's *Science, Women, and Revolution in Russia* analyzes a conflict between two different impassioned aspirations of Russian women of the 1860s–70s: the choice between science and revolution.[5] The remaining slim literature concerns women's roles in higher education.

Of course, women interested in mathematics and the natural sciences appeared in Russia much earlier than the 1860s. Their tracks can be found in memoirs and classical literature in the late eighteenth century. For example, the famous Russian memoirist Philip Vigel (1786–1856) wrote about one such woman, Anna Alexandrovna Turchaninova (1774–1848), whom he met in his childhood. He wrote:

> At less then 20 years old she avoided society, dressed in a slovenly manner, was advantageously engaged in mathematics, knew Latin and Greek, was going to study Yiddish and even wrote verses from time to time—though very unsuccessfully. She was known under the name of Philosopher [*Filosofka*]. All Kiev scholarship hid at that time under monastic cloaks at the Brothers' monastery [*Bratskii monastir'*]. She discovered it and, being free from worldly weakness, was not afraid to make friends with some of the monks teaching sciences in the ecclesiastical academy. . . . Her conversation was very attractive for me. She readily told me about her relations with the honorable scientists, with the professors of Moscow University.[6]

Anna Turchaninova descended from a landowning family that had an estate in the Kiev region, a house in Moscow, and properties elsewhere. She published her poems in several magazines in 1798 and in a book in 1803. More relevant for us is that she translated from Latin and published in verse form a book called *Natural Ethics or the Laws of Morality, Directly Drawn from the Contemplation of Nature*.[7] In 1817, she produced another book, *Lettres philosophiques de Mr. Fontaine et de m-lle Tourtchaniniff*, published in Paris.[8]

Another example was the princess Eudoxia Ivanovna Golitsyna (1780–1850), a woman whom the poet Aleksandr Pushkin ironically dubbed "an academician in a cap." One of the most beautiful women of the time, Golitsyna also had the nicknames *Princesse Nocturne* and *Princesse Minuit*, and because young Aleksandr Pushkin dedicated some poems to her, she is often mentioned in the history of Russian literature. It well known that she conducted an open house at which the cream of society gathered nightly. Not so well known is the fact that her guests also included a member of the St. Petersburg Academy of Sciences, Michail Vasil'evich Ostrogradskii (1801–61), an outstanding Russian geometrician; professor of Moscow University

---

[4] Among many other sources, see, e.g., Toby W. Clyman and Diana Green, eds., *Women Writers in Russian Literature* (Westport, Conn., 1994); Catriona Kelly, *A History of Russian Women's Writing, 1820–1992* (Oxford, 1994). Of course, there simply *were* more women writers than scientists: writers were the first women to struggle against men's monopoly on intellectual labor and the first to achieve success.

[5] Ann Hibner Koblitz, *Science, Women, and Revolution in Russia* (Newark, N.J., 2000). One should also note her biographical study *A Convergence of Lives: So'fia Kovalevskaia, Scientist, Writer, Revolutionary* (New Brunswick, N.J., 1993).

[6] Philip Vigel, *Zapiski*, vol. 1 (Moscow, 2003), 107.

[7] Anna Aleksandrovna Turchaninova, *Natural'naia etika ili zakony nravstvennosti, ot sozertsaniia prirody neposredstvenno proistekaiushchie* (St. Petersburg, 1803).

[8] Anna Aleksandrovna Turchaninova, *Lettres philosophiques de Mr. Fontaine et de m-lle Tourtchaniniff* (Paris, 1817).

and mathematician Nikolai Dmitrievich Brashman (1796–1866); Moris Destrem (1787–1855), a famous engineer; and mathematician, professor, and director of the Institute of Communications, Lieutenant General Pierre-Dominique Bazaine (1783–1838). This woman of fashion was fond of mathematics. She wrote *De l'analyse de la force* (St. Petersburg, 1835, 1837; Paris, 1844). A prestigious journal noted at the time that this book was a "remarkable act of reflection."[9] Famous Russian poet Fedor Glinka (1786–1880) said about this work: "In this book Princess Golitsyna showed such an opinion so exactly hers that it couldn't but appear very new and at the same time correct in its serious conclusions."[10]

But these exceptions only emphasize the rule. A rich woman could afford an eccentric hobby especially if she was unmarried, was married but living separately from her husband, or was a widow. These cases of incidental interest are not the subject of this essay. Although the existence of such women was important, they could hardly be considered scientists and still less professionals. So what did it mean to be a "professional scientist" in Russia during this period? Might it be a person who was self-taught and never held a chair in any official scientific organization but had substantial knowledge in her chosen science, spending the majority of her time in scientific efforts, publishing the results of those investigations, and winning recognition from the scientific community?[11] The problem of Russian women integrating themselves into professional science has simply not been adequately explored. In this article, I shall make some preliminary efforts to fill this gap. I say "preliminary" because my investigation is not yet complete. I have been collecting data on this question, published and archival, for several years, but it is a very large project, especially for the Soviet period. Nevertheless, in the article I shall try to reconstruct the history of Russian women integrating themselves into professional science from 1860 to 1940, to elucidate some of the main characteristics of this process.

HOW DID WOMEN BECOME INTERESTED IN SCIENCE?

The first half of the 1860s was a unique time in Russian history. It was the age of the Great Reforms: serfdom was abolished (1861), and, among other very significant changes in Russian governance, jury trial and local government were promulgated (1864). For young male students, this was a time of freedom, energy, and hope. It was also a time when an interest in natural sciences spread in society, closely connected in their minds with their general situation. Thanks to their brothers, school teachers, and private tutors, and to the periodical press, young ladies were deeply involved with this newborn interest. It became fashionable to attend popular-science lectures, to be a part of a small youth circle that discussed natural sciences (as well as philosophy

---

[9] "L'analyse de la force, par M-me la princesse Eudoxie Galitzine, née Izmaïloff. St.-Petersbourg, chez Hintze, 1837, 8-vo., pp. VI et 12," *Biblioteka dlia chteniia* 25, nos. 11–12 (Moscow, 1837), 74.

[10] Fedor Glinka, "Kniga kniagini Avdot'i Ivanovni Golitsiniy i vecherniaia beseda ee v Moskve," in *Moskvitianin*, vol. 12 (Moscow, 1843), 538.

[11] This sounds a lot like the definition of an amateur. I argue that there is a small but important difference. If a person obtained a level that was strongly associated with the necessary professional skills, is it so important that he (or she) did not have an official document? If a person published a monograph that was acknowledged by his (or her) colleagues, is it important that the author did not have academic status? If exploration of nature was taking plenty of his (or her) time, is it important that a person was not receiving a salary? The answer to all of the questions: it might be vitally important to the person, but it made little difference for the sciences.

and politics) and read books on the natural sciences. In her famous memoirs, Elena Nikolaevna Vodovozova (1844–1923) wrote about the first months of 1862, when she had just graduated from Smol'nyi Institute for Young Ladies:

> Studying the natural sciences was considered the first instrument for the self-education, for preparation for any activities and a really useful social life. They were taken as a necessary foundation of all occupations, without exception. . . . In the '60s, reverence for natural history spread among overwhelming sections of Russian society and had a special character. Extraordinarily useful results were awaited not only from the research activities of scientists but also from every popular book; it made no difference which field of knowledge it belonged to. It was thought that an educated person must draw his attainments first of all from this source. . . . Now it is difficult to imagine with what total enthusiasm the publishing of the translation of *Brehm's Life of Animals* was met. Not to read this book meant to invite reproaches and derision. But people were interested not only in zoology but in the others areas of knowledge as well: mineralogy, botany, physiology, chemistry, partly even in anatomy.[12]

Vodovozova was perhaps less than delighted with all this because she herself was a humanist. She described several situations in which a girl gifted in music or art forced herself to study chemistry and zoology without any success and sometimes with very sad consequences.[13] The fashion persisted, supported by a strong European influence. The majority of Russian noble girls were taught from childhood to speak and read foreign languages. As a rule these languages were French, German, Italian, and sometimes English. Their knowledge of languages was much better than that of their male peers, so women could read almost any western European book in which they were interested without wasting time waiting for someone to translate it. And as male relatives had a habit of seeking their help with the translation of the most important books, girls had access to scientific literature. In addition, Russian noble families used to spend some time in Europe, and young girls and women often accompanied their fathers or husbands on these sojourns. Sometimes the family lived abroad long enough for the daughters to take classes in Italy or elsewhere. Returning home, they brought an acquired interest in natural sciences, antiquity, or fossils with them.

One such woman was Anna Mikhailovna Raevskaia (1820–83). As a lieutenant general's daughter, she received a very good education at home. Among her teachers was, for example, Ostrogradskii, who had a high opinion of her mathematical abilities. When she was eighteen, she married General Nikolay Nikolaevich Raevskii, commander of the Black Sea shoreline. After four years of marriage and the birth of two sons, Raevskaia was widowed. She spent the next five years traveling across Italy viewing antiquities. She was fond of antiquity, archaeology, ethnography, and anthropology and became acquainted with many experts in these fields.[14]

After returning home, Raevskaia corresponded with her European friends who were helping buy different artifacts for her. During her life, she assembled several very interesting anthropological and archaeological collections. For example, in Naples she bought a good mineralogical collection from Vesuvius and in Germany gathered a

---

[12] Elena N. Vodovozova, *Na zare zhizni: Memuarnye ocherki i portrety*, vol. 2 (Moscow, 1987), 80.
[13] Ibid., 82–3.
[14] One of them was Professor Charles Adolphe Morlot (1820–1867), formerly a curator of the Bern Archaeological Museum in 1851 and later appointed professor of geology and mineralogy in Lausanne. He wrote on the Tertiary and Quaternary geology of Austria, Switzerland, and Denmark and acquainted Raevskaia with some details of prehistory.

valuable collection of fossilized ammonites. She also ordered copies of famous items from European museums and private collections. She maintained close relations with the St. Petersburg Academy of Sciences and several Russian museums, as she often presented her collections to them. A friend of academician Karl Ernst von Baer (1792–1876), one of the most outstanding naturalists of the time, she availed herself of his supervision during her anthropological and archaeological expeditions to the Baltic lands and Finland. She organized archaeological digs in the Petersburg region as well. On February 14, 1872, she was elected a corresponding member of the Imperial Moscow Archaeological Society in spite of the fact that its president, Count Alexei Uvarov (1828–84), strongly objected to the idea of mixing women and science and had never before made such an exception.[15] Anna Raevskaia donated her collections to the Imperial Amateurs' Society for Nature, Anthropology and Ethnography. Contemporary anthropologist Dmitrii Anuchin wrote about her: "She was a master-spirit of a woman and had an analytical mind which is so rare among females."[16]

Even some women from the imperial family began to demonstrate interest in the sciences in the 1860s. One of them, grand duchess Maria Nikolaevna (1819–76), daughter of the emperor Nicholas I, for example, sent the Imperial Moscow Society for Naturalists the lower jaw of a fossilized rhinoceros in 1855. This jaw had been discovered near Tyumen' by someone named Shmotin.[17] How it came to Maria Nikolaevna's hands was never mentioned. The society was particularly honored by this gesture because the event coincided with its fiftieth anniversary, so it asked for permission to "beautify the list of its honorable members by including the name of Her Highness."[18] Whether permission was granted is an open question, but the name of Maria Nikolaevna can today be found in the card index of members of the Imperial Moscow Society for Naturalists.[19]

HOW TO BEGIN A CAREER?

In 1859–60, riding the wave of public enthusiasm for the natural sciences and political emancipation, several women asked for and received permission from the authorities to attend lectures at St. Petersburg University as irregular students, as well as at Kiev and Kharkiv universities. Simultaneously, a few young girls began to study at the St. Petersburg Medico-Surgical Academy. There is evidence that neither professors nor male students were opposed to women's presence in the universities. In 1861, Liudmila Ozhigina, who attended lectures at Kharkiv University (in medicine) and Mar'ia Mihailovna Korkunova at St. Petersburg University (in philology) made requests to the Ministry for Popular Enlightenment to permit them to pass state exams.[20] This action brought the whole question to the government's attention.

---

[15] *Imperatorskoe Moskovskoe arkheologicheskoe obshchestvo v pervoe piatidesiatiletie ego sushchestvovaniia (1864–1914 gg.)*, vol. 2 (Moscow, 1915), 297–8.

[16] Dmitrii N. Anuchin, "Anna Mikhailovna Raevskaia," *Izvestiia Obshchestva liubitelei estestvoznaniia, antropologii i etnografii* 90, no. 3 (*Trudy antropologicheskogo otdela* 18, no. 3 [1896], S513).

[17] "Protokol zasedaniia Moskovskogo Obshchestva ispitatelei prirody 20 oktiabria 1855 goda," d. 309, ll. 17–8, Archive of Moscow Society for Naturalists (hereafter cited as MOIP), Moscow.

[18] Ibid., 18.

[19] Society members before 1953 card index in Library of the Imperial Moscow Society for Naturalists, Moscow.

[20] G. A. Tishkin, "Zhenskii vopros i pravitel'stvennaia politika 60–70-h godov XIX v.," in *Voprosy istorii Rossii XIX–nachala XX veka: Mezhvuzovskii sbornik* (Leningrad, 1983), 161–2.

The problem was more difficult than modern historians used to think. It was not only about traditional conservatism—for a conservative a noble woman surrounded by rough and ill-mannered students, including common people, was unacceptable—and the fact that some professors would not tolerate a young lady in the sacred walls of the university. If it were simply a question of allowing women to be present at the universities, the government probably would not have objected. State exams and thus official status were, however, another thing altogether.

The whole system of Russian universities was founded by the state. In the beginning, it was meant to prepare Russian tutors for future Russian specialized schools and more generally educated functionaries for state service.[21] But soon after Moscow University was created in 1755, it became obvious that there were not enough noble sons who wanted to become students. The state had to give university students and future graduates a large assortment of privileges to induce young men from lower estates to enter a university. After more than a century and the foundation of several new universities (Kharkiv, Kazan', Iur'ev, St. Vladimir in Kiev, Novorossiysk in Odessa), and reorganizations (Vilensk), as well as several new unified charters for all Russian universities (1804, 1835), the system of privileges was still operative at the beginning of the 1860s, when a new university charter was discussed all over the country.

By the charter (*ustav*) of the "civil service,"[22] it was open only to the sons from Russian noble families, sons of military officers, and government officials of specific ranks. But if a man successfully graduated from the university, he had a right to enter the civil service despite his origin. Theoretically, it was the only way to obtain the highest managerial positions in the state. With every new rank came new bonuses. For example, the rank of honorable citizen made a person free from taxation, hereditary nobility extended this status to one's heirs, and after twenty-five years of service a person (and after his death his widow and children) had the right to a state pension. Of course women wanted to enter universities to become future doctors, professors, and such, not officials. But sooner or later they would ask for state diplomas (as Ozhigina and Korkunova did) and next for teaching positions in the universities. The problem was that all universities in the country were state ones, and all positions in academia belonged, by statute, to the civil service. Every academic degree was equated with the civil Table of Ranks and the military command hierarchy. To give a woman the right to become a university professor automatically meant to give her access to all positions of the civil service, as there was no way to separate them. This, of course, was unthinkable. In the Russian empire, every subject (*poddannyi*) had a special place and duties depending on estate. Women also had their niche. Social ideals dictated that every girl should be raised to be a good wife and mother. That was her only social role.

After the Ozhigina and Korkunova cases, the Ministry of Popular Enlightenment sent a special letter (in October 1861) to all universities with the demand that their councils answer the three following questions:

---

[21] S. Shevirev, *Istoriia Imperatorskogo Moskovskogo universiteta, napisannaia k stoletnemu ego iubileiu, 1755–1855* (Moscow, 1998), 12.

[22] The so-called civil service was the most prestigious and elitist service in the state—in fact, a ruling bureaucratic hierarchy. Only sons of the nobility, military officers, and government officials could enter and take positions. A person who entered the civil service could achieve the highest administrative positions in the state. University graduates, even those descended from the lowest classes, were given the right (as the highest benefit) to enter the civil service.

1) Is it possible to permit women's presence at university lectures together with [male] students in all departments or not? 2) Which conditions should be put forward for that attendance? 3) Could such persons be admitted to the academic degrees examinations, and what rights should they enjoy if the test were successful?[23]

All university councils, with the exception of those of Moscow and Iur'ev universities, responded affirmatively to the first question.[24] The discussion in the Medical Council of the Ministry of Internal Affairs about the request of Liudmila Ozhigina was also favorable, yet the government's final decision was negative. Although the new university charter (1863) contained no explicit item forbidding women's entering, the circular letter (July 20, 1863) strongly forbade women to attend universities in Russia even as irregular students.[25] This rule was in effect (with some exceptions in the revolutionary period of 1905–7) until the Bolshevik decree in 1918.[26]

Since university doors were closed for women in Russia, young girls interested in the natural sciences had to look for other opportunities. There were three possible routes, and all of them were exploited. The first option was to begin a struggle with the bureaucracy in the hope that sooner or later the permission to enter universities would be granted. The second was to go abroad to European universities. The last, and easiest, was self-education. This way is perhaps the most interesting because it produced the first generation of Russian women scientists, but all three routes played a role in future developments.

Self-education usually was closely connected with the family and thus essentially required a scientifically oriented father or husband. The paternal variant was atypical for Russia, although there were examples of a professional scientist raising his daughter as his assistant. The most notable case of the period was that of Izmail Ivanovich Sreznevskii (1812–80), a specialist in Slavic philology. He was the personal teacher of his daughter Olga Izmailovna Sreznevskaia (1845–1930), whom he trained to be his scientific secretary. For this purpose, her domestic education included such foreign languages as Latin, Italian, French, Provençal, Catalan, Spanish, English, German, Czech, and Serbian. From her youth, she accompanied her father on his travels in Europe, visiting museums, libraries, scientific conferences, and other such institutions and gatherings. She never married. Historian of Russian literature Nikolai Nikol'skii (1863–1936) accurately characterized her mode of life in her obituary:

> For more then 20 years she was the closest and most tireless of her father's assistants in his numerous scientific activities, although her participation was not always noticeable to all at the time, and the list of her published works is very, very short. Bibliographers

---

[23] F. 418, op. 30, d. 630, l. 1, Central Historical Archive of Moscow (Tsentral'nyi Istoricheskii Arhiv Moskvy, hereafter cited as TSIAM), Moscow.

[24] *Zamechaniia na Proekt obshchego ustava imperatorskih rossiiskikh universitetov* (St. Petersburg, 1861), parts 1–2.

[25] Tishkin, "Zhenskii vopros i pravitel'stvennaia politika" (cit. n. 20), 165.

[26] In 1864, all women were expelled from the universities and the St. Petersburg Medico-Surgical Academy (which was under the jurisdiction of the Ministry of War). Only one young woman, Varvara Aleksandrovna Kashevarova-Rudneva (1844–99), was allowed to finish the full course in the academy, and that was only because she had an obligation to serve six years in Bashkiria, as there was great need for treating syphilis there and Muslim women would not see male doctors. E. Likhacheva, *Materialy dlia istorii zhenskogo obrazovaniia v Rossii*, vol. 4 (St. Petersburg, 1901), 479.

would probably be unable to include more than 10–15 of her works in it. And even among them not all were published under her name.[27]

Olga Sreznevskaia's most important works were published under her father's name.[28] The most famous of Izmail Sreznevskii's scientific works is considered the preparation of the *Data for the Old Russian Language Dictionary Based on Written Monuments*.[29] The three volumes of the dictionary were republished several times and are still in use today. Yet before Sreznevskii's death in 1880, data for only the first two letters of the alphabet were ready. It took Olga Sreznevskaia ten years of hard work before the first volume was ready for publication in 1890 and more than twenty years to complete the third volume, published in 1912. But her name is not mentioned anywhere. Contemporaries attempted to make up for the absence by voting her a corresponding member of the St. Petersburg Academy of Sciences in 1896.

In Russia during the second half of the nineteenth century, the cases of women becoming scientists under the guidance of their husbands were more common than those of women doing so under the guidance of their fathers. It was fairly usual for a young wife to want to help her husband in his everyday duties. Most Russian scientists' families were not rich. Ordinarily, a university professor could not afford a professional secretary. His wife—having as a rule a good knowledge of European languages, extensive practice in correspondence, a copybook hand, and skill in drawing (an obligatory part of young ladies' education), as well as being interested in his success, full of goodwill, and never asking for a salary—made for an excellent choice for this position. The level of her education in the natural sciences was not important because she could gain the necessary experience through her work. Being involved in her husband's researches meant a woman sometimes had to communicate with other scientists and even scientific societies; after a time, her own scientific interests might appear. However, without a deep interest in the natural sciences, the wife could find her enthusiasm vanishing with time, as her children dominated her attention. (The average number of children in a Russian high-class family was more then five, for common women closer to nine children—although hardly more than half survived.)

A good example of such a case was that of Elena Vasil'evna Bogdanova, wife of zoologist and Moscow University professor Anatolii Petrovich Bogdanov (1834–96). We do not have much information about her life, but one can find her name as well as a description of her scientific activities in the minutes of the Imperial Amateurs' Society for Nature, Anthropology and Ethnography. Professor Bogdanov was one of the society's founders and among its most active members. Society activity was at a peak from 1860 to 1880. During this time, the society (created in 1864) organized three large exhibitions in Moscow, attracting the attention of thousands of people: an Ethnographical Exhibition (1867), simultaneously with the Slavonic Congress; a

---

[27] N. Nikol'skii, "O. I. Sreznevskaia: Nekrolog," *Izvestiia,* USSR Academy of Sciences (Otdeleniia Obshchestvennikh Nauk), ser. 7, 1931, no. 7:776–7.

[28] For example, she translated into Russian and published with commentary a very interesting medieval text: Rjui Gonsales de Klavixo, *A Journal of a Trip to Timur's Court in Samarkand in 1403–1406* (St. Petersburg, 1881). In the actual publication, Izmail Sreznevskii was named as editor and commentator and the name of the translator is absent, although all biographers affirm that this publication was Olga Sreznevskaia's work.

[29] Izmail Sreznevskii, *Materialy dlia slovaria drevnerusskogo iazika po pis'mennim pamiatnikam,* vols. 1–3 (St. Petersburg, 1890–1912; repr., Moscow, 1958).

Polytechnical Exhibition (1872); and an Anthropological Exhibition (1879). Considerable effort resulted in extensive collections of scientific items, and a few museums were founded as a consequence (among them the famous Moscow Polytechnical Museum). Bogdanov took part in all this, and I assume that he involved his wife in the organizational work. Be that as it may, on March 23, 1874, some society members suggested electing Madam Bogdanova to the list of the society's members because:

> Elena Vasil'evna Bogdanova always took part in the work of the Society, beginning when a few members started to explore entomological fauna of the regions belonging to the Moscow educational district soon after the Society's foundation. She helped to gather collections for the Zoological Museum. During the Ethnographical Exhibition and after it, Bogdanova put a lot of energy into making pictures from university's craniological collection, and later when the Polytechnical Exhibition was prepared, she was a constant participant in composition and organization of the different collections.[30]

Elena Bogdanova was unanimously voted an ordinary (*nepremennyi*) member of the Imperial Amateurs' Society for Nature, Anthropology and Ethnography.[31] But later one cannot locate her name among its active members or find scientific papers published under her name or with her participation; nor did she attend scientific congresses. It seems that her interest faded with time.

To avoid such a case, Elie (Il'ya Il'ich) Metchnikov (1845–1916) created a whole theory about how a scientist should choose a future wife. From early youth, he dreamed of meeting a school age girl, marrying her, and then tutoring her according to his scientific ideals. At twenty-two, he became fond of a thirteen-year-old girl but soon realized that she did not love him. So he married a woman of his own age, but she was already fatally ill and died shortly afterward. A little later, he returned to his original plan. He found a girl of school age from a neighbor's family and became her tutor in zoology and soon proposed marriage. Olga Nikolaevna Metchnikova (1858–1944) remembered that her father was against these lessons and was happy to hear a proposal of marriage. Her mother was a little anxious about the young age of her daughter, but Metchnikov managed to persuade her. It seems that nobody asked Olga. She wrote:

> I had no suspicions about my teacher's feelings and was very embarrassed when I learned about them. I absolutely could not understand how so clever and educated a man could marry a paltry girl. The thought that he was mistaken in me frightened me, and it seemed to me that I was going to sit for an exam for which I was not at all ready.[32]

The morning after the wedding, she spent preparing her zoological work in order to surprise to her husband. After passing secondary school exams, she began to study biology under her husband's direction. They worked together during many long years, but later (after his death) she wrote: "Though I was always interested in science, art was my life's passion."[33] In spite of all Metchnikov's diligence, his second wife never became a scientist.

---

[30] 73rd Session of the Natural Sciences Amateurs Society [Records], 23 March 1874, f. 455, op. 1, d. 12., ll. 151–2, TSIAM.
[31] Voting list for session 23 March 1874, f. 455, op. 1, d. 9, ll. 66–7, TSIAM.
[32] O. N. Metchnikova, *Zhizn' Il'i Il'icha Metchnikova* (Moscow, 1926), 72.
[33] Ibid., 75.

But sometimes a successful marriage opened before a woman a real path toward scientific investigations. There were cases in which a woman, after beginning as her husband's assistant, developed independent explorations; there were also cases in which a woman, when choosing her husband, took into account the possibility that he could assist her scientific interests and work. One such woman was Alexandra Viktorovna Potanina (1843–93), wife of Grigorii Nikolaevich Potanin (1835–1920), a renowned Russian traveler and explorer of Central Asia and Siberia. Practically uneducated and the daughter of a clergyman without any fortune, she understood that her prospects were very poor. She met her future husband when she was visiting her brother, who was serving a sentence in the little town Nikolsk, in the Vologodskaya region. Potanin also lived in exile there. He had taken part in the 1861 student disorders in St. Petersburg, was exiled to Siberia, and later became a member of the so-called Society for Siberian Independence. In a break between these political activities, in 1863–64 he took part in the expedition to the South Altai and Tarbagatai organized by the Russian Geographical Society and made a good showing. In 1874, upon a petition from the Russian Geographical Society, he was pardoned and got married. Together, Potanina and Potanin organized expeditions to northwestern Mongolia and Tuva in 1876–77 and into northern China, eastern Tibet, and central Mongolia in 1884–86 and 1892–93. They collected a great deal of geographic data about unknown regions of Central Asia, herbaria, and zoological collections, as well as data about the culture of Turkic and Mongolian nations. At first, Alexandra Viktorovna played the role of her husband's assistant, but later she began her own investigations. She had a unique opportunity to explore women's lives among the nations they came across. Potanina published several articles on the ethnography of the peoples of Siberia and Central Asia. Living in Irkutsk between journeys, Potanina played the role of salon hostess. When she and her husband were preparing for their last expedition in 1891, she was already seriously ill; she refused to stay home, however, and died on the way to Shanghai in 1893.[34]

Other young ladies began their lives in better circumstances than Potanina had. One of them was Olga Armfeld (1845–1921), daughter of Alexander Armfeld, professor of medicine at Moscow University. A graduate from the Nikolaevskii Sirotskii Institute (a Moscow secondary school for young ladies), she was seeking an opportunity to continue her scientific education. In 1864–68, she spent time at the Zoological Museum of Moscow University ordering its collections, assisting experiments, helping with the translation of the biological books, and corresponding with foreign scientists. She befriended a group of young naturalists, university graduates dreaming about scientific careers. In 1867, one of them, Alexei Pavlovich Fedchenko (1844–72), became her husband. The following year, the Imperial Amateurs' Society for Nature, Anthropology and Ethnography recommended him to a Turkestan governor-general, Konstantin Petrovich von Kaufman (1818–82), as a scientist needed to explore Turkestan—a new and almost unexplored acquisition for Russia. It was clear from the very beginning that Olga Fedchenko would accompany her husband. But, unlike Potanina, she had her own individual tasks. Scientists from the Imperial Amateurs' Society for Nature, Anthropology and Ethnography treated her as an equal member of the expedition and made her responsible for all botanical

---

[34] A. V. Potanina, *Iz puteshestviy po Vostochnoy Sibiri, Mongolii, Tibetu i Kitaiu: Cbornik statey* (Moscow, 1895); V. M. Zarin and E. A. Zarina, *Puteshestviia A. V. Potaninoy* (Moscow, 1950).

aspects of it. One can find confirmation of her official status in a letter von Kaufman sent on October 4, 1868, to General Abramov, commander of the Zeravshnskii region (the starting point of the expedition): "I propose, dear Sir, to provide any assistance to Mr. Fedchenko and his wife who is also to accompany him as a scientist in a commission given to him."[35]

Of course, in reality no official status (and no salary) was stated in any official document, but the work and the responsibility were real. She was not only a plant collector but also the only painter for the expedition. During the famous Turkestan expedition (1868–71), Olga Fedchenko also found time to help her husband make maps, collect insects, correspond with fellow scientists, and manage accounts. Participation in such an important expedition would be considered a very successful beginning for the scientific career of any researcher.

Thus the early 1860s produced a few young women with strong enough interests in the natural sciences to choose it as their main occupation. Under the pressure of circumstances or because of practicality or traditional thinking, they preferred marriage as the easiest way to fulfill their plans. But while Fedchenko and Potanina were looking for suitable husbands according to the old rules, some of their peers were trying to change those rules. Being cut off from Russian universities, they turned to Europe—another typical path for a segment of the Russian intelligentsia.

One of the first on that road was Nadezhda Prokofievna Suslova (1843–1918), the daughter of a serf (but a rich one, a steward at Count Sheremet'ev's estate who was later freed and became the owner of a textile factory). Suslova attended a Moscow private school for young ladies and later (in 1860) lectures at St. Petersburg University and at the St. Petersburg Medico-Surgical Academy. In 1862, she published her first original scientific paper. But in 1864, after expulsion from both St. Petersburg University and the St. Petersburg Medico-Surgical Academy under the order of the Russian government, Suslova and one of her friends, Maria Aleksandrovna Obrucheva (1839–1919) (in the future Bokova-Sechenova, wife of the famous physiologist Ivan Sechenov), went to Zurich to attend school there. In 1867, Suslova passed her doctoral exam at Zurich University and became a Doctor of Medicine.[36] This event generated a big response in liberal circles of Russian society, as many magazines published reports about it, and famous people, beginning with Aleksandr Herzen, greeted Suslova when she returned home in 1867. In 1868, her doctoral dissertation was published in Russia. Suslova's example was very attractive to some other young women, including Sofia Vasil'evna Kovalevskaia.

The first problem on this new path was parental permission. By Russian law, a woman (even one of full legal age) could not travel anywhere without a special *vid* (passport)—permission given by her parents, husband, or a state functionary (if she was a widow). The second problem was money. Parents were rarely supportive, so ruses such as pro forma marriages were developed to secure the passport. This was the choice made by Sofia Korwin-Krukovskaia (1850–91) when she married Vladimir Kovalevskii in 1868 (although later this marriage became real). Having the status of a married woman, she induced the parents of her cousin Iuliia Vsevolodovna

---

[35] "1868 г. October 4. Pis'mo Turkestanskogo general-gubernatora K. P. fon Kaufmana k nachal'niky Zeravshanskogo okruga A. K. Abramovu," in *A. P. Fedchenko: Sbornik dokumentov* (Tashkent, 1956), 53.

[36] E. A. Pavliuchenko, *Zhenshchiny v russkom osvoboditel'nom dvizhenii ot Marii Volkonskoi do Very Figner* (Moscow, 1988), 152.

Lermontova (1846–1919) to allow their daughter to accompany her to Heidelberg. Kovalevskaia and her husband left for Heidelberg in spring 1869; in autumn 1869, Lermontova arrived in Heidelberg. Their other companion, Anna Mikhailovna Evreinova (1844–1919), had to run away from home with neither passport nor money and cross the border illegally; she reached Heidelberg on January 10, 1869.

The Heidelberg women's commune of 1869 was very important in the history of higher education for women as three of its members achieved doctoral degrees. In 1872, Anna Evreinova received permission from Leipzig University and in 1873 prepared a dissertation, passed the necessary exams, and attained the rank of Doctor of Laws, the first among Russian women to do so. In spring 1874, Göttingen University granted Sofia Kovalevskaia the rank of a Doctor of Philosophy *honoris causa* for her mathematical studies. In autumn 1874, Iuliia Lermontova completed her dissertation and passed her exams at Göttingen University, becoming a Doctor of Chemistry. After returning home, all of them were at the center of public attention.

From the early 1870s on, more and more Russian girls chose the same path. In 1872, Sofia Mikhailovna Pereiaslavtseva (1849–1903) arrived in Zurich and in four short years earned a Doctor of Philosophy, with a specialization in zoology and embryology. Elizaveta Fedorovna Litvinova (1845–1919/1922) also arrived in Zurich in 1872[37] and received a doctoral degree in 1876, hers in mathematics, philosophy, and mineralogy from Berne University. All the women mentioned above were born in the mid-1840s, but already women born in the 1850s—five or ten years younger— were even more active. Vera Figner (1852–1942), who studied in Zurich in 1872, wrote that if earlier there had been 15–20 Russian girls in Zurich during a year, in 1873 at Zurich University and the Polytechnic School alone 103 Russian women were studying.[38] Some modern historians consider this number to be correct, others believe it was much larger.[39] The well-known Soviet historian of biology Leonid Iakovlevich Bliakher agued that in 1873 there were no fewer than 130 Russian women studying in Zurich.[40] One should not forget the other European universities: German, northern European, French. Thus from the late 1860s and early 1870s onward, an educational tour through European universities became very popular among Russian young women. This popularity did not diminish until the First World War, in 1914.

Yet one should not confuse the history of the higher education for women with the history of women scientists. Only a minority among these women wanted to become scientists. For others it was an opportunity to have a full, sensible life or to work for humankind's benefit (that is why there were a lot of future physicians), practicality, or even fashion. Obtaining permission to attend lectures was not easy, studies were long and difficult, not to mention the doctoral dissertation and exams. There was a potentially easier way. Nearly half from this very first cohort of female students I mentioned above put aside their studies and turned to revolutionary activity. Vera Figner herself was the best example of such behavior, as she left the university a few months

---

[37] Litvinova was already a widow; her maiden name was Ivashkina. She married in 1866, but her husband, a doctor, died in 1872.
[38] V. N. Figner, "Ocherki avtobiograficheskie. Zurich," in *Polnoe sobranie sochinenii*, 2nd ed., vol. 5 (Moscow, 1932), 47.
[39] F. E. Ivanov, *Studenchestvo v Rossii kontsa XIX–nachala XX veka: Sotsial'no-istoricheskaia sud'ba* (Moscow, 1999), 105.
[40] L. Ia. Bliakher, "Sofia Mikhailovna Pereiaslavtseva i ee rol' v razvitii otechestvennoi zoologii i embriologii," *Trudy Instituta istorii estestvoznaniia i tekhniki* 4 (Moscow, 1955): 170.

before receiving her doctoral diploma in medicine and returned to Russia because her comrade revolutionists needed her. So women's urge for the highest education should not necessarily be equated with the urge for professional scientific work. One should also grant that a doctoral degree itself did not guarantee the possibility of scientific work in Russia, not only in the nineteenth but also in the twentieth century.

HOW TO GET A RESEARCH POSITION IN RUSSIA

After obtaining an appropriate education, professional skills, and in some cases academic status, a person who was really interested in the natural sciences looked for a position in some university or equivalent institution. Nadezhda Suslova returned to Russia in 1867 with a degree from a foreign university and was not allowed to practice medicine immediately. As was the case for any foreign doctor in Russia, she had to ask for permission to confirm her status. In 1868, Suslova successfully passed an examination before a special medical commission and won the right to a private medical practice.[41] This served as a strong example for future Russian women doctors, but there was no such thing as a private practice for mathematicians, zoologists, botanists, or chemists in Russia. If a woman was married to a scientist, then his work opened doors for her own investigations, as well as providing an income. But in this case, a woman depended completely on her husband's goodwill, as her first role was as his assistant. This was not easy. For example, the famous archaeologist Countess Praskovia Sergeevna Uvarova (1840–1924) always spoke about her husband, Count Alexei Sergeevich Uvarov (1825–84), who was already an authoritative archaeologist before their wedding, with great respect, as if he were her sun and moon: her teacher, companion, and colleague. Yet we also have Dmitrii Anuchin's account:

> The life of Countess Praskovia Sergeevna Uvarova naturally divides in two parts. The first—from 1840 until 1885—is a period of development, education, participation in the *beau monde*, marriage, assisting her husband, preparing for future work, and the second—the next thirty and we hope a long line of further years—serving the Moscow Archaeological Society as its president, working on the collection, preservation, exploration, and publication of old documents, and in the whole tireless work for the understanding of antiquity and for the success of Russian archaeology. For all of us . . . , the second part of the Countess's life is more interesting.[42]

The first part of her life (before her husband died in 1885) was counted as "preparation for the public service"—a preparation that took forty-five years. At the age of forty-five she became a widow and was well known in the scientific community because during many previous years she was a chief organizer not only of the regular meetings of the Moscow Archaeological Society (founded by Count Uvarov) but of the All-Russian Archaeological Congresses. In spite of all this, before her husband's death she was not even a member of the society, as he disapproved of learning for women and did not want to make an exception for his wife. After his death, however, society members elected her not only a member of the Moscow Archaeological Society but its president. So in addition to renown, she obtained official status. After

---

[41] Pavliuchenko, *Zhenshchiny v russkom osvoboditel' nom dvizhenii* (cit. n. 36), 153.
[42] Dmitrii N. Anuchin, "Grafinia P. S. Uvarova v ee sluzhenii nauke o drevnostiakh na postu predsedatelia Imperatorskogo Moskovskogo arkheologicheskogo obshchestva," in *Sbornik statei v chest' grafini Praskov'ii Sergeevny Uvarovoi* (Moscow, 1916), ix.

that, Uvarova published more than eighty scientific works and was the editor for many more, organized nine All-Russian Archaeological Congresses, and was voted an honorary member of the St. Petersburg Academy of Sciences and an honorary professor of Moscow University. Of course, after her husband's death she was a very rich woman and could permit herself anything she wanted. And she wanted to be a professional archaeologist.

In general, newly founded Russian scientific societies (the majority of them were created after 1863) were glad to give some space to women scientists. In 1864, Olga Armfeld became a founding member (*chlen-osnovatel'*) of the Imperial Amateurs' Society for Nature, Anthropology and Ethnography. On January 16, 1868, Varvara Kashevarova-Rudneva read a paper at a session of the Society of Russian Physicians.[43] On October 17, 1874, one of the oldest and most respected Russian societies, the Imperial Moscow Society for Naturalists, elected Olga Fedchenko (Armfeld) as a corresponding member.[44] On October 15, 1875, she was elected as an honorary member of the Imperial Amateurs' Society for Nature, Anthropology and Ethnography.[45] In 1875, Lermontova became a member of the Russian Chemical Society in St. Petersburg.[46] The Russian Geographical Society elected its two first women as collaborating members on May 20, 1877.[47] In 1880, Sophia Pereiaslavtseva was elected a head of the Sevastopol biological station, which belonged to the Novorossiysk Society of Naturalists. During the next ten years, she fulfilled these duties; it was also a time of intensive exploration work for her.[48] To my knowledge, this was the first time a woman headed a scientific institution in Russia. In 1889, the Eighth Congress of Russian Naturalists and Physicians chose Pereiaslavtseva as the chair of the zoological section.[49]

Thus the Russian scientific community welcomed women; membership in the societies meant the possibility of attending sessions and giving papers, having access to the societies' libraries, to their natural collections, and (even more important) to their periodicals. Already in the late 1870s, one can see women's names in the pages of scientific journals, and during the 1890s their number rose noticeably. Women's presence at scientific meetings also stopped being unusual. But the one thing the societies could not give women (nor men) was a salary. (Pereiaslavtseva was an exception.) If one looks at the membership lists of the Congresses of Russian Naturalists and Physicians of the time, one notes that the majority were *gymnasium* teachers or local government (*zemskii*) physicians. Although the second positions were open to women with doctoral degrees (the payment was paltry) during the late nineteenth century, the first were barred to women until the twentieth century. One reason for this was that gymnasiums for girls were established much later than those for boys. Initially, women were allowed there only as form masters (*klassnaia dama*), then after several years as teachers for the junior class. All other positions

---

[43] S. M. Dionesov, *V.A. Kashevarova-Rudneva—pervaia russkaia zhenshchina—doktor meditsiny* (Moscow, 1965), 38–9.
[44] Record [Protokol] of Moscow Society for Naturalists, 17 Oct. 1874, d. 482, l. 25, MOIP.
[45] "Protokoly zasedanii Moskovskogo Obshchestva liubitelei estestvoznaniia, antropologii i etnographii s sentiabria 1874 po oktiabr' 1876 g.," *Izvestiia Obshchestva liubitelei estestvoznaniia, antropologii i etnographii* 24 (Moscow, 1876), 57–8.
[46] Iu. S. Musabekov, *Iuliia Vsevolodovna Lermontova, 1846–1919* (Moscow, 1967), 47.
[47] L. S. Berg, *Vsesoiuznoe geograficheskoe obshchestvo za sto let* (Moscow, 1946), 203.
[48] Bliakher, "Sophia Mikhailovna Pereiaslavtseva" (cit. n. 40), 181–4.
[49] Ibid, 186.

were for men with university degrees. Boys' gymnasiums were closed to women. Even before 1914, there was a struggle for women's right to teach the highest classes at boys' gymnasiums. Salaries in all other primary and secondary schools were inadequate.

Sometimes it was possible for a woman to get into the state system with the help of a close friend or her husband. For example, Maria Vasil'evna Pavlova (1854–1938) after her marriage (1886) was allowed to work in the Geological Office of Moscow University "only thanks to a personal authority of her husband, Professor A. P. Pavlov (1854–1929), and Professor V. I. Vernadskii."[50] After returning from Europe, Iuliia Lermontova was invited by Aleksandr Mikhailovich Butlerov (1828–86), in 1877, to work in his small private laboratory at St. Petersburg University. In 1880, she was working in the laboratory of Moscow chemist V. V. Markovnikov (1837–1904).[51] But all these jobs were on a volunteer basis. When Lermontova's father died, she had to manage her almost ruined family property in order to salvage what was left, so she had no time for a job and put aside her scientific interests.

When in 1891, after a tremendous confrontation with Aleksandr Kovalevskii, a professor at Novorossiysk University and the secretary of the Novorossiysk Society of Naturalists, Sofia Pereiaslavtseva sent in her resignation and lost her position at the Sevastopol biological station, she found herself in a very difficult situation. Without any means of support, she spent some time with her relatives in St. Petersburg trying translation jobs (the only way, besides teaching, for educated women to earn some money) but could not make a living at it. The following year, however, the Imperial Moscow Society for Naturalists gave her a grant for a foreign journey, and she spent a year at the Naples zoological station. Then in 1893, the Ninth Congress of Russian Naturalists and Physicians set up a collection for one of her monographs, thanks to which she was able to spend a year or more in Paris working in its Museum of Natural History.[52] But when these funds ran out, she was right back where she started, with no means of support. In 1903, she went to Odessa hoping to continue her explorations but fell seriously ill. The situation was so tragic that her old friend, Novorossiysk University professor of botany Liudvig Al'bertovich Rishavi (1851–?), published a note in the city paper *Odesskii listok*:

> Without any means for living, earning her slender bread by translations from foreign languages, among unbelievable pecuniary destitution, sometimes half-starving, Sofia Mikhailovna continued her scientific work, continued publishing her scientific articles, which gave her a European reputation . . . As an old friend of S. M. Pereiaslavtseva's, I believe that my duty is to inform by this letter all her acquaintances and well wishers and also all educated women aspiring to higher education about Pereiaslavtseva's hard, almost hopeless situation.[53]

The St. Petersburg Academy of Sciences sent some money for her as well as a "Literary Fund." But it was too late—she died on December 1, 1903. Some of her friends insisted it was from starvation.

---

[50] "Pamiati M. V. Pavlovoi," *Paleontologicheskoe obozrenie*, 1939, no. 1:1. Mariia Vasil'evna Pavlova graduated from the Sorbonne in 1884 as a paleozoologist. In 1886, she married Aleksei Petrovich Pavlov, geologist, professor of Moscow University, and head of the university's Geological Office.

[51] Musabekov, *Iuliia Vsevolodovna Lermontova* (cit. n. 46), 39–47.

[52] Bliakher, "Sofia Mikhailovna Pereiaslavtseva" (cit. n. 40), 194–5.

[53] F. 575, op. 2, d. 14, l. 9, Manuscript Department of the Russian State Library, Moscow.

Thus the first generation of Russian women scientists were mainly young ladies from noble families or (more rarely) clerical daughters. They achieved education, professional skills, even academic status, authority in the scientific community, and reputations. They became professional scientists in every respect except one very important one: they had no right to take a position at a university or at any other scientific institution. Without this opportunity, the majority of them had to put aside scientific investigations and look for other work (if they could find it) or live in poverty. Actually only a rich or fanatical woman could allow herself such a hobby. And as time passed, more and more Russian noble families became impoverished, and their daughters had to be practical.

THE THIRD WAY

When I mentioned the possibilities open to Russian women after they were expelled from the universities in 1864, I named three different ways they could realize their goals. The first two were described above. While some women were taking root in the scientific community with the help of their husbands and others sought degrees in European universities, a group of women struggled with the Russian bureaucracy for the possibility of higher education for women in Russia. It was a long and difficult dance. In 1861, when the possibility of women's studying in universities was being discussed, the idea of a separate women's higher school emerged. From the "Opinion" written by Alexander Armfeld to the Professors' Council of Moscow University, one can see that professors discussed such an option during their meeting on September 23, 1861.[54] The majority believed that such a course would better satisfy morality and tradition than would allowing women to study at the present universities. Armfeld thought that the there was no need to wait while new universities were organized, as there were already old ones: "One cannot but support such a generous and modern conception . . . But between the moment ideas arise and the moment they can be realized a long time can pass, while our university auditoriums are ready and waiting only for a single word to open to everyone wanting to come in."[55]

Armfeld was certainly right about one thing: living his whole life in Russia, he knew the way business was done—slowly, very slowly. Seven years passed without any movement. Then on December 28, 1867, when the First Congress of Russian Naturalists was opened, Evgeniia Ivanovna Konradi (1838–98) sent a letter asking the delegates to support the concept of higher education for women. She argued that women as future mothers and mentors of their children determined the education of future generations. To teach their children, women should be educated, and education in the natural sciences was impossible without special equipment and professional guidance accessible only in the universities. After reading the letter, the session's chairman, professor of botany Andrei Nikolaevich Beketov (1825–1902), answered that although the congress completely sympathized with these thoughts it was not able to discuss them.[56] The idea was revived the following year. In autumn 1868, several women—led by Anna Pavlovna Filosofova, Nadezhda Vasil'evna Stasova, and

---

[54] A. O. Armfeld, "An Opinion Written by Professor Alexander Armfeld to the Professors' Council of Moscow University," f. 418, op. 30, d. 630, ll. 11, TSIAM.
[55] Ibid.
[56] *Trudy pervogo s"ezda russkikh estestvoispytatelei v Peterburge, proiskhodivshem s 28 dekabria 1867 goda po 4 ianvaria 1868* (St. Petersburg, 1868), 29–30.

Maria Vasil'evna Trubnikova (all three already active figures in the nascent Russian women's movement)—collected nearly 400 signatures on a petition to create a higher school for women.

In 1869, after several maneuvers, governmental permission was granted, and on April 1, 1869, the first Russian educational courses for women opened in St. Petersburg. The school was not a university, of course; the primary aim was to prepare women for university studies to the level of a boys' gymnasium curriculum. In 1870, more public courses opened in St. Petersburg with the same aim, but they invited men as well as women (the so-called Vladimirskie Courses). Simultaneously, a group of Moscow ladies tried to organize something similar in Moscow (Lubianskie Courses). At last in 1872, the first Higher Women's Courses were opened in Moscow (they are known as Courses of Professor Gerie after their founder). They were opened in Kazan in 1876 and in Kiev in 1878, but the Ministry of Popular Enlightenment barred them in Odessa in 1879 and in Warsaw and Kharkov in 1881. In 1878, in St. Petersburg, Higher Women's Courses (Bestuzhevskie) were created with a systematic university curriculum. All these institutions were collective projects without governmental support. On the contrary, the government forbade even publishing announcements with the call for donations for the courses. Furthermore, courses were allowed only in so-called university cities, and only university graduates with academic status could teach there. For example, Sofia Kovalevskaia, who took an active part in the organization of Bestuzhevskie Courses, was not invited to teach there.[57] But the process had begun. Then suddenly in 1886, all courses except the Bestuzhevskie were closed by governmental decree under the pretence of the necessity of a new policy in this area. And even in St. Petersburg, the admission of new students was closed until 1889.

Only in 1900 did the State Council confirm the law about the reconstruction of the Moscow Higher Women's Courses under the control of the Minister of Popular Enlightenment, and in 1901, they reopened. From this time until 1917, several Higher Women's Courses were founded in different cities of the Russian empire, as well as a Women's Medical Institute and a Women's Pedagogical Institute in St. Petersburg. For a short period during the revolution of 1905–7, women were allowed into the universities as irregular students. In 1906–7, there were 1,949 women in attendance at all the imperial universities (except Warsaw), including the St. Petersburg and Kiev Polytechnical Institutions and the Tomsk Technological Institute. After several transformations (in 1908 the permission to attend was rescinded), in 1911 there were 960 women in Russian universities.[58]

At the very beginning, the Higher Women's Courses could not compete with the universities. They had no buildings, laboratories, libraries, or equipment. The educational level of the women students was lower than that of the men. But as the time passed, buildings were erected, laboratories were organized, equipment was purchased, and young girls arrived with better preparation. In the first decade of the twentieth century, the Bestuzhevskie Courses were already equivalent to a university. In

---

[57] E. F. Litvinova, *S. V. Kovalevskaia (zhenshchina-matematik): Ee zhizn' i deiatel' nost'* (St. Petersburg, 1894), 48.

[58] N. I. Shilova, "Zhenshchiny v russkikh universitetakh i tekhnicheskikh uchebnikh zavedeniiakh v 1906–1912," in *Trudy 1 Vserossiiskogo s"ezda po obrazovaniiu zhenshchin, organizovannogo Rossiiskoi Ligoi Ravnopraviia Zhenshchin v S.-Peterburge*, vol. 1 (St. Petersburg, 1914), 31–2.

1912, the Moscow Higher Women's Courses admitted 2,000 girls every year, and the total number of students was 7,000.[59]

But there was one thing that ruined this educational picture of success for women. The Higher Women's Courses in Russia had no right to grant any degrees to their graduates. All they could offer was a certificate, but unlike the universities' degrees, it conferred no rights. A woman with the certificate of the Higher Women's Courses had the same rights as a gymnasium graduate—the right to work as a teacher in a secondary school. So there were no practical benefits for a woman's spending up to four expensive years only to satisfy her curiosity. Nevertheless, the Higher Women's Courses in St. Petersburg from 1878–1912 (twenty-eight graduating classes) gave 3,995 certificates to those who passed all exams, and many more students took the Courses but left without certificates.

For this essay, the most important question is what influence the development of higher education for women had on women's entering science. In 1909, a special poll was taken among Moscow students of the Courses. One question concerned their future plans. In the Department of Physics and Mathematics, of the 481 girls who answered, only 4.1 percent said they were planning to look for a scientific job, and 14.7 percent said that they would do so if there were real opportunities to find such a job with an acceptable salary.[60] There is statistical data about employment of graduates of the St. Petersburg Courses from the first year until 1912 (3,171 out of 3,995, including people from the last class of 1912 who had not yet had time to find a job). The most typical occupation was teaching in different types of secondary schools: 1,567 did this, and 76 of them were working as assistants or teachers in the Higher Schools of different cities. One hundred seventy-six worked in medicine (141 physicians, 35 paramedical personnel), 147 made literary careers, 118 were engaged in "private service" in different offices, 16 turned to art, 191 were continuing their education elsewhere, and 84 served in factories, observatories, and zoological stations.[61] So 160 women who graduated from Bestuzhevskie Higher Women Courses had positions equivalent to those in the higher schools or scientific investigations.

Some industries found that it was very profitable to hire women for calculating or technical work or as a laboratory assistants. They were quite competent and more assiduous than men, and their salary was very low. University professors teaching at the Higher Women's Courses found the same thing. Women trying to build scientific careers in a masculine mold sometimes could stay at the Courses "for preparation for a professor's title." They worked as assistants for their professors much harder than men would. But nevertheless they had no chances. After several years, the majority of such women came to understand that they would remain eternal assistants and left. But as always there were some who wanted to become scientists. The whole generation was caught in this trap, the fate of the second generation of Russian women scientists. The luckiest among them married their professors. After that they could work at the Courses, although usually without payment. A few gained master's degrees in the 1900s, when that became possible. A good example is the case of mathematician

---

[59] "Rech' professora A.N. Reformatskogo," in *Trudy 1 Vserossiiskogo s"ezda po obrazovaniiu zhenshchin* (cit. n. 58), xxxiii.

[60] *Slushatel'nitsi St. Peterburgskikh visshih jenskih (Besstujevskih) kursov: Po dannim perepisi* (St. Petersburg, 1912), 140.

[61] E. Shepkina, "Deiatel'nost' okonchivshikh S.-Peterburgskie Vysshie Zhenskie Kursy," *Vestnik evropy*, Aug. 1913, 342–54.

Liubov' Nikolaevna Zapol'skaia (1871–1943). She was a daughter of a noble family, but her father had to teach. She graduated from the girls' gymnasium in 1887 with a gold medal, completed three years of pedagogical courses, and at last entered Bestuzhevskie Higher Women's Courses in the Physical and Mathematical Department. Then after graduating from the Bestuzhevskie Courses in 1894, she went to Göttingen University in 1895 and in 1902 became a Doctor of Philosophy. After returning home in 1903, she was invited to teach in the Moscow Higher Women's Courses. Subsequently, she published a mathematical book and in 1906 defended a thesis at Moscow University and received a master's degree in abstract mathematics (the first woman to do so in Russia).[62] But a more usual example was the fate of Anna Boleslavovna Missuna (1868–1922), who was a well-known geologist and author of seventeen published scientific works and who worked as only an assistant in the Geological Office of the Moscow Higher Women's Courses (from 1907 until her death). By 1914, almost all assistants and a few professors at the Women's Courses were women. In 1903, a booklet was published by the Bestuzhevskie Higher Women's Courses with an appendix containing a bibliography of their graduates' published works: there were ninety-four names in the list and several hundred items.[63]

In the period 1860–1917, Russian women interested in scientific investigations explored many possible ways to include themselves in the profession. They tried "scientific marriages," European universities, and the establishment of women's education in Russia. And they were successful.

A NEW OLD LIFE

Soviet historiography usually has argued that 1917 suddenly changed the whole life of the country and that the scientific community was no exception. At first glance this appears to be correct, but careful exploration shows another picture. From May 1918 until October 1919, the Bolshevik government produced several decrees connected to the Higher Women's Schools. The first, from May 31, 1918, declared in its first point: "Co-education of students belonging to both sexes is introduced in all educational institutions."[64] The second one, from August 2, 1918, read: "Every person irrespective of his (her) citizenship and sex can become a student of any academy without showing a diploma and a secondary (or other) certificate of completion of school."[65] These two sentences were the result of more than fifty years of struggle. The Provisional Government of 1917 had come close to this decision but did not cross the line. Vladimir Lenin was a strategist. In this case, he wanted to have the students' support, or at least their loyalty, and the demand of coeducation was a way of harnessing active forces.

The third decree, from October 1, 1919, had another target. If young people were potential allies, university professors were potential enemies. They were a privileged

---

[62] A. M. Pavlov, "Pervaia russkaia zhenshchina magistr matematiki," *Istoriko-matematicheskie issledovaniia* 32–3 (1990): 235–41.

[63] *Pamiatnaia knizhka okonchivshikh kurs na S.-Peterburgskikh Vysshikh Zhenskikh Kursakh: 1882–1889; 1893–1903* (St. Petersburg, 1903), 217–42.

[64] "Postanovleniia Narodnogo Komissariata Prosveshcheniia: O vvedenii obiazatel'nogo sovmestnogo obucheniia," in *Sobranie uzakonenii i rasporiazhenii rabochego i krest'ianskogo pravitel'stva*, no. 38 (Moscow, 1918), otdel 1, 473.

[65] "2 Avgusta. Dekret o pravilakh priema v vysshie uchebnye zavedeniia RSFSR," in *Dekrety sovetskoi vlasti*, vol. 3 (Moscow, 1964), 141.

class, an integral part of the ruling elite, with their own ideas about state structure. The strategy on this front was to train a new generation of professors from young people who supported the new authorities. Of course, it was not possible to fulfill such a plan immediately. The new decree rescinded all special academic privileges, and as the Bolsheviks strongly associated academic status with doctoral degrees, master's degrees were abolished. Now the right to become a professor and to head a university department belonged to any persons known for their scientific investigations. All professors having ten years of service as of November 25, 1917, were dismissed and had to be reelected.[66]

In winter 1918–19, the Higher Women's Courses in Petrograd and Moscow were united with the universities or turned into universities. Those women who had a position at the Courses at once became university professors. Of course, the right to pass state examinations, to obtain academic status, and to teach had been granted to women already in 1911,[67] and some of them had hurried to take advantage. For example, Aleksandra Andreevna Glagoleva-Arkadieva (1881–1945) graduated from Moscow Higher Women's Courses in 1910 and remained as an assistant in the Department of Physics. In her free time, she prepared for the state examinations and passed them in 1914.[68] Maria Aleksandrovna Bolkhovitinova (1877–1957) entered Moscow Higher Women's Courses in 1912 and passed state examinations in geology in 1917.[69] Lidiia Karlovna Lepin' (1891–1985) graduated from the Physico-Mathematical Department of Moscow Higher Women Courses in 1917 and passed examinations in November 1917.[70] But there were few opportunities to become a university professor if female students were still not allowed in the university. So now all the women mentioned immediately above and those who were a little older instantly became full members of university faculties.

After this first interference, however, the new authorities did not pay much attention to the "women in science" question. Why would they? If women students or women who would be students were already a serious power, especially in the capitals, before 1917, women in science were still a tiny group with no organization. They could become neither strong allies nor important enemies. Thus women pressing toward careers as scientists were left alone to pave the way in the labyrinths of the universities, research institutes, and other scientific institutions. The beginning of a scientific career (getting higher education, entering graduate school, passing state examinations, participating in scientific projects or expeditions) became much easier after 1918. But women soon discovered that the attitude of the scientific community could frustrate any career even better then the old laws had.

Those women who were in the mainstream of scientific life very soon realized

---

[66] "1 Oktiabria. Dekret o nekotorikh izmeneniiakh v sostave i ustroistve gosudarstvennikh uchenykh i vysshikh uchebnykh zavedenii," in *Dekrety sovetskoi vlasti* (cit. n. 65), 3:381–2.

[67] "Statia 36226. Dekabria 19. Ob ispytaniiakh lits zhenskogo pola v znanii imi kursa vysshikh uchebnykh zavedenii i o poriadke priobreteniia imi uchenykh stepenei i zvaniia uchitel'nitsy srednikh uchebnykh zavedenii," in *Polnoe sobranie zakonov Rossiiskoi imperii, sobranie 3*, vol. 31, pt. 1 (St. Petersburg, 1914), 1297–1300.

[68] F. 641, op. 6, d. 122, l. 19, Archive of Russian Academy of Sciences (hereafter cited as ARAN), Moscow.

[69] F. 311, op. 1, d. 98, l. 85, ARAN.

[70] Interview with Lidiia Karlovna Lepin' [1938], in the private collection of Olga Valkova, 4. About Lepin', see *Professora Moskovskogo Universiteta, 1755–2004: Biograficheskii slovar'*, vol. 1 (Moscow, 2005), 745–6.

that with the changes of their status the treatment of the community changed, too. Maria Pavlova, now a professor in the Geology and Mineralogy Department of Moscow University (as of March 13, 1919), described one painful but relatively common situation. In 1920, she began training a talented student named Esfir Falkova, who successfully graduated from Moscow University in 1924. During her studies, Falkova took part in four geologic and paleontological field expeditions, presented a few papers, and passed all necessary exams. She wanted to continue her scientific studies, and Pavlova recommended her for the position of junior research officer at the Research Geological Scientific Institute of Moscow University. The first session of the Subject Commission (where such decisions were made) rejected the entire question because Pavlova had not sent them the application beforehand. The next session (on February 25, 1925) first decided that every candidate for such a position as Falkova had applied for should have at least one scientific publication and then turned Falkova down. After another university professor, Vera Aleksandrovna Varsanofieva, spoke in support of Falkova, whom she knew personally, some people began saying that Falkova was untalented, rude with students, and indifferent to social questions. During voting, eleven members of the Subject Commission were against (two professors and nine students), five in favor (among them all four women present who were professors and researchers), and four abstained from voting (all four were male professors). Pavlova was strongly insulted, especially because all this intrigue was organized by her old fellow professor Andrei Arkhangelskii (1879–1940) and because her colleagues were too cowardly to support her.[71] Describing the voting results in her memoirs, Pavlova wrote: "Here is the result of the appreciation of my scientific work from those closest to my scientific activities. Here are the indignities A. D. Arkhangelskii wanted to put upon me after 25 years of my friendly trust in him. Yes, today I'm feeling as if I returned from a very hard funeral."[72]

Thus, after the Bolshevik decrees of 1918–19, women were fully included into academic life and had to learn that the success of a scientific career depended not only on their success in investigations. They also had to understand that to secure the possibility of scientific research and necessary funding, they needed to gain enough influence and to have a commanding position in the hierarchy of the scientific bureaucracy. With this came an understanding (perhaps an unconscious one) that women were the weakest pieces on the chessboard. Although there were more women scientists in the 1920s than there had been before, they were still a minority among the scientific community; the old opinion that a woman was less clever than a man did not disappear; and women in Russia were still much more occupied than men with household chores and care for parents, children, and spouses, thus having less time for the work.

Those women scientists who were not connected with the educational system mainly continued working as they had before. Analysis of such women's curricula during 1914–25 shows that the majority of them did not stop expeditionary researches or laboratory investigations. For example, Olga Evertovna Neustroeva-Knorring (1887–1978), who participated in botanical expeditions organized by the Emigrant

---

[71] "Memories of M. V. Pavlova," f. 311, op. 1a, d. 99, ll. 1–7, ARAN.
[72] Ibid., 7.

Table 1. Neustroeva-Knorring Field Expeditions

Date	Location
1914	Khodzhenskii district, Samarkand region
1916	Mountainous Bukhara
1917–18	Orenburgskaya province (*guberniia*)
1919	Kanskii district (Kargatskaia dubrava)
1920	Eastern part of the Omskii district
1924	Leningrad district; Northern Caucasus
1925	Kara-Kalpak autonomous region

Administration (*Pereselencheskoe upravlenie*), spent every summer of 1908–14 in expeditions.[73] In table 1, I have listed her field expeditions during 1914–25.

From 1926 until 1943, annual expeditions took place (with the exception of 1934).[74] Geologist Vera Alexandrovna Varsanofieva (1889–1976),[75] Olga Knorring's contemporary, who had been a student of the Moscow Higher Women Courses in 1907–15 and became a graduate student in 1916, participated in geologic expeditions every summer in 1911–16, in winter 1917, and again in summer 1918. Between 1919 and 1922, the serious illness of her father and her own injuries prevented her from joining any expeditions. Then from 1923 until 1936 (except 1929 and 1935), she made geologic surveys every summer for a 124-page USSR general geological map.[76] Knorring's and Varsanofieva's expeditionary experiences were typical for women scientists in their positions.

In general, the state of affairs in science did not change much during 1914–25 for both women and men. In spite of the fact that the First World War brought some inconveniences (one of the greatest being the rupture of international communications) before 1917, everyday life remained little altered. However the winters of 1918 and 1919 were different. It was a time of civil war, and communications broke down within the country. Many scientists working or studying in Moscow and St. Petersburg lost contact with near relatives. For women this was more important than for men, as in Russia the woman was usually the person taking care of elder family members. In summer 1917, Vera Varsanofieva received a letter from the geologist Maria

---

[73] The Emigrant Administration was founded in 1896 under the Main Administration of Land Management and Agriculture (*Glavnoe Upravlenie Zemleustroistva i zemledeliia*). Between 1908 and 1914, it organized eighty-four botanical expeditions. A. A. Shcherbakova, N. A. Bazilevskaia, and K. F. Kalmikov, *Istoriia botaniki v Rossii (Darvinovskii period, 1861–1917)* (Novosibirsk, 1983), 277–81.

[74] S. Iu. Lipshits, *Russkie botaniki (Botaniki Rossii–SSSR): Biografo-bibliograficheskii slovar'*, vol. 4 (Moscow, 1952), 209–10.

[75] Vera Aleksandrovna Varsanofieva (1889–1976) was an outstanding geologist, explorer of the northern lands, historian of science, and science popularizer. She was the first Russian woman to obtain a Doctor of Geology and Mineralogy (1936) and served as vice president of the Moscow Society for Naturalists (from 1943), as well as corresponding member of USSR Academy of Pedagogical Sciences (from 1945).

[76] Curriculum vitae of V. A. Varsanofieva, f. 3, op. 1, d. 461, ll. 13–4, Russian State Archive of Economics, Moscow.

Ivanovna Shulga-Nesterenko (1891–1964), a close friend and colleague who had returned to Kiev to support her mother, father, and sister—all seriously ill:

> A horrible time, some unknown epidemic falls on the people of the past, on the generation of our fathers. You know it is everywhere, everyone, all families are being wrecked because of peculiar diseases, peculiar disasters fall on the people of the old century. And if we are somehow struggling, trying to stand in the violent storm, they no longer have the necessary strength. And the most horrible thing is our weakness, the impossibility of helping them.[77]

In Moscow and Petrograd, times were hard, too. Many women, especially those who had children, had left for the provinces and returned to the cities only after 1920.[78] Many women scientists from the oldest generation and even from the second one did not survive. There were some well-known names in this mournful list, such as Nadezhda Suslova (1918), Elizaveta Litvinova (1919/1922), Olga Fedchenko (1921), Vera Iosifovna Shiff (?–1919)—graduate of the first Bestuzhevskie Higher Women's Courses (1882), Doctor of Abstract Mathematics of the Göttingen University (1901), and from 1901 a junior teacher of the Bestuzhevskie Courses, and author of a large number of mathematical books—Anna Missuna (1922). But those who were young and free from family and had some professional or personal support never interrupted their investigations. From the notes of Moscow physicist Vera Aleksandrovna Glagoleva-Arkadieva, one learns that there was no central heating in her house until winter 1927–28. Before then, beginning with winter 1919–20, she and her husband had to make do with one and later three stoves.[79] From the same notes, it is clear that all this time she was busy with teaching and research.

After 1921, life returned to normal little by little: the New Economic Policy was proclaimed in March 1921; the civil war was ending at last; connections between different regions were reestablished; cultural, intellectual, and artistic life flourished in the capitals. At the same time, men were demobilizing from the army, and horrible unemployment ensued. Since 1917, women and men had equal political and economic rights, but mass ideology never changes as quickly, and women found themselves in a much worse situation than men did. The Bolsheviks clearly understood that women amounted to half the country's population, and even without the electoral franchise, they were an important labor force; after obtaining the franchise, they became a political force, too.[80] So the new government worked hard to attract women's support, especially that of working-class women. Among other arrangements, the government organized a large propagandistic campaign in which an item concerning the new rights and opportunities women received after the revolution was one of the

---

[77] M. I. Shulga-Nesterenko to V. A. Varsonofieva, 25 June 1917, Kiev, f. 3, op. 1, d. 286, ll. 7–8, Russian State Archive of Economics, Moscow.

[78] For example, that was the decision of Evgenia Vasil'evna Krakau (1886–1977). She graduated from the Bestuzhevskie Higher Women's Courses in 1914, passed the state examinations at St. Petersburg University in 1915, and then married. In 1918, when she left Petrograd for her native Smolensk district looking for pedagogical work, she already had two daughters. She and her family returned to Petrograd in 1922, and she began working at the Main Physical Observatory. In 1938, she became a PhD in physical and mathematical sciences. Selezneva, *Pervye zhenshchiny geofiziki i meteorologi* (cit. n. 1), 68–70.

[79] F. 641, op. 6, d. 122a, ll. 19–20, ARAN.

[80] According to the Population Census made on January 28, 1897, there were 99.8 women for every 100 men in the Russian empire (*Rossiia: Entsiklopedicheskii slovar'* [Leningrad, 1991], 86–90), and after two wars this ratio obviously would have shifted in favor of women.

most important. All women in traditionally male professions became a part of this campaign; women scientists were no exception. For example, one of the most popular Soviet journals, *Ogonek*, with a circulation of more than 3 million in the 1920s, published articles about women scientists regularly alongside articles about women pilots, sea captains, and so on.[81] The author of one such article told his readers about visiting the Institute for Applied Botany in Leningrad and a conversation with its director, N. I. Vavilov. He wrote about how Vavilov introduced the most successful institute employees to him, among whom there were many women:

> "Didn't you know," N. I. Vavilov said, "that thirty percent of all research fellows of the experimental agricultural stations of the USSR are women? Didn't you know that in the fight of the humanity with nature, in this fight for the renovated earth, women in the USSR total up to one third?"....
>
> "Can you tell me," I asked one young woman scientist, "is it true that women usually choose secondary crops for investigations and in whole are better as doers than as leaders?... They are good as the scientists' 'right hands' but weaker in independent work?"
>
> "It is not true," answered the young professor."We women rushed into all aspects of the agricultural experimental work."...
>
> I looked at the young woman scientist, at the certainty with which the girl was classifying and arranging new varieties of humanity's corn, at her eyes, in which one could see the happiness of her future investigations and triumphs and silently apologized for my silly thoughts about bluestockings, silently delighted with the wise fire in this beautiful person, silently bowed to women fighters for the better of the future days.
>
> 30 percent of the fighters....
>
> Bravo![82]

And so on. Such publications were typical. Of course they had an influence on public opinion and on young girls planning their futures, but I cannot argue that they had a noticeable influence on the scientific community. Nevertheless, the official position of the government proclaimed so widely gave women some confidence in their position. During the first decades of the Soviet state, the number of women scientists and students in higher education grew. Statistics of this growth became a part of the propagandistic strategy intended not only for the Russian people but also for foreigners. For example, in the album *USSR—Country of Women's Equality* (Moscow, 1938), one can read: "There is no single area of scientific exploration in which woman couldn't take part like a leader and organizer, like a teacher, researcher and scientific worker."[83] Then the following numbers are given:[84] 1929—5,100 women worked as researches in scientific institutions (institutes and branches); 1936—11,800 women did so.

There is no information about how these numbers were generated, so they remain suspect. But we can assume that they were not entirely fabricated. And if this assumption is correct, then the number of women scientists doubled in less than ten years. Fortunately, we have data on American women scientists for almost the same period for

---

[81] For example, in 1927 alone the following were published: M. Sandomirskii, "Vuzovka," *Ogonek*, 20 Feb. 1927, no. 8 (204):9; A. Bragin, "Tridtsat' protsentov boitsov," *Ogonek*, 6 March 1927, no. 10 (206):3; M. G., "Zhenshchina-konstruktor," *Ogonek*, 6 March 1927, no. 10 (206):4; A. Shabanova, "Pervaia Zhenshchina-Vrach," *Ogonek*, 9 Oct. 1927, no. 41 (237):14; E. V. Kozlova, "On Horseback across Mongolia: Impressions of an Expedition Member," *Ogonek*, 9 Oct. 1927, no. 41 (237):12; V. Komarova, "Avtobiografii V. D. Stasovoi-Komarovoi," *Ogonek*, 16 Oct. 1927, no. 42 (238):12.

[82] Bragin, "Tridtsat' protsentov boitsov" (cit. n. 81), 3.

[83] *SSSR—Strana Ravnopraviia Zhenshchin: Al'bom-vystavka* (Moscow, 1938), 23.

[84] Ibid.

comparison. Margaret W. Rossiter gave the following values in her well-known book *Women Scientists in America. Struggles and Strategies to 1940*: 1921—450 female scientists; 1938—1,912 female scientists.[85] We can assume that such a difference in the number of Soviet women and the number of American women was partly the result of the publicity campaign provided by the Soviet government and its policy in the field.

At the same time, scientific organizations and their professional unions were collecting statistical data, too, although information is also lacking about their methods. Tables 2–5 present information I have compiled from several sources. Table 2 lists information on the numbers of men versus women in three regions.[86]

Information in the USSR as a whole (in scientific institutions—and it seems that the Higher Educational Institutions were not included) is shown in table 3.[87] I should also mention that it is possible that this data includes also specialists in humanities. Sometimes one finds information about the women's professions,[88] but very rarely are there data about women's positions inside institutions. We do have such information about Irkutsk in 1927, as shown in table 4.[89]

There is also information about the women staff of the Higher Educational Institutions in April 1935 (percentage from the total occupying the given positions), which is shown in table 5.[90]

And finally we have data about girl students in higher education (percentage of all students): 1928—28.1 percent female; 1935—38.0 percent female; 1937—41.0 percent female.[91]

Even if this data is not representative, one notices a growing trend. Yet the majority of these women occupied positions as junior members of teaching and research staffs. Even without any formal barriers, they could not advance. One of the reasons was that the large group of women scientists of the time was very young and at the beginning of their careers.

One also notes a change in the character of the women. In the first two generations (and chiefly in the second one), they used to be diligent, assiduous, almost invisible, never looking for anything for themselves except the opportunity to work—or so many of their obituaries characterized them. But some women—representatives of the third generation (by my count)—were not of the same mettle. They were young, active, full of force and ambition, and believed in the political support of the new authorities. They did not want to stay invisible. They were ready to learn new rules, to change them if necessary, and to take an active part in the very old game played by the scientific community. It is possible to surmise that in such conditions, women scientists had to create some new strategies to advance their careers.

The first strategy that could lead to high-ranking positions and influence on the development of the sciences consisted in supporting the new authorities by disseminating the new Marxist ideology and in becoming active members of the ruling Bol-

---

[85] Margaret W. Rossiter, *Women Scientists in America: Struggles and Strategies to 1940* (Baltimore, 1981), 136.
[86] *Nauchnye Rabotniki Irkutska* (Irkutsk, 1927), 41; *Nauka i Nauchnye Rabotniki SSSR*, vol. 4, *Nauchnye Rabotniki SSSR bez Moskvy i Leningrada* (Leningrad, 1928), 801; *Nauchnye Rabotniki Moskvy* (Leningrad, 1930), 6.
[87] I. A. Kraval', ed., *Zhenshchina v SSSR: Statisticheskii sbornik* (Moscow, 1936), 98.
[88] *Nauchnye Rabotniki Krima: Spravochnik* (Simferopol, 1927).
[89] *Nauchnye Rabotniki Irkutska* (cit. n. 86), 41.
[90] Kraval', *Zhenshchina v SSSR* (cit. n. 87), 97.
[91] *SSSR—Strana ravnopraviia zhenshchin* (cit. n. 83), 34; Kraval', *Zhenshchina v SSSR* (cit. n. 87), 109.

Table 2. Scientists by Region

Year	Region	Number of men scientists	Number of women scientists
1927	Irkutsk	235	63
1928	USSR (except Moscow and Leningrad)	9,609	1,588
1930	Moscow	8,056	1,484

Table 3. Researchers and Graduate Students in USSR

Date	Total number of researchers	Number of women	Total number of graduate students	Number of women graduate students
1 April 1929	22,600	5,153	1,000	233
1 Jan. 1933	47,900	12,358	6,400	1,480
1 Jan. 1935	38,200	11,116	4,300	1,157

Table 4. Researchers in Irkutsk (1927)

Position	Total number	Men	Women
All researchers in the city	235	172	63
Professors	40	40	0
Senior lecturers, teaching main courses	15	14	1
Readers and junior members of teaching or research staff	105	74	31
Graduate students	16	10	6
Attending physicians	11	7	4

Table 5. Higher Educational Institutions (April 1935)

Position	Women staff (%)
Total number of women	15.0
Directors and deputy directors	3.1
Professors	2.9
Senior lecturers, teaching main courses	11.3
Readers and junior members of teaching staff	22.3

shevik Party. The latter strategy, however, was not very popular in the 1920s. It is possible that the reason for this was the traditional indifference towards politics adopted by Russian women scientists of the older generations or in the foreignness of the very notion that political activities could assist the success of a scientific career. There are no complete statistics, but from data collected by the Moscow Regional Committee of the USSR Higher School and Scientific Institutions Professional Union during

February–March 1940 (lists of women scientists who were professors, doctors and PhDs working in nineteen Moscow scientific and educational institutes), one can see that the majority of women born in the 1880s and 1890s had no relations with the Communist Party, at the same time those born in 1900s were party members (or candidate members or members of Leninist Young Communist League).[92] Nevertheless, among women born in the nineteenth century, there were some who made their careers working with the new authorities. Take the case of the mathematician and professor of Moscow University Sofia Aleksandrovna Ianovskaia (1896–1966). She was born in Odessa, studied there first in a women's gymnasium, then at the faculty of mathematics of the Higher Women's Courses in Odessa (1914–18). She began to take part in revolutionary activity in 1916. In 1918, during German occupation of the city, she joined the Bolshevik Party, fulfilling a variety of missions, including editing revolutionary periodicals. In 1919 she served in the Red Army. As she wrote in her autobiography:

> From 1920 until 1923 I worked in the District Party Committee [Gubkom] in Odessa mastering the communications and statistical department [*Informatsionno-statisticheskii otdel*] . . . In 1923 (at the insistence of the Department of Culture and Propaganda (*Kul'tprop*) of the Central Committee) I was sent to the natural sciences department of the Red Professors' Institute in Moscow. But since that opened only in 1924, for a year I studied at Moscow State University . . . In 1924 I entered the natural sciences department of the Red Professors' Institute planning to specialize in the philosophy of mathematics. Simultaneously I taught a course of natural dialectics at the physico-mathematical department of Moscow University for mathematics students and graduates. From 1927 I was taken on in the mathematical section of the Communist Academy as a senior staff scientist . . . In 1929 I graduated from the Red Professors' Institute and from the Communist Academy. From 1930 I was confirmed in the position of professor both at Moscow State University and at the Red Professors' Institute where I have been working until recently.[93]

In 1935, she became a Doctor of Physico-Mathematical Sciences. In the 1930s, she was one of the most notorious "red professors" of Moscow University, writing for such journals as *Under the Banner of Marxism* (*Pod znamenem marksizma*), *Natural Sciences and Marxism* (*Estestvoznanie i marksizm*), and so on. She was the first to publish the mathematical manuscripts of Karl Marx with extensive commentary. Later, however, she became a well-known historian of mathematics and an active propagandist for mathematical logic in Russia.[94] So she began her career in politics and turned to science with the blessing of her political supervisors.

Two other successful strategies shared one similarity: a woman should choose for her investigations a discipline so essential for the government that all specialists were welcomed (e.g., geology), or she should work in a region whose development was a priority (e.g., Central Asia), or both. Geologic exploration was developing rapidly in this period. As a result, there was a whole constellation of women geologists who became doctors or professors. One example is Elizaveta Dmitirievna Soshkina (1889–1963), a graduate of the Moscow Higher Women's Courses (1915), a geologist and

---

[92] F. 6733, op. 2, d. 14., Central State Archive of Moscow Region, Moscow.
[93] "Autobiography of S. A. Ianovskaia," f. 641, op. 6, d. 161, l. 56, ARAN.
[94] For more about S. A. Ianovskaia, see: I. G. Zenkevich, *Sud'ba Talanta (Ocherki o zhenshchinakh-matematikakh)* (Briansk, 1968), 71–8; I. G. Bashmakova, S. S. Demidov and V. A. Uspenskii, "Zhazhda Iasnosti," *Voprosi istorii estestvoznaniia i tekhniki*, 1999, no. 4:108–19.

paleontologist who received her master's in 1937 and doctoral degree in 1946. In 1948, she became a professor, and for several years she was the head of the laboratory in Paleontology Institute of the USSR Academy of Sciences.

There were other scientific institutions where one could find women among the junior research staff in the 1920s–30s, but the percentage of women doctors, professors, and heads of departments was much lower. For example, in the State Astronomical Institute of P. K. Shternberg in 1936–37, there were five women among senior staff scientists (four PhDs) and four women research assistants. Pavel Karlovich Shternberg (1865–1920), who took the position of observatory director in 1916, was a supporter of the women's higher education and started teaching as a professor in the Moscow Higher Women's Courses in 1901. He invited some of his students to work in the observatory (for example, Anna Sergeevna Miroliubova [1886–1978] and Maria Aleksandrovna Smirnova [1892–1986], who spent their entire lives at the observatory and then at the institute). But it seems that among all the women who worked at the institute in the 1920s–1930s, only one later became a professor and received her doctoral degree: Evgeniia Iakovlevna Boguslavskaia (1899–1960).[95]

The final way Soviet women became scientists was the most traditional, the most successful, and the most common: "scientific marriages." Marriages between female students and university professors were common in Russia from the very beginning of higher education for women. With the advent of coeducation, marriages between students continued to be frequent occurrences. As the outstanding physiologist and full member of the USSR Academy of Sciences Lina Solomonovna Shtern (1878–1968) wrote in her unpublished article "Woman's Role and Significance in Science" (1957):

> A woman has not only rights but obligations to develop all her gifts in full measure in order to promote humanity's progress. But unfortunately a woman satisfies her potential in accordance with that of her spouse, holding herself back to give him headway. Even today one can come across situations when a woman holds herself back to create suitable conditions for the man standing near her, and this lies very deeply in one's consciousness; an urge to push him forward even becomes unconscious. When business concerns a man, she puts a man forward before herself and above other talented women. From my own experience from the time when I directed a large collective of scientists, I had to struggle with this . . . I always aspired to put women up to the mark and met opposition from those women whom I considered necessary to move forward. They held themselves back somehow.
>
> The equal rights given to us mostly have a declarative character. A woman has to give more time and strength to her family then a man does, so he can give all his strength to science. And science as is well known is very jealous and demands the whole human being. It is impossible to serve the Lord and Mammon at the same time. A woman trying to do that doesn't employ her equality in full measure. Cases when women find it permissible to place her family's cares on her husband are very rare.[96]

But sometimes when a woman was energetic and talented, her career was made much easier with the support of her husband, and as he advanced in position her capabilities grew. There are examples of very successful couples, such as Aleksandr Fedorovich Kots (1880–1964) and his wife, Nadezhda Nikolaevna Ladigina-Kots

---

[95] This list may be incomplete. See: "List of women—senior staff scientists and list of women research assistants of the State Astronomical Institute of P. K. Shternberg," f. 641, op. 6, d. 161, ll. 46–7, ARAN.

[96] Personal Collection of L. S. Shtern, f. 1565, op. 1, d. 367, ll. 4–5, ARAN.

(1889–1963). He was a doctor in biological sciences, professor and founder of the State Darwinian Museum (1913) in Moscow. They married in 1911, when she was a third-year student of the Moscow Higher Women's Coursers and he was one of her teachers. She assisted Kots in all stages of the foundation of the museum while simultaneously exploring animal psychology. In 1913, she organized a laboratory for animal psychology at the museum. In 1917, Ladigina-Kots became a full member of the Institute of Psychology of Moscow University. She published several interesting works in this field and in 1941 received her doctorate in biology, becoming one of the most well-known animal psychologists in the USSR. It was Kots who had roused her interest in the doctrine of evolution, took her along on foreign trips, and gave her an opportunity to work in his museum. In turn, Ladigina-Kots helped her husband with technical work at the museum (not to mention providing him with a family).[97] Surveying the most noted Russian women scientists of the first half of the twentieth century, one finds that the majority of them not only were married but had husbands who occupied high positions in the scientific hierarchy (although the data is only preliminary and not statistical).

CONCLUSION

The article analyzes the history of Russian women entering the sciences from the 1860s to the 1940s. Of course, it was a slow process. It began with the complete unavailability of higher education for women in Russia, not to mention academic positions, and it ended with a state of formal equality (although obviously this was quite far from true equality). I argue that there were three "generations" of women scientists during this period.

The first generation followed the wave of general interest in the natural sciences in the early 1860s, typically women born in the 1840s or early 1850s. This generation may have been small numerically, but they nevertheless made significant contributions. Without formal education or academic status but with energy and persistence, they not only demonstrated the possibility of becoming a woman scientist but gained support from the scientific community. Some of these women were the first to receive doctoral degrees (all except one of them in Europe). Together they demonstrated that women were able both to study successfully in universities and to participate in the sciences. They became an example to a large number of women. Simultaneously, Russian women began a struggle for access to higher education.

The second generation consists of women ten to twenty years younger than the first. On the one hand, they already had examples of successful female scientific careers before them; on the other hand, they hoped that they would get an opportunity to build traditional scientific careers and to take research positions at scientific institutions, especially after Higher Women's Courses were permitted in several Russian cities. The majority of women belonging to this generation began their careers successfully at the Higher Women's Courses and continued them at European universities. A European doctoral degree became the standard path for this generation. But they underestimated the conservatism of the Russian bureaucracy. So those who still wanted to work as scientists had either to become volunteers (as their predecessors

---

[97] N. N. Ladigina-Kots, "Avtobiografiia," in *Gosudarstvennii Darvinovskii muzei. Stranitsy istorii: Osnovateli muzeia* (Mozhaisk, 1993), 73–83.

did) or to stay in a role of assistant or laboratory technician their entire lives. Yet when young women from the *third* generation arrived in auditoriums of the Higher Women's Courses, they met women who had already worked there on a regular basis. They might occupy positions much lower than their professional skills would indicate, but their presence had become ordinary. It was clear that sooner or later the barrier would fall.

In the 1890s and 1900s, some limitations were abolished. By 1917, women who had graduated from the Higher Women's Courses received the right to take state exams in the universities and to obtain university degrees with the right to prepare doctoral dissertations. After the doctorate, they could obtain positions at scientific institutions. Certainly, they were still excluded from the benefits of the civil service, and many of these new rights did not exist in practice. But this changed with the October Revolution of 1917, when the new government equalized all rights for women in the course of looking for allies. It was a long-awaited moment, and many women hurried to make use of it.

But very soon they found out that equality produced new problems. Living standards in academia collapsed after the revolution, and the treatment of women in the scientific community changed now that they had become real competitors. Women had to learn new rules and create new ways of successfully participating in science. Teaching and other duties often left too little time for investigations. In fact, teaching often left no time as the majority of academics worked in several institutions just to eke out a living. So scientific investigations remained an individual initiative for women.

Since the Soviet government had declared that there was no "woman's question" in the USSR, it did not specifically address the issue of "women in science." Nevertheless, in 1929 the Women's Department (*Zhenotdel*) of the CPSU Central Committee collected information about women scientists.[98] Unfortunately, right at that time the department closed, and documentation of the last period of its activity did not survive, so it is difficult to uncover the purpose of this endeavor. On November 19–20, 1936, the Moscow Regional Union of the Higher School and Scientific Institutions Workers organized a Creative Conference of Women Scientists, whose purpose was to "report for the Congress of Soviets about the achievements of women scientists in the development of Soviet sciences."[99] In November 1937, a similar conference was held in Leningrad.[100] It was shown that not only were the number of women in science growing but among them some had reached positions of very high academic status.

The number of women in Russian science seriously began to grow from the late 1880s onward. When it became easier for women to become scientists, it turned into a mass profession, and many undertook this path for whom science was not a high mission but simply a potential route to a secure life. And they brought a very old mindset with them: a man (a husband) as a worker was much more important to a family so his career must be develop first and women should take care of him, his children, and his home and support him in any way (at home and at the scientific institution). But it was impossible to change mass psychology immediately. That was a task for future generations of Russian women scientists.

---

[98] For example, see the letter from Rudzit (a functionary in the Women's Department [*Zhenotdel*] of the CPSU Central Committee) to M. V. Pavlova. 2 Dec. 1929, f. 311, op. 1a, d. 98, l. 70, ARAN.

[99] Letter from Rudzit to A. A. Glagoleva-Arkadieva, 2 Nov. 1936, f. 641, op. 6, d. 98, l. 147, ARAN.

[100] Papers of L. S. Shtern, f. 1565, op. 1, d. 324, l. 3, ARAN.

Wishful Science:
The Persistence of T. D. Lysenko's Agrobiology in the Politics of Science

By Nils Roll-Hansen*

ABSTRACT

The suppression of genetics in Soviet Russia was the big scandal of twentieth-century science. It was also a test case for the role of scientists in a liberal democracy. The intellectual's perennial dilemma between scientific truthfulness and political loyalty was sharpened by acute ideological conflicts. The central topic of this essay is how the conflict was played out in Soviet agricultural and biological science in the 1930s and 1940s. The account is focused on the role of the then current Soviet science policy and its basic epistemic principles, the "unity of theory and practice" and the "practice criterion of truth."

INTRODUCTION

The Soviet alternative to international genetic science was called Lysenkoism after its leading figure, the agronomist Trofim Denisovich Lysenko. He called it agrobiology or Michurinism, after a gardener with ambitions to become a Russian Luther Burbank.[1] During the cold war ideological standoff, Lysenkoism was the central topic in the history of Soviet science. Stalinist tyranny combined with scientific ignorance, opportunism, and moral deviousness among the Lysenkoists, was depicted as primary causes of the tragedy. With the fall of the Soviet system well behind us, the time

---

* Department of Philosophy, Classics, History of Art and Ideas, University of Oslo, Box 1020 Blindern, 0315 Oslo, Norway; nils.roll-hansen@ifikk.uio.no.

A first version of this paper was presented in the history of science and medicine seminar at Yale University in September 2002. Its further development, especially concerning scientific freedom and the political role of intellectuals, has benefited much from input and comments by the editors of the present issue of *Osiris*.

Abbreviations used in this article: ARAN—Archive of the Russian Academy of Sciences; *BV*—*Biulleten' VASKhNIL* (bulletin published by the Lenin Academy of Agricultural Science); SGS—Stenographic report from the conference on genetics and selection organized by the editorial board of the journal *Under the Banner of Marxism*, 7–14 Oct. 1939; SRSKh—*Sotsialisticheskaia Rekontruktsiia Sel'skogo Khoziaistva* (journal published jointly by the Lenin Academy of Agricultural Science and the Ministry of Agriculture); TsGANKh—Tsentral'nyi Gosudarstvenuyi Arkhiv Narodnogo Khoziaistva SSR (Central State Archive for the National Economy), Moscow; VL—(Vavilov Letters, 1929–40) *Nauchnoe nasledstvo*, vol. 10, *Nikolai Ivanovich Vavilov, iz epistoliarnogo naslediia 1929–1940 gg.* (Moscow, 1987).

[1] *Agrobiology* is a selection of Lysenko's works, first published in Russian in 1948, with a number of new editions and translations into foreign languages.

is ripe for a reconsideration of the nature and significance of Lysenkoism, paying more attention to the scientific and science policy issues.[2]

Science policy is the area in which science and politics overlap. It faces both ways and coordinates the two kinds of activity. The Soviet Union was the first country in the world to provide generous state support for science and technology. With the first five-year plan and the collectivization of agriculture, starting in 1928, a systematic and centralized science policy was formed. In natural and technological sciences, funds were increased, researchers recruited, and new institutions formed at breathtaking speed.[3] As late as the 1970s, the Soviet Union was far ahead of any other country in sheer numbers of scientists and engineers. This all-out effort laid the foundation of successes such as the Soviet thermonuclear bomb and the first Earth satellite. It also created the research system in which Lysenko made his career.

The Bolshevik enthusiasm for science as a motor of social progress appealed to liberal and left-wing scientists and intellectuals in the West. The British physical chemist and Communist J. D. Bernal set Soviet science and technology policy up as a paradigm for the rest of the world.[4] But there was also grave concern that such a centralized, state-governed organization of science would, in the long run, suppress essential intellectual freedom and undermine scientific as well as social progress. In 1940 the Oxford zoologist John Baker, together with the Hungarian-born physical chemist Michael Polanyi and the prominent Oxford ecological botanist Arthur Tansley, formed the Society for Freedom in Science in direct response to Bernal's ideas.[5]

Dividing resources between basic and applied research, between theoretical and practical science, is a persistent plight for the politics of science. One apparent escape is to say pragmatically that practical effect is what counts in the end and thus there is no important distinction to be made. This was the spirit of the slogan "unity of theory and practice," which dominated the politics of Soviet science in the 1920s and 1930s. The implication was that "pure" theoretical research is an abstract and useless activity, "cut off" from the working masses.[6] However, the dilemma between basic and applied science was nothing special to Soviet science. It still persists, as does the popularity of the pragmatic solution.[7]

The difference between theoretical and practical science points to a broader issue of cultural politics concerning the role of intellectuals. Is it important for rational politics to distinguish between, on the one hand, politically neutral theoretical science that is valid for all and serves common goals and, on the other, practical science engaged in solving specific social and technological problems and thus directly bound to certain political and economic agendas? Attempts at sharp separation would undoubtedly be harmful; the interaction between theory and practice has always been

---

[2] An attempt in this direction is Nils Roll-Hansen, *The Lysenko Effect: The Politics of Science* (Amherst, N.Y., 2005).

[3] A. G. Korol, *Soviet Research and Development: Its Organization, Personnel, and Funds* (Cambridge, Mass., 1965), 21.

[4] J. D. Bernal, *The Social Function of Science* (London, 1939).

[5] William McGucken, "On Freedom and Planning in Science: The Society for Freedom in Science, 1940–46," *Minerva* 16 (1978): 42–72.

[6] N. I. Bucharin et al., *Science at the Cross Roads: Papers Presented to the International Congress of the History of Science and Technology Held in London from June 29 to July 31, 1931, by the Delegates of the USSR*, 2nd ed. (1931; London, 1971).

[7] See, e.g., Michael Gibbons et al., *The New Production of Knowledge* (London, 1994); D. E. Stokes, *Pasteur's Quadrant: Basic Science and Technological Innovation* (Washington, D.C., 1997).

a major source of scientific progress. Nevertheless, isn't an understanding of the difference between theoretical understanding and practical action essential for proper organization of their interaction?

A corresponding distinction applies to scholarship and cultural activity in general. Throughout the twentieth century, liberal democrats tried to avoid radical politicization of cultural activities, seeing themselves as defending classical Enlightenment ideals against attacks from the Communist Left as well as the Fascist Right. In addition to Michael Polanyi, mentioned above, the Jewish-French philosopher Julien Benda, the American sociologist Robert Merton, and the Austrian-British philosopher Karl Popper are examples of well-known defenders of a nexus between scientific freedom and liberal democracy.

The Lenin Academy of Agricultural Science plays a central role in the following account because it epitomized the revolutionary Soviet science policy. Established in June 1929, it was to be "the academy of the general staff of the agricultural revolution"—the general staff being the Ministry of Agriculture.[8] Collectivization was conceived as a technoscientific as well as a social revolution. Working in tandem, a new science and a new social organization were to bring unprecedented progress in agriculture. Production would increase at the same time that excess labor would be transferred to industry. Agricultural science thus appeared as the key to solving two main problems: feeding the population and manning the new industries.

HISTORICAL INTERPRETATIONS OF LYSENKOISM

With the official ban on genetics in the Soviet Union following the August 1948 congress of the Lenin Academy, "Lysenko" and "Lysenkoism" became household words in the West. They referred to the paradigmic example of how the Soviet regime had betrayed scientific truth and intellectual freedom.

Julian Huxley, evolutionary biologist and the first general secretary of UNESCO, gave his diagnosis in 1949: "the major issue at stake was not the truth or falsity of Lysenko's claims, but the overriding of science by ideological and political authority."[9] In 1948–49, a flood of articles revealed the faults of Lysenkoism. Up to the 1960s, most of the literature on Lysenkoism was written by biologists and other natural scientists. In addition to Stalinist tyranny, Marxist theory of science and Lamarckian ideas of heredity were central factors of explanation for Lysenkoism.[10] The first comprehensive historical account of Lysenkoism, by Soviet biologist Zhores Medvedev, was published in 1969.[11] Written during the 1960s as part of the internal struggle to get rid of Lysenko, the account had circulated widely as samizdat before being translated into English and published in the United States. This balanced and insightful overview remains the most readable general account of the Lysenko story.

The publication in 1970 of *The Lysenko Affair* by the American historian of science

---

[8] *Sotsialisticheskoe zemledelie* [Socialist agriculture], 22 Jan. 1930, 2; article on the basic tasks of the Lenin Academy of Agricultural Science written by Nikolai Vavilov, its first president.

[9] Julian Huxley, *Soviet Genetics and World Science: Lysenko and the Meaning of Heredity* (London, 1949), ix.

[10] See, e.g., H. J. Muller, "The Destruction of Science in the USSR" and "Back to Barbarism—Scientifically," *Saturday Review of Literature*, 4 and 11 Dec. 1948; R. B. Goldschmidt, "Research and Politics," *Nature* 109 (4 March 1949): 219–27; Conway Zirkle, *Evolution, Marxian Biology, and the Social Scene* (Philadelphia, 1956).

[11] Zhores Medvedev, *The Rise and Fall of T. D. Lysenko* (New York, 1969).

David Joravsky marked a new epoch in the historiography of Lysenkoism. He started his research on the assumption that Lysenko had at least boosted farm yields—"why else would commissars of agriculture repeatedly say so?" But even this minimal expectation of rationality was disappointed. In the end, Joravsky saw no role for rational scientific or science policy arguments in Lysenkoism. He concluded that it was a "romantic" Western myth that Marxist theory in alliance with scientifically outdated Lamarckian theories of heredity were important causes of Lysenkoism. In his view, Lysenkoism "rebelled against science altogether. Farming was the problem, not theoretical ideology. Not only genetics but all the sciences that impinge on agriculture were tyrannically abused by quacks and time-servers for thirty-five years."[12] In spite of such romantically exaggerated conclusions, Joravsky's book remains a classic, a treasury of interesting facts and sharp analyses. Another leading American historian of Soviet science, Loren Graham, agreed that Lysenkoism had little to do with either Marxist theories about science or serious issues in biology. The "Lysenko episode was a chapter in the history of pseudoscience rather than the history of science."[13] I will argue that it is true that Lysenko ended up by rejecting sound science. But many leading Russian biologists approved of his early work. Its theoretical interest and potential practical importance were recognized internationally. The interesting problem is what took Lysenko and his followers into the sphere that Graham calls "pseudoscience."

An alternative explanation, taking Marxist theory of science seriously, was sketched in the mid-1970s by supporters of the radical science movement. Richard Lewontin and Richard Levins saw Lysenkoism as "an attempt at scientific revolution"—a genuine attempt to transform science into a better instrument for social justice and progress. In their view, Lysenkoism raised important unsolved problems about the relationship of theoretical science to practical work. As practicing biologists, they also found Marxist philosophy to be fruitful in their own research. In other words, the issues of scientific method and science policy raised by Lysenkoism could not be so easily dismissed. Similar views were argued by left-wing intellectuals in France.[14]

This neo-Marxist "dialectical" perspective on Lysenkoism was not developed beyond the level of a sketch. But it serves to remind us that well beyond the Second World War the sociopolitical project of the Soviet Union had a broad appeal not based in repression and terror. Genuine enthusiasm also carried the project. For instance, the leading physicist and later dissident Andrei Sakharov wrote in a private letter about his grief at Stalin's death in 1953: "I am under the influence of this great man's death. I am thinking of his humanity."[15] Not until after Gorbachev made a last attempt to save the Soviet project did the vision completely fade away.

Valery Soyfer's *Lysenko and the Tragedy of Soviet Science* (1994) is valuable both for its wealth of detailed information and for the intense participant perspective.[16] As a student at the Timiriazev Agricultural Academy in the 1960s, and later faculty

---

[12] David Joravsky, *The Lysenko Affair* (Cambridge, Mass., 1970), ix.
[13] Loren Graham, *Science and Philosophy in the Soviet Union* (New York, 1972), 195.
[14] Richard Lewontin and Richard Levins, "The Problem of Lysenkoism," in *The Radicalisation of Science: Ideology of/in the Natural Sciences*, ed. Hilary Rose and Steven Rose (London, 1976), 32–64; Dominique Lecourt, *Proletarian Science? The Case of Lysenko* (London, 1977).
[15] Andrei Sakharov, *Memoirs* (New York, 1990), 164.
[16] Valery Soyfer, *Lysenko and the Tragedy of Soviet Science* (New Brunswick, N.J., 1994). Russian edition: Valerii Soyfer, *Vlast' i nauka: Istoriia razgroma genetiki v SSSR* [Power and science: The history of the rout of genetics in the USSR] (Tenafly, N.J., 1987).

member, he became involved in the last phase of the struggle against Lysenkoism—and its aftermath. The fall of Lysenko in 1964 did not bring immediate change in the governance of Soviet science. Soyfer's book was written in a dissident mood in the early 1980s, when the regime still suppressed historical investigations of Lysenkoism. His dissident late Soviet perspective still labors under the yoke of the cold war. The genuine dilemmas and hard choices that always face an uncertain science play a secondary role in his account. During the rise of Lysenko, from first national fame at the end of the 1920s to Lysenko's presidency of the Lenin Academy in 1938, the best road to fruitful development and sound application of agricultural biological science was not as clear as it appeared half a century later.

A new generation of Russian historians of science writing in the post-Soviet period have emphasized the normal social features of science under Stalin's regime. The complex integration of scientific and political establishments implied an intimate two-way relationship rather than simple subordination of science to politics.[17] Scientists were under political control but were also indispensable advisers to the government. Their political leverage could be used in competing with each other. This was nothing unique to Soviet science, although the centralized and brutal nature of the regime made stakes higher. In this institutional perspective, the Lysenko episode was due to a fashion that affected all of Soviet science. The social mechanisms driving "Stalinist science" are much the same in the traditional totalitarian and the new institutional account. Scientific issues play a minor role. The early success as well as the later fall of Lysenko is mainly described and explained in terms of ritualized public "discussions," lobbying political bosses, and so on. But strong moral condemnation has given way to a more relaxed and social relativist attitude.[18]

Alexei Kojevnikov's account of the 1948 debacle demonstrates the fruitfulness of the new institutional approach. Under mature Stalinism, the rules of intraparty democracy were extended to science, and scientific issues decided accordingly. Higher (political) authorities defined the problem and set the stage for a "free" discussion followed by a "vote." The result was binding and the losing side was obliged to make self-criticism (repent). Kojevnikov shows how Lysenko provoked the top leadership to set up such a game at a time when he was still in a key position to pick participants.[19] This explanatory model makes clearer why the fate of genuine science and scholarship could differ so much between disciplines. In 1948 a monopoly was given to Lysenko's obscure agrobiological alternative to genetics. In 1950 the followers of Nikolai Marr were denied a similar monopoly in linguistics. In 1949 plans for a corresponding "discussion" in physics was simply called off. The intraparty style of democratic centralism, as well as the strong confidence of the top political leaders

---

[17] See, e.g., Alexei Kojevnikov, "President of Stalin's Academy: The Mask and Responsibility of Sergei Vavilov," *Isis* 87 (1996): 18–50; Nikolai Krementsov, *Stalinist Science* (Princeton, N.J., 1997); Alexei Kojevnikov, "Dialogues about Knowledge and Power in Totalitarian Political Culture," *Historical Studies in the Physical and Biological Sciences* 30 (1999): 227–47; Kojevnikov, *Stalin's Great Science* (London, 2004); Nikolai Krementsov, *International Science between the World Wars: The Case of Genetics* (London, 2005).

[18] See, e.g., Krementsov, *Stalinist Science* (cit. n. 17), 58–60. This view was to a considerable extent anticipated by Mark Adams, who stressed the adaptability of scientific institutes to different political regimes and demands without much change in the substance their research. See, e.g., his "Science, Ideology, and Structure: The Kol'tsov Institute, 1900–1970," in *The Social Context of Soviet Science*, ed. L. L. Lubrano and S. G. Salomon (Boulder, Colo., 1980), 173–204.

[19] Kojevnikov, *Stalin's Great Science* (cit. n. 17), 207–14.

in their own scientific judgment, made for erratic results. However, broad consensus among the representatives of the scientific community was respected.[20]

VAVILOV AND LYSENKO

Nikolai Vavilov is usually described as the staunch defender of science who stood up against Lysenko's pseudoscience, was struck down by political intervention, and finally suffered martyrdom in the cause of genetics. He can also be described as a fellow traveler who understood too late the threatening nature of the science policy he was involved in. Before discussing this view, it is important to lay out a few facts about the economic setting and the two main actors.

Before the First World War, Russia was a big exporter of grain and other foodstuffs. With the civil war came food shortage and hunger. From then on, agricultural production and food supply proved a chronic problem for the Soviet government. Agriculture symbolized by grain production was at the center of public attention. Great efforts to improve the situation had little success. This painful economic failure is an ominous backdrop to the story of Lysenkoism.

Trofim Lysenko was born 1898 in a peasant family. He received poor and late basic education. His scientific training was extramural. Undoubtedly a gifted, intelligent, and inspiring person, he emerged as the leader of a school that rejected standard genetics. In 1938 he became president of the Lenin Academy of Agricultural Science, and he organized the 1948 congress that condemned classical genetics. Only in 1965 was he finally deposed from his dominant position in agricultural science and biology.

Nikolai Vavilov was ten years older than Lysenko and the son of a rich self-made merchant of peasant origin. He received thorough scientific training in Russia and abroad. From the early 1920s into the mid-1930s, he was the main entrepreneur of Soviet agricultural science and president of the Lenin Academy from 1929 to 1935. After first supporting and protecting Lysenko, he turned in the late 1930s to sharp opposition. Vavilov was arrested in 1940 and perished in prison three years later.

Vavilov was a liberal progressive, not a Marxist or a socialist. His view of science and its social role was similar to that of the British physicist and longtime editor of *Nature* Richard A. Gregory. Gregory's *Discovery, or the Spirit and Service of Science*, first published in 1916, was translated into Russian in 1923, edited and prefaced by Vavilov.[21] Gregory shared the demands of radicals such as Bernal, J. B. S. Haldane, and Joseph Needham that science must become more relevant and active in solving pressing social problems but did not share their socialist ideology. Vavilov similarly shared the utopian sentiments that motivated the policy of forced collectivization and did not heed warnings that collectivization would destroy the social structure of the agricultural community. His model was the United States, where more than 17 million people had left the farms between 1910 and 1920.[22] Lenin also revered the American agricultural revolution. He is said to have had *The New Earth*,[23] a glorification of the recent American agricultural revolution, as bedside reading.

---

[20] Ibid., 216.

[21] R. A. Gregory, *Otkrytiia, tseli i znachenie nauka* (Petrograd 1923).

[22] Mark Popovskii, *Delo akademika Vavilova* [The file of academician Vavilov] (Ann Arbor, Mich., 1983), 38–9.

[23] W. S. Harwood, *The New Earth* (New York, 1906). Russian translation: V. S. Garvud, *Obnovlenie zemlia* (Moscow, 1909).

The world collection of plants was Vavilov's big project. Cultivated plants and their wild relatives were collected from all over the world to serve as material for breeding. This was the world's first large-scale gene bank, a grand and foresighted idea. But as the country became caught up in a crash program of modernization, there was diminishing understanding of and support for such long-term scientific investments.

Lysenko was a junior researcher in the already vast system of Soviet agricultural research when he first attracted public attention through a presentation in *Pravda* in 1927: a young "barefoot professor" at the experimental station of Gandzha in Azerbaidzhan had found a way to extend the period of growth and "make the fields green in winter," quite independently of academic science.[24] The impressed journalist did not consider how much milder winters were south of the Caucasus.

Lysenko's research at Gandzha was published in a small monograph, *Effects of the Thermal Factor on the Duration of Phases in the Development of Plants* (1928).[25] The central idea is that a certain "sum of heat" measured as "degree-days" is needed for a plant to pass through each of its developmental phases, from germination to flowering and ripe fruit. Eventually, this idea developed into Lysenko's theory of stages in the development of plants. He built on earlier work by the leading Soviet cotton specialist Gavril Zaitsev, a friend and colleague of Vavilov's working in Tashkent.[26]

The developmental physiology of plants, in particular the effects of light and temperature, was a new area in research. The intense interest was due not least to potential agricultural applications. Theoretical thinking and experimental methods in Lysenko's work had weak aspects and received deserved criticism, but its methods and problems were typical of the state of the art.[27]

In January 1929 Lysenko presented a paper on the effects of cold treatment on germinating wheat at a big national conference on biological agricultural science in Leningrad.[28] The leading Soviet expert in this field was Nikolai Maksimov, head of the plant physiology lab in Vavilov's Institute for Plant Cultivation. Maksimov had been critical of Lysenko, not least of his unwillingness to listen and learn, and he wanted to reject Lysenko's paper to the congress. But Vavilov found the paper sufficiently original and promising.[29] Maksimov attended Lysenko's lecture with critical remarks, but he also gave Lysenko's ideas ample room in the paper that summed up this section of the congress.[30]

---

[24] Vitaly Fyodorovich, "Polia zimoi" [The fields in winter], *Pravda*, 7 Aug. 1927, 6.

[25] T. D. Lysenko, *Vliianie termicheskoge faktora na prodolzhitel' nost' faz razvitiia rastenii; opyt co zlakami i khlopchatnikom* [Effects of the thermal factor on the duration of phases in the development of plants: Experiments with grasses and cotton], *Trudy Azerbaidzhanskoi tsentral' noi opytno-selktsionnoi stantsii, im .tov. Ordzhonikidze v gandzhe* [Works of the Central Experimental Station of Azerbaidzhan named Comrade Ordzhonikidze] (Baku), 1928, no. 3.

[26] See S. Reznik, *Zaveshchanie Gavrila Zaitseva* [The testament of Gavril Zaitsev] (Moscow, 1983).

[27] See, e.g., Roll-Hansen, *The Lysenko Effect* (cit. n. 2), 58–64.

[28] While normal annual plants germinate and produce seeds within the same year, winter annual plants germinate in the late autumn, pass the winter as small seedlings, and then set flower and seed the following spring or summer. If a winter annual plant is sown in the spring, it will not flower, or will flower poorly, during that same year. Winter rye and winter wheat are examples of winter annual plants. Spring rye and spring wheat are normal annual plants completing a whole life cycle within the same year (growth season).

[29] Semion Reznik, *Nikolai Vavilov* (Moscow, 1968), 268.

[30] N. A. Maksimov and M. A. Krotkina, "Issledovaniia nad posledeistviem ponizhionnoi temperatury na dlinu vegetatsionnogo perioda," *Trudy po prikladnoi botanike, genetike i selktsii* [Works on applied botany, genetics, and selection] 23, no. 2 (1929/30): 427–78.

VERNALIZATION—PROMISES AND FAILURES

The success of Lysenko's vernalization research had a serious economic background. Grain was the key agricultural product, and harsh winters that killed the seedlings of winter sowings became a critical problem. This happened extensively in both 1927 and 1928, on the eve of the first five-year plan.

The ambitious young Lysenko was not quite satisfied with the attention he had received at the Leningrad conference and decided to prove the importance of his work with a striking public demonstration. He instructed his peasant father to soak some sacks of winter wheat seed grain, bury them in a snowdrift, and then sow at the ordinary time for spring wheat. As summer came, the field of Lysenko's father stood out as a wonder to the peasant community, according to local and national newspapers. After sensational presentations in the press and on-site inspections from agricultural specialists and government authorities, Lysenko was given a new job at the Ukrainian Institute for Selection and Genetics at Odessa. He became head of a new laboratory for *iarovizatsiia*. It was well known that winter grain could be sown in the spring and produce a normal harvest if slightly germinated seed had been subjected to a period of low temperature. The task of Lysenko's group was to further investigate this phenomenon and see whether useful practical methods could be developed.

A school of followers grew around Lysenko and his theory of stages in the development of plants. Their work soon attracted international attention. From 1929 on, the two Imperial Bureaux of Plant Genetics, British institutions established in 1928 to serve communication of new knowledge in plant breeding and applied plant physiology, was instrumental in making Lysenko's work known.[31] A scientist of the bureaux introduced the term "vernalization," a Latinized version of the Russian iarovizatsiia, which remains the standard scientific term today. As late as the 1970s, Lysenko's name was also routinely in textbook accounts of vernalization, as a token of his status as a founding father in this field.

In the 1920s and 1930s, the physiology of plant development was internationally perceived as an unexplored field with high potential for practical applications—in other words, a promising field, one vulnerable to wishful thinking. When the world's leading cotton specialist met Lysenko in 1933 and characterized him as "a biological circle squarer," Vavilov answered liberally that an "angry species" such as Lysenko who "walked by faith and not by sight" might make some useful discovery, perhaps even "how to grow bananas in Moscow." It did no harm and might do some good to let him go on working.[32]

It soon turned out that the vernalization of winter seed grain was not practical, and Lysenko shifted to vernalization of spring grain. He claimed that this would counteract troublesome summer droughts by speeding up ripening. The method was introduced on a mass scale without much testing. When leading agricultural experts asked

---

[31] Roll-Hansen, *The Lysenko Effect* (cit. n. 2), 64–8, 142–8. The system of Imperial Agricultural Bureaux was expanded in 1928. Among the new bureaus established was an Imperial Bureau for Plant Genetics in Cambridge and another one in Aberystwyth, for "crops other than herbage plants" and for "herbage plants," respectively. See W. R. Black, "Imperial Agricultural Bureaux," *Journal of the Ministry of Agriculture*, 1929, 461–7, on 465.

[32] S. C. Harland on "The Lysenko Controversy," *The Listener*, 9 Dec. 1948, 873.

for more precise experimental trials,[33] Lysenko angrily suggested that his opponents were "wreckers" trying to sabotage the agricultural revolution.[34]

Vavilov was reviewer for the Commissariat of Agriculture. As late as July 1935, he was vigorously defending Lysenko's vernalization of seed grain as "a great achievement" in general agrotechnology.[35] But after rapid expansion in the mid-1930s,[36] the procedure was quietly reduced to empty ceremony and disappeared.[37] Only Lysenko's reputation for practical achievements remained due to the positive propaganda of the mass media and the suppression of public scientific criticism.

VERNALIZATION OF THE WORLD COLLECTION

The main reason for Vavilov's enthusiastic view of vernalization was its applications to plant breeding. An impatient government decree of July 1931 demanded that new varieties of grain be produced in four to five years instead of the former ten to twelve years. Vernalization was a promising tool in responding to this pressure by making effective use of the world collection of plants. Many foreign plant varieties would not develop normally in the Russian climate but could be made to do so with the help of vernalization. This was essential both to investigate their hereditary properties and for manipulating flowering to make hybridization possible.

In March 1932 Vavilov approached the Odessa institute and Lysenko for cooperation on vernalization of the world collection.[38] At the Sixth International Congress of Genetics, held in Ithaca, New York, in August 1932 he praised the "remarkable discoveries" of Lysenko and the "enormous new possibilities" they opened to plant breeders.[39] In a lecture at the Leningrad House of Scholars in April 1933, Vavilov told how plant breeding "last year unexpectedly received help from physiology."[40]

At this time Vavilov's Institute of Plant Industry (Vsesoiuzny Institut Rastenievodstva, VIR) in Leningrad was seething with political trouble. Graduate students wanted revolution in science. During Vavilov's half-year travel to the United States, a number of his leading collaborators were arrested or forced to leave the institute. Without experienced older scientists, the research would not be effective, Vavilov tried to explain to the Young Turks in January 1934. In particular, he worried that they were not sufficiently interested in vernalization. In May, he once more asked Lysenko for help.[41]

---

[33] P. N. Konstantinov, P. I. Lisitsyn, and D. Kostov, "Neskol'ko slov o rabotakh Odesskogo instituta selektsii i genetiki" [Some words on the works of the Odessa institute of selection and genetics], *SRSKh*, 1936, no. 11:121–30.

[34] T. D. Lysenko, "Otvet na statiu 'Neskol'ko slov o rabotakh odesskogo instituta selektsii i genetiki' akad: Konstantinova P. N., akad. Lisitsyna P. I., i Doncho Kostova" [Reply to the article "Some words on . . ."], *SRSKh*, 1936, no. 11:131–8.

[35] F. 8390, op. 1, e. 604, ll. 85–9, TsGANKh.

[36] R. O. Whyte, "History of Research in Vernalization," in *Vernalization and Photoperiodism*, ed. A. E. Murneek and R. O. Whyte (Waltham, Mass., 1948), 1–38, 9.

[37] See Eric Ashby, *Scientist in Russia* (London, 1947), 115.

[38] Vavilov to F. S. Stepanenko and T. D. Lysenko, 29 March 1932, VL, 165.

[39] Nikolai Vavilov, "The Process of Evolution in Cultivated Plants," in *Proceedings of the Sixth International Congress of Genetics, Ithaca, New York, 1932*, vol. 1 (New York, 1933), 331–42, 340.

[40] N. I. Vavilov, "Problema selektsii v SSSR" [Problems of selection in the USSR] (manuscript of lecture to be given at Doma Uchionykh in Leningrad, 28 April 1933), f. 8390, op. 1, ed. khr. 284, ll. 48ff., TsGANKh.

[41] See various letters from Vavilov to Russian colleagues, including Lysenko, in the period August 1932 to May 1934, VL.

RESEARCH PLANNING

Planning became a fashion in Soviet science in the early 1930s, under the leadership of Nikolai Bukharin. And Vavilov, president of the Lenin Academy, was an active participant.[42]

Bukharin's general view of science was strikingly presented at the 1931 International Congress for History of Science in London. A delegation of prominent Soviet scientists arrived spectacularly in a special airplane at the last minute. Bukharin, head of the delegation, was the most scholarly and intellectual of the top Soviet leaders; among the participants was Nikolai Vavilov. Bukharin explained the scientific revolution happening in the Soviet Union: "the rupture between intellectual and physical labour" was being eliminated and scientific research was rising to a new level of efficiency. His prime example was plant breeding.[43]

In spite of the great scientific efforts, agricultural production declined, and there was widespread hunger in the early 1930s. A government decree of July 1934 stated that the Lenin Academy "had not fulfilled the basic task to which it had been assigned." Inadequate organization and narrow specialization were among the basic faults. Among the few bright spots were Lysenko's vernalization and Vavilov's world collection of plants.[44]

In June 1935, the Lenin Academy was radically reorganized. Lysenko was one of the fifty new academicians (members) appointed. Many were bureaucrats rather than scientists, some without any scientific training. An old Bolshevik and former vice minister of agriculture, A. I. Muralov, took over as president. Vavilov became a vice president.

LYSENKO MOVES INTO GENETICS: NEW WHEAT VARIETIES IN RECORD TIME

Vavilov's interest in vernalization of the world collection stimulated Lysenko's shift to plant breeding and genetics. But although many established experts received his early research in plant physiology well, the reactions to his ideas on breeding and genetics were generally negative. That environmental conditions could somehow influence heredity in a direct and adaptive manner represented an interesting possibility that could not be discarded out of hand. Such neo-Lamarckian ideas were, in fact, pursued extensively by one of Lysenko's staunchest opponents, the plant geneticist Anton Zhebrak. But Lysenko's experiments and arguments indicated scientific incompetence rather than originality.

Lysenko and his co-workers had a holistic approach to plant development, seeking a unified theory of development and heredity. They stressed the interaction of environmental and hereditary, internal and external, factors. Heredity was a property of the organism as a whole. The chromosomes had no special role. Thus the theory of stages in the development of plants was extended to include heredity.[45] This biological ho-

---

[42] *Trudy vsesoiyznoi konferentsii po planirovaniio genetiko-selektsionnykh issledovanii* [All-union conference on the planning of genetics and breeding research] (Leningrad, 1933).

[43] N. I. Bucharin, "Theory and Practice from the Standpoint of Dialectical Materialism," in Bucharin et al., *Science at the Cross Roads* (cit. n. 6), 11–33, 15–6.

[44] *Izvestiia*, 20 July 1934, 1.

[45] A. Favorov, "Theoretical and Practical Significance of Lyssenko's Research on the Vernalization of Agricultural Plants," *Herbage Reviews* 1 (1933): 9–14.

lism provided Lysenko with a bridge from developmental physiology to genetics. Lysenko presented his ideas in a small book coauthored with his ideological and political adviser, I. I. Prezent.[46] As proof of the truth of his statements, Lysenko announced the creation of new and superior early ripening spring wheat varieties in only two and a half years. In early June 1935, members of the Lenin Academy reviewed the work of Lysenko's team in Odessa.

Two weeks later, the presidium of the Lenin Academy met in Moscow. Lysenko was not present. G. K. Meister, a prominent plant breeder and party member who had just been appointed as one of the vice presidents, argued that Lysenko simply did not understand what genetics was about, and even Muralov wondered whether Lysenko, in his laudable boldness, had not crossed the border to unhealthy extremism. But Vavilov defended him passionately. Why had the critics not spoken up in Odessa? Of course the new strains had not yet been fully tested, said Vavilov, but "[w]e were singing the praise of the method," not of an "accidental new variety." If Meister and some of the other speakers had been at Odessa, they would have understood how hard it was to find a way to confront youthful enthusiasm and "impatience" with tactful corrections. Vavilov also reminded the meeting that a worldwide critical revision of the principles of genetics was taking place—for example, the Morgan school of Drosophila genetics, founders of the chromosome theory of heredity, had been criticized for lack of "dialectical depth" by another prominent Soviet breeder and geneticist, A. A. Sapegin.[47]

THE INTERNATIONAL CONGRESS AND THE "TWO DIRECTIONS IN GENETICS"

In 1932, Vavilov asked the Soviet government for permission to invite the Seventh International Congress of Genetics, scheduled for 1937, to convene in Moscow. But the ministry was cold about the idea. In fact, it only allowed a couple of scientists to go to the Sixth Congress, held in New York State, in 1932. Vavilov urged Lysenko to go, but without success. Sweden was chosen as the place for the next congress. But when the Swedes withdrew their offer in the summer of 1935, Vavilov seized the opportunity and was able to obtain permission from the government as well as assent from the Permanent International Committee on Genetics to hold the next congress in Moscow.

Vavilov acknowledged, in his letter to the chairman of the International Committee, Norwegian medical doctor and geneticist Otto Mohr, that Soviet genetics was young and inexperienced, but he counted on help from people such as American geneticist H. J. Muller, then working in the Soviet Union. (Yet Muller, in a letter to his friend Mohr, had just expressed doubt that Soviet genetics was ready for the demanding task of holding a congress and worried that it would divert energy from research.) By the end of 1935 a local organizing committee had been set up with Muralov as president. Among the members were Vavilov, Lysenko, the botanical ecologists Boris Keller and Vladimir Komarov, and the experimental biologist Nikolai Kol'tsov.

---

[46] T. D. Lysenko and I. I. Prezent, *Selektsiia i teoriia stadiinogo razvitiia rasteniia* [Selection and the theory of the development of plants through stages] (Moscow, 1935). Prezent was a political activist with some training in law. He played a central role in developing a theory of "creative Darwinism" for education and other popularization of science in the 1930s. After Vavilov had turned down Prezent's offer of advisory services, he joined Lysenko in 1932.

[47] F. 8390, op. 1, ed. khr. 604, ll. 85–93, TsGANKh.

The human geneticist Solomon Levit was secretary, and Muller was in charge of the program. Both Keller and Komarov had Lamarckian sympathies and were skeptical of classical genetics. Keller was an active player in Soviet science politics and a supporter of Lysenkoist ideas. Komarov was to become president of the Academy of Science in the following year, 1936.

Through 1935–36, the Lenin Academy and the Ministry of Agriculture staged a broad debate on dialectical method in the science of biological heredity. A primary purpose was to sort out differences between classical genetics and Lysenkoist agrobiology. There were sharp disagreements, for instance, on the handling of seed production in highbred plants, and the ministry demanded a resolution. Growing political tension on the eve of the Great Terror (1936–38) made open scientific debate difficult, especially on ideologically sensitive topics such as dialectical method. Nevertheless, a number of leading biologists and agricultural scientists contributed. Vavilov, however, remained conspicuously silent. To conclude the debate, the Lenin Academy organized a conference on "the two directions in genetics" in December 1936. At the opening, Muralov proclaimed that based on "the Marxist-Leninist-Stalinist worldview" and a rejection of "fascist 'theories' of race," a comprehensive examination of different genetic theories was to "provide unity of method" for practical breeding work.[48] This tall order expressed bureaucratic voluntarism rather than scientific rationality.

The coming international congress was an underlying issue for the conference delegates. Vavilov and the geneticists saw such a congress as a possible way to stem the Lysenkoist tide and promote genuine genetic science. Not surprisingly, Lysenko objected to the congress, which threatened to undermine his scientific standing. The tense and ominous situation was underscored by the arrest of Levit before the December conference. As an active party member Levit was vulnerable. He had also spent a year in Muller's lab in Texas on Rockefeller money and was director of the Institute for Medical Genetics, an institution suspected of promoting eugenic ideas.[49]

Hard-hitting scientific criticism of Lysenko's genetic ideas from Muller and the leading Soviet geneticist, A. S. Serebrovskii, could not prevent the congress from turning into a public relations catastrophe for genetics. The defiant foreigner Muller, himself a proponent of eugenics on socialist premises, made the fateful step of arguing that Lamarckism rather than Mendelism was the eugenically obnoxious theory. This unleashed a violent campaign tarnishing Serebrovskii and other geneticists for earlier eugenic utterances. Vavilov kept a low profile, primarily defending his own Institute of Plant Industry. Lysenko launched a frontal attack on classical genetics. Geneticists and plant breeders were ranged against him, but a number of other influential leaders in biological science supported many of his views. In the Soviet mass media, genetics came off as a theory full of bourgeois metaphysical prejudice, one holding up the practical advance of agriculture and harboring dubious sympathies with eugenics.

In the meantime, a special commission had reported to the Soviet government that preparations for the congress were lagging and that on the preliminary program there

---

[48] A. I. Muralov, "Zadachi dekabr'skoi sessii" [Tasks of the December session], *BV*, 1936, no. 12:1–3, on 2.

[49] V. V. Babkov, "Medical Genetics in the Soviet Union," *Herald of the Russian Academy of Sciences* 71 (2001): 553–61.

were many "Fascists."⁵⁰ On November 17, the Council of Ministers (the Soviet cabinet) decided to cancel the congress for 1937.⁵¹

But only gradually through rumor, unreliable mass media messages, and private communication did information about the canceling of the congress reach the Permanent International Committee on Genetics. On December 14, 1936, the *New York Times* reported that Vavilov had been arrested, that other Soviet geneticists were being threatened with arrest, and that the 1937 congress in Moscow had been canceled. A week later, an editorial in *Izvestiia* reacted to this "slander": Vavilov had not been arrested, and the congress had only been postponed at the request of Soviet geneticists. Furthermore: "Real freedom for research work, real intellectual freedom exists only in the USSR," where science works for the benefit of the people and not a small group of capitalists, claimed the editorial. As was later revealed, Stalin himself had heavily edited the article.⁵²

On January 7, 1937, Mohr wrote to Muralov, with copies to Vavilov and Levit: "it is urgently needed that you send me immediately detailed information on the situation." On February 13, Muralov and Vavilov sent an answer with the brief explanation that many scientists and institutions had wanted a postponement, adding that the government had now permitted the international congress to be held in 1937.⁵³ Included, however, were bulletins from the December 1936 genetics conference. Mohr and others could now read for themselves the arguments used against classical genetics.

For Mohr and most other members of the international committee, the clandestine political interventions to control the program of the congress were irreconcilable with traditional liberal ideals of science. Increasing political terror and the beginning of Moscow prosecutions soon made it clear that the Soviet Union was no longer a suitable place. A congress in Moscow would threaten to undermine genuine science by obscuring the differences between science and politics. Through the spring of 1937, Mohr maneuvered diplomatically to gain time. He needed to form for himself, and communicate to the members of the international committee, a trustworthy picture of the situation. By July 1937, he had become convinced that holding the congress in Moscow would be a grave mistake and sent a new memorandum to the members of the international committee recommending that the congress be moved from Moscow to Edinburgh, to be held in 1939.⁵⁴ Some left-leaning Western geneticists, such as Muller, J. B. S. Haldane, and the American representative to the committee, Leslie Dunn, continued to support a congress in Moscow. It would be a betrayal of Soviet genetics to move it, they argued. To Mohr it was a relief when Muller admitted in November 1937 that Mohr had been right and that now even the Russian geneticists agreed.⁵⁵

⁵⁰ Memo dated 8 Oct. 1936, f. 201, op. 3, d. 10, ll. 1–3, ARAN.

⁵¹ A more detailed version of the following account of the cancellation of the Moscow genetics congress is found in Roll-Hansen, *The Lysenko Effect* (cit. n. 2), 230–43. A different interpretation with detailed descriptions of the workings of Soviet institutions is found in Krementsov, *International Science* (cit. n. 17), 42–72.

⁵² For a detailed account, see Krementsov, *International Science* (cit. n. 17), 45–52.

⁵³ Mohr to Muralov, 7 Jan. 1937. Muralov and Vavilov to Mohr, 13 Feb. 1937. Copies of these letters were later sent to all members of the International Committee of Genetics together with a memorandum in which Mohr briefly described what had happened, expressed grave concern for free scientific debate, and stated that he was investigating alternative countries for the congress. See, e.g., Federley Papers, University Library, University of Helsinki, Helsinki.

⁵⁴ Federley Papers, University Library, University of Helsinki.

⁵⁵ Mohr to Muller, 26 Nov. 1937, Muller Archives, Lilly Library, Indiana University, Bloomington.

SUPPRESSING FREEDOM OF SPEECH

As the international committee struggled to get reliable information about the state of affairs in the USSR, events moved quickly and dramatically in Soviet genetics. On January 7, 1937, at a meeting for authors and staff of the state publishing house for agricultural literature, the newly elected president of the Academy of Sciences, Komarov, complained that knowledge of Darwin was scandalously low.[56] And the minister of agriculture, I. A. Iakovlev, explained in a crassly polemical speech how classical Mendelian genetics was incompatible with Darwinism.[57]

In an atmosphere full of fear, one prominent biologist dared to challenge the political control over science. Although Nikolai Kol'tsov had actively opposed the October coup of the Communists in 1917, he had been able to preserve and expand his school of research. By 1939 he was the grand old man of Russian experimental biology. He was the one who read out Muller's speech to the December 1936 conference, and he saw clearly how fatal the image of genetics communicated by the mass media could become to the future of genetics in the USSR. Therefore he wrote to Muralov asking for the publication in *Pravda* and *Izvestiia* of "extensive articles written by genuine geneticists in defense of their science." But in the presidium of the Lenin Academy, the scientists—Vavilov, M. M. Zavadovskii, and G. K. Meister—feared a political confrontation and chose not to support Kol'tsov.[58]

Toward the end of March 1937, the Lenin Academy summoned a special session for all staff and members to discuss its own troubled affairs in the light of the new Soviet constitution, officially declared to be "the most democratic in the world." The curiously distorted debates at this meeting, called an aktiv, vividly demonstrate how political conformism, fear for one's own career, and narrow administrative logic can cooperate to suppress scientific freedom and autonomy.

Muralov accused Kol'tsov of eugenic views unacceptable under the new democratic constitution. But Kol'tsov defiantly repeated that the academy had failed to correct misleading and highly tendentious press reports. An open and democratic discussion in the spirit of the new constitution was just what he wanted. His former statements on eugenics had been quoted out of context and misinterpreted. For their time and context, his claims were legitimate and scientifically well founded. Kol'tsov found no reason to take back a word of what he had said.[59] His opponents were infuriated by this unwillingness to repent and perform self-criticism. How could this man, Kol'tsov, dare to pose proudly as a modern Galileo in defense of science? asked the scientific secretary.[60] The scientists were silent.

The difficult balance between political-bureaucratic and scientific authority was the central topic of the aktiv. Vavilov reminded the audience that the 1935 reorganization had been aimed at getting rid of an inefficient bureaucratic system, but all Muralov's selfless efforts had just made the situation worse.[61] D. N. Prianishnikov,

---

[56] V. L. Komarov, "Izdavat' Timiriazeva i Darvina" [Publishing Timiriazev and Darwin], *SRSKh*, 1937, no. 4:27–9.

[57] Ia. A. Iakovlev, "O darvinisme i nekotorykh antidarvinistakh" [On Darwinism and some anti-Darwinians], *SRSKh*, 1937, no.4:17–26, 24–5.

[58] A. E. Gaissinovich and K. O. Rossianov, "'Ia gluboko ubezhdion chto ia prav . . .'" [I am deeply convinced that I am right . . .], *Priroda,* 1989, no. 5:86–95; no. 6:95–103.

[59] F. 8390, op. 1, ed. khr. 954, ll. 82–4, TsGANKh.

[60] Ibid., ll. 37–9.

[61] F. 8390, op. 1, ed. khr. 956, ll. 42–8, TsGANKh.

an expert on fertilization and a staunch liberal defender of genuine science, pointed to the central bureaucracy's lack of contact with active scientific research as a problem. The reorganized Lenin Academy of Agricultural Science's preoccupation with short-term practical results made the academy fail in its primary task of planning and organizing scientific research.[62]

In his closing speech, Muralov challenged Vavilov on his specialty of rust-resistant wheat. Muralov had actively directed such research through instructions to the research stations. "Should I just keep away from this?" he asked rhetorically. "Absolutely," replied Vavilov from his seat. "Is it not the task of the presidium to organize research so that new varieties can be produced as quickly as possible?" retorted Muralov. "[Y]ou call this bureaucratic interventionism, but we call it organization of research." Vavilov answered, "You should take advice from the best specialists," giving a couple of names. "But you did not mention Lysenko and Tsitsin,"[63] continued Muralov. Vavilov then held his tongue. "I will fight to produce new varieties as quickly as possible," concluded Muralov.[64]

VICTORY AND CONSOLIDATION OF LYSENKOISM

Soon after this aktiv, the leadership of the Lenin Academy was hit by the terror. As elsewhere in Russia, active Communist Party members were particularly at risk. During the summer of 1937, Muralov and the scientific secretary were arrested. Meister, also a party member, became president, but a few months later he was arrested. For a period in the autumn and winter of 1937–38, Vavilov acted as academy president. Thus the Great Terror had the unintended effect of removing the last hurdles to Lysenko's ascent. On February 28, 1938, he was appointed president of the Lenin Academy, the top position in Soviet agricultural science.

In his new position, Lysenko promoted agrobiology and harassed genetics. The official government policy, however, still supported open competition between the two directions in genetics to allow the theory that proved most successful in "practice" to triumph. Keeping Vavilov and Zavadovskii as vice presidents of the academy was an expression of this balancing policy.

In March 1939, the Eighteenth Congress of the Communist Party marked the end of the Great Terror, and it again became possible to challenge the hegemony of Lysenkoism. A group of Leningrad biologists wrote to Andrei Zhdanov, the party secretary responsible for science, about the lack of open public debate on genetics. Free competition between the two directions in genetics was suppressed by administrative power, they claimed. On Zhdanov's initiative, a second conference "On the Controversy in Genetics and Breeding" was held in October 1939 under the auspices of the party's theoretical journal, *Under the Banner of Marxism*.

The philosophical staging consolidated Lysenkoism. While the 1936 conference had been summarized and evaluated by a genuine geneticist and breeder, Meister, the 1939 conference was summarized and evaluated by a philosopher, Mark Mitin. He mildly criticized Lysenko for not taking seriously enough chromosomes and other facts produced by classical genetics and his supporters for occasionally failing to

---

[62] F. 8390, op. 1, ed. khr. 954, ll. 97–100, TsGANKh.
[63] A leading plant breeder who gradually turned away from Lysenko in the 1940s and 1950s.
[64] F. 8390, op. 1, ed. khr. 956, ll. 84–6, TsGANKh.

observe rules of proper scholarly conduct. But, Mitin stated, Lysenko's practical achievements were beyond doubt: "Against this nobody has said, or could say anything from the rostrum of our meeting, because these things have been introduced into practice, into life, and have been widely disseminated."[65] Mitin disregarded Vavilov's sharp criticism that the Ministry of Agriculture had introduced new methods on Lysenko's recommendation with no support from "experimental data whatever."[66] And when Zhebrak asked for a commission to evaluate Lysenko's new, useless varieties of wheat, he was quickly put down by Chairman Mitin.[67] Yet Mitin's praise of Lysenko was not without a kernel of truth: the practical usefulness of the vernalization of seed grain was taken for granted. Vavilov did not criticize Lysenko on this point, and the two most knowledgeable experts, P. I. Lisitsyn and P. N. Konstantinov, who had raised the issue in 1936, were not present.

A certain redressing of the balance between Lysenkoism and classical genetics followed. Some of Lysenko's attempts to interfere in internal affairs of Vavilov's Institute of Plant Industry and the academy's Institute of Genetics were stopped. But the general effect was to strengthen the ideology of "unity of theory and practice" that had helped Lysenko all along. Many Western left-wing scientists continued to accept the official Soviet claim to an open scientific debate subject to the practice criterion of truth. The theories of both "the schools of Lysenko and Vavilov alike are subjected to the acid test of practice," the British embryologist Joseph Needham had written in 1938.[68]

But in the tense international situation the balance was precarious. The Hitler-Stalin pact of 1939 obscured the difference between enemies and friends. Vavilov had been under surveillance by the secret police since the early 1930s. By 1940, there was a large file on him, including numerous denunciations for political disloyalty, betrayal, and sabotage; in August 1940, he was arrested and charged with spying for the British. He was indeed working with his British contacts to achieve the publication of *Theoretical Basis of Plant Breeding*[69] in English.[70] This was a three-volume handbook, edited by Vavilov, summing up the main results of Soviet plant-breeding research. Ironically, it contained a broad and very positive presentation of Lysenko's contributions.

A regular rout of classical genetics followed Vavilov's arrest. A number of his colleagues in Leningrad were also arrested. A Lysenko sympathizer took over the Institute of Plant Industry, and Lysenko himself became director of the academy's Institute of Genetics, in Moscow. Kol'tsov was removed as director of the Institute for Experimental Biology and died soon after. A Lysenkoist became his successor. In spite of this dramatic setback, the research and teaching of classical genetics did survive in a number of institutes and university departments. But most important, the events of 1939 and 1940, symbolized by Vavilov's fate, made it clear to the scientific

---

[65] Mark B. Mitin, "Za peredovuiu sovetskuiu nauku" [For a leading Soviet science], *Pod znaeniem Marksxizma* [Under the banner of Marxism] 10 (1939): 149–76, on 150.

[66] "Genetics in the Soviet Union: Three Speeches from the 1939 Conference on Genetics and Selection," *Science and Society: A Marxian Quarterly* 4 (1939): 183–233, on 187–8.

[67] SGS, vol. 1, 221–2.

[68] Joseph Needham, "Genetics in the USSR," *Modern Quarterly* 1 (1938): 370–4. Needham used the pseudonym "Helix and Helianthus."

[69] Nikolai Vavilov, ed., *Teoreticheskie osnovy selektsii rastenii* [Theoretical basis for the selection of plants], 3 vols. (Moscow, 1935–37).

[70] See Darlington Papers, Bodleian Library, University of Oxford, Oxford, C 114.

community in the Soviet Union that no constructive compromise or free scientific competition with Lysenkoist agrobiology was possible. From now on it was war—within the constraints set by the institutional culture of Stalinist science.[71]

THE 1948 GENETICS CONFERENCE

After Nazi Germany attacked the Soviet Union in June 1941, ideological orthodoxy was toned down. Western democracies and the Soviet Union were now comrades in arms. By the end of World War II, there was growing hope for classical genetics. Criticism of Lysenko's theories as well as his leadership in the Lenin Academy was reviving. Anton Zhebrak, as head of a section in the Central Committee secretariat for a period in 1945–46, worked actively to establish new contacts with Western geneticists. But the emerging cold war, with its campaigns against internationalism and subservience to Western capitalism, soon provided Lysenko with a platform for counterattack.[72]

The dramatic events around the August 1948 conference of the Lenin Academy were precipitated by a young official in the Central Committee, Iurii Zhdanov. He was trained as a chemist and belonged to the inner circle of the regime. He was the son of Andrei Zhdanov, the most influential leader on cultural questions after Stalin. Iurii Zhdanov had even discussed principles of science policy personally with Stalin. In April 1948, he spoke to an audience of party officials on "controversial questions of contemporary Darwinism."

Following the official policy line of open and fair scientific debate and competition, young Zhdanov evaluated the two schools of genetics. He repeated the standard criticism of classical genetics: it lacked practical results, was obsessed with fruit flies, believed in unchanging genes, and generally suffered from a divorce of theory from practice. Zhdanov praised Lysenko for his great practical achievements but then expanded on the criticism of Lysenkoism. Lysenko had neglected recent discoveries about polyploidy of chromosomes,[73] rejected the use of hybrid corn,[74] not fulfilled the promise of new useful varieties of cereals in two to three years, and so on. I. Zhdanov also warned against illegitimate ways of suppressing other schools of thought and against the way philosophers had intervened in favor of Lysenko from the 1930s on.[75]

Iurii Zhdanov's lecture indicated the growing influence of genuine genetic science in the central party apparatus. But Lysenko adroitly used the budding criticism to force a showdown while he still had the upper hand. He used his political connections and administrative powers as president of the Lenin Academy to organize a special session of the academy with a majority biased in favor of agrobiology. A few geneticists spoke courageously for their science, but others just sat quietly, expecting the worst. At the end of the session, Lysenko dealt a final blow to his opponents. He announced that his keynote address on "The Situation in Biological Science" had

---

[71] Vividly described in publications by Krementsov and Kojevnikov (cit. n. 17).

[72] These developments have been described in detail by Krementsov, *Stalinist Science* (cit. n. 17).

[73] Methods for multiplication and manipulations of the number and composition of chromosomes were seen internationally as important techniques for creating genetic variation and progress in breeding.

[74] By 1948, this was a great agricultural success in the United States, often seen as the first big practical result of classical genetics.

[75] Iu. Zhdanov, "Vo mgle protivorechii" [In the darkness of contradiction], *Voprosy filosofii* 7 (1993): 65–92.

already been approved by the Politburo. In fact, it had been carefully edited by Stalin himself.[76] Those who had spoken against him had the choice between public repentance and leaving their jobs. A ban on teaching and research in classical genetics was soon confirmed by the Ministry of Education and the Academy of Sciences.

The August 1948 congress and the official Soviet ban on genuine genetic science was a striking example of self-defeat by wishful thinking—or shooting oneself in the foot. Translations of the verbatim report of the proceedings were distributed internationally. All over the world, scientists could read in detail the pseudoscientific Lysenkoist argument. The result was a major defeat for the Soviet Union on the intellectual front of the cold war.

There was strong opposition to Lysenko in the central party organs in 1948, and without Stalin's personal support he would most likely have been deposed. A number of idiosyncrasies made him sympathetic. Like many left-leaning politicians and intellectuals, Stalin had a soft spot for Lamarckian ideas about the malleability of heredity under environmental influence.[77] The mechanistic and somewhat inhumane deterministic taste of classical genetics did not suit his romantic and holistic tendencies. Stalin was also a passionate hobby gardener who felt he, not unlike Lysenko, had an intimate practical knowledge of plants. Branched wheat was a characteristic Stalin hang-up. Many spikes on each straw suggest a manifold increase in yields. In December 1946, Stalin had presented Lysenko with a sample of branched wheat from his homeland, Georgia. Lysenko dutifully started a project, although branched varieties of wheat had long been known to breeders and repeated attempts to make practical use of this property had failed. But, of course, Stalin's utopian hope was not completely wild. Sometimes new attempts succeed where many have failed before.

THE PREDICAMENT OF INTELLECTUALS

What most shocked the international scientific community in 1948 was the undisguised suppression of intellectual freedom.

From the 1930s, liberal intellectuals had cooperated with Communists in the fight against Fascism. In a critical political situation, this appeared the only viable alternative in spite of the authoritarian and antiliberal tendencies of Communist ideology. With the common victory over Nazi Germany, the Soviet Union emerged in a positive light, and hopes were strong that freedom and social equality could now be realized together in the West and the East.

At the end of World War II, it was a widespread view among Western biologists that both Lysenko and Vavilov were genuine scientists with important contributions. There were rumors but no reliable knowledge about Vavilov's death. A thoroughly respectful overview of the new genetics in the Soviet Union by two scientists at the Commonwealth Bureau of Plant Breeding and Genetics concluded that Lysenko's genetic ideas were highly problematic. But they took for granted that his practical and theoretical achievements in vernalization merited a general scientific standing comparable to that of Vavilov.[78] In a review of their book, the plant physiologist Eric Ashby noted that it

---

[76] Kirill Rossianov, "Editing Nature: Joseph Stalin and the 'New' Soviet Biology," *Isis* 84 (1993): 728–45.

[77] See, e.g., Stalin's letter to Lysenko, 7 Oct. 1947, published by Iurii Vavilov in "Avgust 1948: Predistoriia" [August 1948: The prehistory], *Chelovek*, 1998, no. 4:104–11, 109–10.

[78] P. S. Hudson and R. H. Richens, *The New Genetics in the Soviet Union* (Cambridge, UK, 1946), 4.

was "almost orientally apologetic" in its efforts to be fair to Lysenko. But Ashby did not remark on its respect for Lysenko's achievements in vernalization.[79] Ashby's own report from his period as a diplomat in Moscow during the war described Lysenko as an honest and sincere scientist, no charlatan or showman, in spite of his unfounded ideas about genetics.[80] At this point, the prospects looked good for genetics in the Soviet Union. Western geneticists believed that it would soon flourish[81] and organized a campaign to support the efforts of their Soviet colleagues to get rid of the Lysenkoist yoke. They were careful, however, to keep to purely scientific criticism and avoid political issues that could make the situation difficult for their Soviet colleagues.[82]

Other intellectuals were more explicit about the cultural threat from the Soviet Union. Karl Popper's *The Open Society and Its Enemies* targeted the sociology of knowledge inspired by Marxism as a main source of the revolt against reason that was threatening liberal democracy after the Second World War.[83] For Michael Polanyi and other leaders of the Society for Freedom in Science, the standard example of what Bernalism might lead to was the suppression of genetics in the Soviet Union.[84] When the society was established in 1940, its ideas met with little approval, but by the end of the war this had changed. There was broad acclaim of its ideals in the journal *Nature* by once skeptical scientists, and even the left-leaning British Association of Scientific Workers found the ideas germane; the society's membership soared.[85] The French watchdog of intellectual freedom Julien Benda, who had joined the Popular Front in the 1930s with a commitment that brought him close to membership in the Communist Party, also expressed concern about the threat of the Soviet system to intellectual freedom.[86] In a new preface for the 1947 edition of *La trahison des clercs* (The treason of the intellectuals), he stressed traditional scientific ideals of adherence to truth and neutrality in political matters as a counterforce to totalitarianism whether from right or left, which was not unlike Robert Merton's ethos of science with universal validity and disinterestedness as central ideals.[87]

In 1948 there was still widespread sympathy in the West both for the ideal of a science that would truly serve the people and for some of Lysenko's general biological ideas. The ban on genuine genetic science, however, undermined remaining sympathy. It dawned on many left-wing intellectuals that Communism and freedom were more at odds than they had hoped. The valiant defense of Lysenko's Lamarckism and

[79] Eric Ashby, "Genetics in the U.S.S.R.," *Nature* 158 (31 Aug. 1946): 286–7.

[80] Ashby, *Scientist in Russia* (cit. n. 37), 116.

[81] See, e.g., L. C. Dunn, "Science in the U.S.S.R.: Soviet Biology," *Science* 99 (28 Jan. 1944): 65–7; Julian Huxley, "Science in the U.S.S.R.: Evolutionary Biology and Related Subjects," *Nature* 156 (1 Sept. 1945): 254–6.

[82] Nikolai Krementsov, "A 'Second Front' in Soviet Genetics: The International Dimension of the Lysenko Controversy, 1944–1947," *Journal of the History of Biology* 29 (1996): 229–50.

[83] Karl Popper, *The Open Society and Its Enemies* (Princeton, N.J., 1950); see also Popper, *The Poverty of Historicism* (London, 1957).

[84] Michael Polanyi, "The Autonomy of Science," *Memoirs and Proceedings of the Manchester Literary and Philosophical Society* 85 (1943): 19–38.

[85] McGucken, "On Freedom and Planning in Science" (cit. n. 5).

[86] David Schalk, *The Spectrum of Engagement: Mounier, Benda, Nizan, Brasillach, Sartre* (Princeton, N.J., 1979), 43.

[87] Julien Benda, *La trahison des clercs* (Paris, 1947), 83–92. The book was first published in Paris in 1927; the first English translation was published in New York in 1928. R. K. Merton, "The Normative Structure of Science," in *The Sociology of Science* (Chicago, 1970), 267–78. Originally published as "Science and Technology in a Democratic Order," *Journal of Legal and Political Sociology* 1 (1942): 115–26.

science for the people by a cultural icon such as George Bernard Shaw simply highlighted its futility.[88]

In Britain, the brilliant population geneticist and scientific popularizer J. B. S. Haldane had argued insistently that some claims of Lysenko's might well be right and that his agrobiology deserved support as a competing research program. But after it was proclaimed as the truth in 1948, Haldane quietly faded out of the Communist movement.[89] Nevertheless Haldane retained a predilection for unorthodox genetic ideas,[90] and in his auto-obituary he coyly suggested that if he had had Lysenko's power over science, he would probably have committed similar mistakes.[91] The physicist Bernal, however, continued his defense of Lysenko well beyond 1948. In the first edition of *Science in History* (1954), Bernal still presented agrobiology as a valid exemplar of socialist science. But in the second edition of 1957, he admitted that political loyalty to the Soviet Union had led him "to pass over the inadmissible way in which the controversy was conducted in the Soviet Union."[92] Yet the ideals of science that fueled Lysenkoism retained their attractiveness into the 1970s. How to separate the scientific mistakes of Lysenkoism from its valid sociopolitical and epistemic principles was an important topic for the radical (left-wing) science movement.[93]

The influence of Lysenkoism was stronger and more lasting in France than in the Anglo-American cultural sphere, reflecting the cultural basis of a strong Communist Party. In French agricultural botany, there was a lasting tradition, organized as the Friends of Michurin, pursuing ecological studies in the spirit of Lysenko and Vavilov.[94] Writings of Nobel Prize geneticists Jacques Monod and André Lwoff witness how important this struggle was in the French scientific community. Dialectical materialism was the main target of Monod's slim 1970 volume of essays in the philosophy of biology, *Chance and Necessity*. Lysenko was quite right and Soviet geneticists mistaken, declared Monod; the theory of a gene that is stable through generations is completely irreconcilable with the principles of dialectics.[95] Monod's own epistemic alternative was "the ethics of knowledge," in harmony with the existentialist mood of his youth, respecting objectivity and avoiding the traps of vitalism and animism. This ethics expresses a code of behavior for scientific intellectuals similar to the ideals of Benda, Merton, and Popper. For Monod, the ethics of knowledge was "the only attitude at the same time rational and deliberately idealist on which a true socialism can perhaps be built."[96]

Lwoff in the preface of a new 1975 edition of *La trahison des clercs* vigorously defended the nonconformist Benda against common misinterpretations. For instance, Lwoff said, the Great Larousse encyclopedia described Benda's book as "a pamphlet against the intellectuals." To the contrary, Lwoff stated, the book exhorted truth, justice, and reason as the three universal values that intellectuals should pursue and

---

[88] George Bernard Shaw, "The Lysenko Muddle," *Labour Monthly*, Jan. 1949, 18–20.

[89] A sympathetic and informative account and analysis is given by Diane Paul, "A War on Two Fronts: J. B. S. Haldane and the Response to Lysenkoism in Britain," *J. Hist. Biol.* 16 (1983): 1–37.

[90] J. B. S. Haldane, *Biochemistry of Genetics* (London, 1954).

[91] Soyfer, *Lysenko and the Tragedy* (cit. n. 16), 282.

[92] J. D. Bernal, *Science in History*, 2nd ed. (London, 1957), 827.

[93] Lewontin and Levins, "The Problem of Lysenkoism"; Lecourt, *Proletarian Science?* (Both cit. n. 14.)

[94] A central person was Claude-Charles Mathon, working for many years at the University of Poitiers.

[95] Jacques Monod, *Le hasard et la nécessité* (Paris, 1970), 52.

[96] Ibid., 193.

chastised those who did not. Like much other contemporary philosophy Marxism, said Lwoff, drawing on Benda, was a philosophy of action that "does not admit of stable truths but only of truths determined by the moment." Lwoff parodied the "logic of contradiction" inherent in dialectical materialism as the right to contradict truth. His testimony was the suppression of genetics in the Soviet Union, starting with Lysenko's announcement of a new genetics in 1935. In 1975, there still had been no sanctions against the scientists and politicians responsible. Instead a scientist who published a book exposing the affaire ended up in a psychiatric hospital, wrote Lwoff.[97]

Monod and Lwoff were defending scientific autonomy in a new period of political radicalization. They felt that in the 1970s Marxist ideas about the social nature of science had once more become popular and threatening in universities and cultural life.

THE PERSISTENCE OF LYSENKOISM IN THE SOVIET UNION

After the climax of 1948, Lysenko enjoyed a few years of supreme power in Russian biology, experiencing a cult of personality only a step below that of Stalin himself. But accumulating failures of his practical advice gradually undermined Lysenko's authority. One spectacular failure was cluster planting of trees as part of the grand Stalin plan for the transformation of nature.[98] Assuming altruism and mutual sacrifice rather than competition between individuals of the same species, Lysenko advised the planting of clusters, rather than isolated individuals, to increase survival under harsh conditions. After a few years, there could be no denying that cluster planting in the steppe had been a great waste of resources. The final showdown between Lysenko and his academic critics in 1965 also focused on questions of practical economic results. A commission of the Academy of Sciences found that claims by Lysenko about the milk yields of his cows at the Lenin Hills Farm were at least unfounded, if not based on directly falsified data.[99]

Although practical failure was the most effective political argument, inconsistency with established biological theory was the basis of the scientists' campaign. Even Stalin was aware that science must observe standards of objectivity valid across cultural and political divides. In editing Lysenko's 1948 speech, Stalin had deleted references to "bourgeois biology" and claims that "all science is class science."[100] In a famous 1950 pamphlet on *Marxism and the Problems of Linguistics*,[101] Stalin rejected the monopolistic ambitions of certain schools, insisting that Soviet science must be open to criticism and observe international standards of objectivity.

Opposition to Lysenko grew after Stalin's death in 1953. He had to resign from the presidency of the Lenin Academy in 1956. But with support of party general secretary Nikita Khrushchev—a peasant son with a background from Ukrainian agricultural politics—Lysenko fought back. In 1958, a new editorial board took charge of

---

[97] André Lwoff, introduction to Julien Benda, *La trahison des clercs* (Paris, 1975), 9–40, 17–9. He probably had Zhores Medvedev and his book *The Rise and Fall of T. D. Lysenko* (cit. n. 11) in mind.

[98] Described in the government resolution as a "Plan for Erosion-Control, Forest Planting, Introduction of Grassland Crop Rotation, and Building of Ponds and Reservoirs to Guarantee Large and Dependable Harvests in the Steppe and Forest-Steppe Regions of the European USSR." Quoted from Soyfer, *Lysenko and the Tragedy* (cit. n. 16), 205.

[99] Ibid., 283–8.

[100] Rossianov, "Editing Nature" (cit. n. 76), 732.

[101] J. V. Stalin, *Markzism i voprosy iazykoznaniia* [Marxism and the problems of linguistics] (Moscow, 1950).

the oppositional *Botanical Journal*, and in 1961, Lysenko returned to his position as president of the Lenin Academy. But by now the Academy of Sciences, led by physical scientists, was heavily behind the drive to oust Lysenko.

In 1963, a new crisis emerged as the magazine *Neva* described the fateful 1948 session as a result of authoritarian Stalinism. This was sharply denounced by high officials, and the editors publicly regretted their "gross error." Soon thereafter, the election of new members to the Academy of Sciences precipitated a confrontation between it and Khrushchev. His candidates were voted down, and Khrushchev was furious. He considered a radical reorganization of the academy but did not have time to retaliate as his career stranded on broad failures in agricultural policy. For the first time in its history, the Soviet Union had had to import grain. In October 1964, Khrushchev was removed from his commanding position and replaced by Leonid Brezhnev, who was an engineer by training, and Lysenko quickly lost his important administrative posts.[102]

By 1965, the direct political suppression of genetics in the Soviet Union was finally at an end. But the new liberty was not matched by openness about the causes of repression. The general ideas and attitudes governing science policy persisted with little change. In 1967, Mark Popovskii's articles about the fate of Nikolai Vavilov were sharply attacked. Such emotionally inflammatory writings would spread uncertainty and stimulate revengefulness among the public and could not be permitted. Party ideologues worried that persons responsible for the mass media had become "less demanding of skilled treatment of modern problems of science and practice."[103]

A typical example of this continuity is the career of agricultural economist and bureaucrat Pavel Lobanov. He chaired the 1948 session and became president of the Lenin Academy when Lysenko had to step down in the period 1956 to 1961. With Lysenko's final retreat in 1965 Lobanov once more took over as president of the Lenin Academy of Agricultural Science—until 1978.

CONCLUSION

My account is focused on the politics of science—the arena in which science and politics overlap and interact—because I believe that provides the best explanatory perspective on the development of Lysenkoism, its rise and fall as well as its protracted aftermath. And I have pointed to the Lenin Academy of Agricultural Science as the key institution at which the pragmatic principles of Soviet science policy, expressed in the practice criterion of truth and the unity of theory and practice, were played out with more consequence than in other Soviet institutions of science policy.

The rise of Lysenkoism from 1929 to 1939 is described in detail, and the rest of the story, up to the 1980s, is only sketched. There are two main reasons for this. First, the early period has so far received less historiographical attention. The scientific content and contemporary standing of Lysenko's early work in plant physiology, in particular, has mostly been superficially treated. Second, this is the period when an ideologically distorted view of the relation between science and politics precipitated an irrational institutional and cultural system of science politics.

This perspective throws new light on the role of Nikolai Vavilov. He was not only

---

[102] For a detailed account, see Soyfer, *Lysenko and the Tragedy* (cit. n. 16), 251–83.
[103] Ibid., 291.

the martyr of genetics but also a main entrepreneur of the research system in which Lysenko made his career. He supported Lysenko's career in its early stages and did not publicly turn against him until the late 1930s. As Polanyi pointed out, it was with some reason that Lysenko turned Vavilov's own pronouncements from 1932 against him at the October 1939 conference: "The divorce of genetics from practical selection" that characterizes the West "must be resolutely removed from genetics-selection work in the U.S.S.R."[104] In the debates over the "two directions in genetics" in 1935–36, it was others who defended the value of genetics as a theo-retical science. To argue that disregard for the different character of theoretical and applied science was a major source of the political appeal of Lysenkoist misconceptions[105] was not popular at that stage. When Kol'tsov stood up for scientific and intellectual freedom in 1936–37, Vavilov gave him no public support. Such outspokenness was, of course, dangerous under the Great Terror. But Vavilov proved his courage later. So perhaps he was too involved with current government science policy to clearly see the threats that Kol'tsov pointed to.

The vagueness and ambiguity of the classical ideal for the political role of the intellectual, proclaimed by Benda and others, has worried the British social anthropologist Ernest Gellner. He has pointed to paradoxical and treacherous consequences of a naturalistic and pragmatic interpretation of the ideal. Jean-Paul Sartre's softness on Stalinism was an example of how easily it happens that "he who preaches against the treason of the clerics commits it in the very same sermon." The task of "*not* committing the treason is far, far more difficult" than most intellectuals have assumed, claimed Gellner.[106] It is not surprising that Nikolai Vavilov, as a scientific technocrat, had difficulty in breaking free from the role of a loyal fellow traveler.[107]

Why did biology and not, for instance, physics, chemistry, or geology, succumb to a primitive and retarded pseudoscience? The answers lie in the important differences that existed in the status of current knowledge and the institutional situation of biology versus those that existed for other sciences. The biological knowledge at stake—in plant physiology as well as genetics—was new, still to a large extent uncertain, and disputed. The Lenin Academy of Agricultural Science was set up in 1929 in the spirit of a socialist revolution in science. This was a spirit with international appeal, as can be seen from the enthusiastic reaction of the British scientific Left to Bukharin at the London 1931 International Congress of the History of Science.[108]

Considering its incompatibility with international biology and its accumulating practical failures, Lysenko's agrobiology could not last. But why did it take so long to disprove it and why was the Soviet government so slow to give it up? The continuity of science policy doctrines, institutions, and personnel prolonged the life of Lysenkoism almost to the end of the Soviet Union. There was no clear abandonment of the practice criterion of truth or the unity of theory and practice, only a slow and partial recognition of their inadequacy.

[104] Polanyi, "The Autonomy of Science" (cit. n. 84), 34.

[105] M. M. Zavadovskii, "Protiv zagibov v napadakh na gentiky" [Against deviations in the attacks on genetics], *SRSKh*, 1936, no. 8:84–96, on 95–6.

[106] Ernest Gellner, "La trahison de la trahison des clercs," in *The Political Responsibility of Intellectuals*, ed. Ian Maclean, Alan Montefiore, and Peter Winch (Cambridge, UK, 1990), 17–27.

[107] For the history of the term, see David Caute, *The Fellow-Travellers: A Postscript to the Enlightenment* (London, 1973), 1–7.

[108] See, e.g., G. Werskey, "Introduction: On the Reception of *Science at the Cross Roads* in England," in Bucharin et al., *Science at the Cross Roads* (cit. n. 6), xi–xxix.

Stalin's Rocket Designers' Leap into Space:
The Technical Intelligentsia Faces the Thaw

By Slava Gerovitch*

ABSTRACT

This article explores the impact of the professional culture of rocket engineering in Stalin's Soviet Union on the engineering and organizational practices of the space program during the Khrushchev era. The Stalinist legacy and the dual military/civilian character of rocket engineers' work profoundly affected the identity of this elite part of Soviet technical intelligentsia. Focusing on such notions as control, authority, and responsibility, this article examines the role of engineering culture in shaping the Soviet approach to the automation of piloted spacecraft control. Through patronage and networking, rocket engineers were able to overcome the inefficiency of Soviet industrial management and to advance their agenda of space exploration.

INTRODUCTION

On September 25, 1938, Joseph Stalin signed a list of seventy-four military specialists and defense engineers, authorizing their execution by firing squad. The rest of the Politburo followed suit, huddling their signatures below Stalin's. This was a routine procedure; in 1937–1938, Stalin signed more than 350 such lists, condemning to death at least 39,000 people whose execution required his personal sanction. Number twenty-nine on the September 25 list was an engineer from a rocket research institute, one Sergei Korolev. He had been arrested in June 1938 on a trumped-up charge of wrecking and sabotage and tortured into confession.[1] Two days after Stalin's approval, a quick trial hearing was held. Fifty-nine people on the list were sentenced to death and immediately executed. Korolev was lucky: after retracting his

---

* Science, Technology and Society Program, E51-165, Massachusetts Institute of Technology, Cambridge, MA 02139; slava@mit.edu.

Earlier versions of this paper were presented in November 2006 at the panel "From Prison Labs to Nuclear Plants: Revisiting the Soviet Technical Intelligentsia," at the annual meeting of the American Association for the Advancement of Slavic Studies, in Washington, D.C., and in March 2007 at the Slavic Department of the University of North Carolina at Chapel Hill. I wish to thank the participants of these forums, the editors of this volume, and an anonymous referee of *Osiris* for their very useful comments. I am especially grateful to Asif Siddiqi, Archie Brown, Ethan Pollock, Christopher Putney, and Donald Raleigh for their penetrating questions. The staff of the Russian State Archive of the Economy and the staff of the Russian Academy of Sciences Archive in Moscow were most helpful in locating relevant documents. Research for this article was supported by the National Science Foundation under Grant No. SES-0549177. Unless otherwise noted, translations are my own.

[1] On the circumstances of Korolev's arrest, see Asif A. Siddiqi, "The Rockets' Red Glare: Technology, Conflict, and Terror in the Soviet Union," *Technology and Culture* 44 (2003): 470–501.

© 2008 by The History of Science Society. All rights reserved. 0369-7827/08/2008-0008$10.00

confession, he received only a ten-year sentence. Another engineer from the same institute, Valentin Glushko, had been arrested three months before Korolev, on the same charge, and was also sentenced to prison time. Glushko was sent to work at a *sharashka*, a prison design bureau. Korolev served the first several months at the notorious Kolyma labor camp, barely survived, and eventually ended up in the same sharashka as Glushko. Both were released in 1944, after successfully completing the design of a new airplane, but the charges against them were not formally dismissed for another ten years.[2]

The names of Korolev and Glushko are now associated with some of the most remarkable technological achievements of the twentieth century. Korolev, the chief designer of rocket technology, and Glushko, the chief designer of rocket engines, played crucial roles in the development of Soviet rocketry, building the first Soviet intercontinental ballistic missile, launching *Sputnik*, and sending the first man into space.[3]

The Stalinist oppressive policies also adversely affected many other rocket engineers. Boris Raushenbakh, a leading control systems engineer, was interned in a labor camp as an ethnic German during World War II.[4] Vasilii Mishin, Korolev's first deputy, who later succeeded Korolev as chief designer, lost his father in the Gulag.[5]

Nevertheless, the Stalin era has been held up as the golden age of rocketry in the historical mythology that permeates the Russian space industry. Space engineers often call the postwar years a "romantic" period, the time of optimism and enthusiasm.[6] I use the term "mythology" here without implying the truth or falsity of any particular historical claims but simply to stress the foundational, identity-shaping character of such claims.[7]

The Soviet space program achieved its greatest successes—*Sputnik*, the first man in space, the first group flight, the first woman's flight, the first multicrew mission—during the Khrushchev era. This was a tumultuous period, which combined many contradictory trends. Khrushchev's "secret speech" at the Twentieth Congress of the Communist Party condemned Stalin's "cult of personality" and opened the gates for de-Stalinization in wider society. Yet the violent Soviet suppression of the Hungarian uprising, the vociferous campaign against the Nobel laureate poet Boris Pasternak, and Khrushchev's public outbursts against liberal intellectuals testified not only to

---

[2] Konstantin Tomilin, "Stalin sanktsioniroval ubiistvo Koroleva," *Sarov* (June 2002); http://russcience.euro.ru/papers/korolev.htm (accessed 30 Oct. 2006). The list with Korolev's name is kept in f. 3, op. 24, d. 419, l. 170, Archive of the President of the Russian Federation, Moscow; http://www.memo.ru/history/vkvs/spiski/pg11170.htm (accessed 30 Oct. 2006).

[3] The most comprehensive history of the Soviet space program is Asif A. Siddiqi's thoroughly researched *Challenge to Apollo: The Soviet Union and the Space Race, 1945–1974* (Washington, D.C., 2000), which includes an excellent bibliographic essay. For Korolev's well-researched biography, see Iaroslav Golovanov, *Korolev: Fakty i mify* (Moscow, 1994). For materials about Glushko, see V. F. Rakhmanin and L. E. Sternin, eds., *Odnazhdy i navsegda* (Moscow, 1998).

[4] Boris V. Raushenbakh, *Postskriptum* (Moscow, 2001).

[5] Deborah Cadbury, *Space Race* (New York, 2006), 95.

[6] E.g., see introductory remarks by Yurii Koptev, then the head of the Russian Aerospace Agency, in Yurii A. Mozzhorin, *Tak eto bylo* (Moscow, 2000); http://epizodsspace.testpilot.ru/bibl/mozjorin/tak/rka.html (accessed 30 Oct. 2006).

[7] On the selectivity of memories woven into various narratives of Soviet space history, see Asif A. Siddiqi, "Privatising Memory: The Soviet Space Programme through Museums and Memoirs," in *Showcasing Space*, ed. Martin Collins and Douglas Millard (London, 2005), 98–115; and Slava Gerovitch, "Creating Memories: Myth, Identity, and Culture in the Russian Space Age," in *Remembering the Space Age*, ed. Steven J. Dick and Roger D. Launius (forthcoming).

Khrushchev's oscillating personality but also to the uncertainty and instability that characterized this period's politics and culture. As the historian Polly Jones has noted, "Characteristic of the Khrushchev period were repeated swings in official policy, as the new leadership attempted to maintain a tense balance between enthusiasm for discarding the past, and uncontrolled iconoclasm, between mobilizing the energy of 'new forces', and giving in to anarchy, between maintaining the Soviet system, and causing its implosion."[8]

Among the confusions and contradictions of the Khrushchev era, the tremendous Soviet technological leap into space had a very specific symbolic meaning. In the public mind, it represented a daring breakthrough into the future—both into the technological utopia of interplanetary travel and into the political utopia of Communism. According to the 1963 poll by a popular youth-oriented Soviet newspaper, Yurii Gagarin's flight was viewed as the greatest human achievement of the century, and *Sputnik* the greatest technological feat.[9] In this sense, space exploration epitomized the thaw as a movement beyond Stalinism into a new and exciting political and cultural territory.[10]

This article attempts to look for the roots of the triumphs of the Khrushchev era in an earlier period by establishing the continuity of professional cultures of rocketry and space engineering. My focus is on two mutually shaping processes: the development of the space industry, and the formation of the identity and the professional culture of space engineers as a specific group of the Soviet technical intelligentsia in the late 1950s–1960s.

Historian Kendall Bailes stressed that the prewar Soviet "technostructure" did not simply follow orders issued by the "power structure." On the contrary, the technical intelligentsia played an active role in reshaping the Soviet social and cultural landscape.[11] Walter McDougall has noted a similar trend in the postwar period. In his study of the U.S. and Soviet space programs, he has emphasized the growing political influence during the cold war of the technocratic elites on both sides of the iron curtain.[12] Recent studies of the inner bureaucratic workings of the Soviet space program reveal a complex picture, in which different groups of space engineers competed for space projects and had to negotiate with multiple power brokers in the Communist Party and government apparatus, in the military, and in the defense industry.[13]

Focusing on the role of professional culture in rocket and space engineering, this

---

[8] Polly Jones, introduction to *The Dilemmas of De-Stalinization: Negotiating Cultural and Social Change in the Khrushchev Era*, ed. Polly Jones (London, 2006), 1–18, on 1. Jones's volume includes an excellent bibliography on the Khrushchev period. On de-Stalinization trends in Soviet science, see Slava Gerovitch, *From Newspeak to Cyberspeak: A History of Soviet Cybernetics* (Cambridge, Mass., 2002).

[9] Boris A. Grushin, *Chetyre zhizni Rossii v zerkale oprosov obshchestvennogo mneniia*, vol. 1, *Zhizn' 1-ia: Epokha Khrushcheva* (Moscow, 2001), 403.

[10] Slava Gerovitch, "'New Soviet Man' Inside Machine: Human Engineering, Spacecraft Design, and the Construction of Communism," *The Self as Project: Politics and the Human Sciences in the Twentieth Century*, Osiris 22 (2007): 135–57.

[11] Kendall Bailes, *Technology and Society under Lenin and Stalin: Origins of the Soviet Technical Intelligentsia, 1917–1941* (Princeton, N.J., 1978).

[12] Walter A. McDougall, . . . *The Heavens and the Earth: A Political History of the Space Age* (New York, 1985).

[13] See Andrew John Aldrin, "Innovation, the Scientists, and the State: Programmatic Innovation and the Creation of the Soviet Space Program" (PhD diss., Univ. of California, Los Angeles, 1996); William Barry, "The Missile Design Bureaux and Soviet Piloted Space Policy, 1953–1974" (PhD diss., Oxford Univ., 1995); and Siddiqi, *Challenge to Apollo* (cit. n. 3).

essay draws on recent studies of patronage, networking, cultural norms, and social identity in Soviet society.[14] These studies have stressed the central role of personal networks in strengthening the Soviet state and at the same time helping individuals overcome bureaucratic bottlenecks. Sheila Fitzpatrick has argued that in Stalinist society "outward conformity to ideology and ritual mattered, but personal ties mattered even more."[15] Kiril Tomoff writes, "Born of inefficiency and encouraged by the Party's longstanding, self-proclaimed right to intervene to correct any bureaucratic shortcoming, unofficial networks permeated the bureaucratic system."[16] Gerald Easter and Jerry Hough have explored how cohesive groups of Soviet functionaries with close personal ties established and maintained the effectiveness and stability of the Soviet state.[17] Barbara Walker has suggested that "the Soviets were able to create such an inefficient means of redistribution at all precisely because effective prior networking and patronage relations mitigated and obscured the profound inadequacy of the bureaucratic system as it took shape."[18]

Professional networks played a particularly important role. Belonging to a professional network not only shaped the identity of engineers, scientists, and managers but also allowed them to consolidate their efforts in furthering their professional agenda. For example, according to Mark Adams, by using personal networks, the scientific community during the Khrushchev era proved "more resourceful at manipulating [the Soviet] system to serve its own agendas than even the most optimistic advocate of academic freedom might have hoped."[19]

Stressing the inconsistencies and uncertainties of ideological discourse and the shifting, unsettled nature of Soviet identities, these studies overturn the stereotypical picture of the Soviet citizen as either blindly supporting or passively resisting specific government policies. This perspective illuminates the complex dynamic of the technical intelligentsia's service to the state: the engineers constantly grappled with their problematic identity and tried to formulate their own technocratic agenda, while negotiating and reinterpreting government policies. Instead of positing an opposition between the technical intelligentsia and the state, it would be more productive to talk about the inner tensions that defined the intelligentsia's identity and about the intelligentsia's involvement in the formulation and implementation of government policies through both official and unofficial means.

---

[14] Major literature reviews in this field are Sheila Fitzpatrick, "Politics as Practice: Thoughts on a New Soviet Political History," *Kritika: Explorations in Russian and Eurasian History* 5 (2004): 27–54; Anna Krylova, "The Tenacious Liberal Subject in Soviet Studies," *Kritika* 1 (2000): 119–46; and Barbara Walker, "(Still) Searching for a Soviet Society: Personalized Political and Economic Ties in Recent Soviet Historiography: A Review Article," *Comparative Studies in Society and History* 433 (July 2001): 631–42.

[15] Sheila Fitzpatrick, *Everyday Stalinism: Ordinary Life in Extraordinary Times; Soviet Russia in the 1930s* (New York, 1999), 227.

[16] Kiril Tomoff, "'Most Respected Comrade . . .': Clients, Patrons, Brokers, and Unofficial Networks in the Stalinist Music World," *Contemporary European History* 11 (2002): 33–65, on 65.

[17] See Gerald M. Easter, *Reconstructing the State: Personal Networks and Elite Identity in Soviet Russia* (Cambridge, UK, 1999); and Jerry F. Hough, *The Soviet Prefects: The Local Party Organs in Industrial Decision-Making* (Cambridge, Mass., 1969).

[18] Walker, "(Still) Searching for a Soviet Society" (cit. n. 14), 635.

[19] Mark B. Adams, "Networks in Action: The Khrushchev Era, the Cold War, and the Transformation of Soviet Science," in *Science, History, and Social Activism: A Tribute to Everett Mendelsohn,* ed. Garland E. Allen and Roy MacLeod (Dordrecht, Netherlands, 2001), 255–76, on 271.

ORGANIZATIONAL PROBLEMS OF THE ROCKET AND SPACE INDUSTRY

The date of May 13, 1946, when Stalin signed a decree establishing a Special Committee for Reactive Technology, is still celebrated today as the birthday of the Russian rocket and space industry. The committee was headed by Stalin's chief lieutenant, Georgii Malenkov, and included several leading defense industry managers. The missile program was organized on the same principles as the atomic project, managed by the Special Committee No. 1: a crash program with direct political support from the top, vast funding, and enormous resources. Key institutions of rocketry development were created, including the Scientific Research Institute No. 88, which at that time included Korolev's design bureau.[20]

The story of Stalin's critical personal involvement in the launching of the Soviet rocket industry has been told many times and acquired mythic proportions. Some of it is clearly based on hearsay and has not been confirmed by documentary record.[21] Yet in the identity-shaping folklore of Soviet rocket designers, the truly significant support that the Soviet government gave to the highest priority missile program in the late Stalinist period is often personified in the figure of Stalin as a great benefactor of Soviet rocketry. After Stalin's death in March 1953, Korolev, clearly unaware of Stalin's personal role in his imprisonment, expressed genuine sorrow in a private letter to his wife: "Our Comrade Stalin passed away . . . My heart hurts so much, my throat is clogged, and there are no thoughts and no words to express the tragedy that befell all of us. This is truly a national, immeasurable tragedy—our beloved Comrade Stalin is no more."[22]

The launch of *Sputnik* in October 1957 on top of the R-7 intercontinental ballistic missile, designed by Korolev, became a highly visible sign of success of the missile program supervised by the Special Committee for Reactive Technology. The successful completion of the three major defense industry projects of late Stalinism—nuclear weapons, the ballistic missile, and radar—engendered a radical reorganization of the defense complex. Coupled with Khrushchev's far-reaching reform of the national system of economic management, this led to the complete dismantling of the old Stalinist system of defense industry management. In December 1957, the three special committees supervising the nuclear weapons, the ballistic missile, and the radar programs were abolished, and a new agency—the Commission on Military-Industrial Issues under the USSR Council of Ministers—was created instead to coordinate the work of the defense industry ministries.[23] That same year, another reform threatened the ministries themselves. Khrushchev proposed a radical management reform of the national economy, replacing the system of central ministries with a system of regional economic councils. Instead of a single ministry controlling all enterprises in a particular branch of industry across the entire Soviet Union, a regional

---

[20] For the full text of the 1946 decree, see Boris E. Chertok, *Rockets and People*, vol. 2, *Creating a Rocket Industry*, ed. Asif A. Siddiqi (Washington, D.C., 2006), 10–5.

[21] Asif A. Siddiqi, series introduction to Boris E. Chertok, *Rockets and People*, vol. 1 (Washington, D.C., 2005), ix–xix, on xvi.

[22] Nataliia Koroleva, *Otets*, 2 vols. (Moscow, 2002), 2:269.

[23] Chertok, *Rockets and People* (cit. n. 20), 2:23. On the Military-Industrial Commission, see Nikolai Stroev, "Voennaia aviatsia," in *Sovetskaia voennaia moshch' ot Stalina do Gorbacheva*, ed. Aleksandr Minaev (Moscow, 1999), 279–82.

economic council would supervise all the industries located on its territory. This threw the economy into chaos.[24]

The defense industry was not immune to the general economic disarray. In a letter to Khrushchev, a group of top managers tried to persuade him to exempt the defense industry from the management reform, but they failed.[25] Like other plants and factories, those producing missiles were transferred to the control of their regional councils. When Khrushchev famously proclaimed in the United Nations that Soviet plants were producing rockets "like sausage," he was indeed correct. After the transfer, many of the problems that plagued the production of sausage now afflicted the production of missiles: broken supply chains across regional borders, poor coordination between central and regional agencies, and overlapping and conflicting spheres of authority among multiple supervising bodies.

The nascent space program had a particularly difficult time adapting to the new administrative regime. In the Soviet Union, there was no single central agency, like NASA, solely responsible for the funding and the supervision of space activities. Space projects were officially authorized by joint resolutions of the Party Central Committee and the Council of Ministers, but these decisions often came with no financial backing. In 1959, Korolev's bureau received no funding for the development of the Vostok spacecraft and for the rockets it used to launch automatic lunar probes. By early 1960, the bureau had a deficit of 95 million rubles; by the end of February, it had exhausted all the funds allocated for the first quarter, and by March 1960, it had no cash at all. Korolev constantly petitioned his superiors for the 95 million rubles that the bureau had already spent fulfilling party and government resolutions. After a month of bureaucratic wrangling, Korolev received a 50 million ruble grant and a 22 million ruble loan, still far short of his needs.[26]

After Gagarin's successful orbital flight on board *Vostok* in April 1961, a euphoric Khrushchev showered the space industry with medals and awards, and he became much more receptive to more ambitious plans of space travel to the Moon, Mars, and beyond. The new, larger projects, however, faced even greater organizational and financial problems than the Vostok program did. The growing complexity of rocket and spacecraft design and production required cooperation and coordination on an unprecedented scale. Korolev's Experimental Design Bureau No. 1, which served as the rocket and space technology integrator, had to deal with 200 to 300 subcontractors. As a deputy chairman of the Military-Industrial Commission recalled, any broken deadline could lead to the "total disorganization of the entire project."[27] Because the space program developed on the basis of ad hoc decisions of the party and the government, space projects often were not included in long-term economic planning. Their implementation required multiple adjustments of production plans for hundreds of enterprises across the Soviet Union. The cumbersome system of central planning had great difficulty managing such fast-paced large-scale technological projects.

---

[24] See Roy A. Medvedev and Zhores A. Medvedev, *Khrushchev: The Years in Power* (New York, 1978), 104–7.

[25] Irina V. Bystrova, *Voenno-promyshlennyi kompleks SSSR v gody kholodnoi voiny* (Moscow, 2000), 250.

[26] F. 4372, op. 79, d. 355, ll. 175–6, 216–7, Russian State Archive of the Economy (Rossiiskii Gosudarstvennyi Arkhiv Ekonomiki, hereafter cited as RGAE), Moscow.

[27] Georgii Pashkov, quoted in Aleksandr Ishlinskii, ed., *Akademik S. P. Korolev: Uchenyi, inzhener, chelovek* (Moscow, 1986), 318.

In July 1963, in an attempt to bring some order to the increasingly chaotic network of supply chains, the Council of Ministers established a system of monetary sanctions for undisciplined suppliers who did not fulfill their assignments on time. The reasons for delays, however, were often referred further and further down the supply chain, which made it nearly impossible to find and punish the "real" culprit. For example, in November 1965 a factory in Sverdlovsk was threatened with hefty fines for its failure to deliver a batch of launchers and missiles to the Ministry of Defense on time. By arguing that subcontractors had caused the delay in production, the factory officials obtained an exemption from the sanctions.[28]

The Soviet lunar program was besieged by inadequate funding and huge organizational problems. A rift between Korolev and Glushko resulted in Glushko's refusal to build engines for Korolev's N-1 lunar rocket and forced Korolev to collaborate with an engine contractor from another ministry. In the meantime, other leading missile designers, such as Vladimir Chelomey and Mikhail Yangel, actively promoted competing proposals for lunar missions. Korolev and his rivals cleverly used their political patronage ties with the top echelon of the Soviet government and lobbied for their own versions of government decrees. The ensuing compromise split the lunar program—including all the funds and resources—between Korolev and Chelomey. This resulted in an unprecedented duplication of effort in the design and production of lunar rockets and spacecraft.[29] The acute shortage of funds forced Korolev to cancel the construction of a ground testing facility for the entire cluster of first-stage engines for the N-1 rocket. This proved to be a fateful decision, one that spelled the ultimate failure of the entire program.[30]

The troubles with the organization of the rocket and space industry continued after Khrushchev's ouster from power in October 1964. In 1965, the Soviet government abolished the regional economic councils and restored the system of industrial branch ministries. The newly created Ministry of General Machine Building gathered under one administrative roof most of the bureaus and factories involved in rocket and space design and production. Korolev tried to seize this moment to transfer to the ministry as many of his subcontractors as possible. For example, in October 1965, he attempted to acquire control over the Balashikha plant that had been assigned the production of a fueling system for the N-1 rocket. He complained that the plant's performance was "exceptionally poor." Less than a year before the deadline, the plant had completed only 1 percent of the total amount of work. Korolev lost this round of bureaucratic power play: the plant was transferred to another ministry.[31]

The restoration of the ministry system did not solve the problem of component supply; in some ways, the system became even more complicated. Every contract between two organizations from different ministries now had to be approved by both ministries. Trying to reduce their overall load, various ministries often refused contracts for complex rocket and space equipment. For example, in February 1966, the Ministry of Electronic Industry flatly rejected a request to start production of ground control equipment for missiles and spacecraft. The head of the ministry declared that the proposal was "totally unrealistic and obviously impractical."[32] In August 1966,

[28] F. 4372, op. 81, d. 1249, ll. 139–40, RGAE.
[29] See Siddiqi, *Challenge to Apollo* (cit. n. 3), chaps. 9 and 11.
[30] Ibid., 392.
[31] F. 4372, op. 81, d. 1239, ll. 25–7, RGAE.
[32] F. 4372, op. 81, d. 1945, l. 16, RGAE.

the Ministry of Heavy Machinery refused to produce girders and support constructions for the N-1 assembly, even though the ministry had been assigned this task by the Party Central Committee, the Council of Ministers, and the Military-Industrial Commission.[33] "Having different ministries is like having different governments," one contemporary observed.[34]

In such circumstances, it was hardly surprising that the leaders of the rocket and space industry looked back at the Stalin years with a bit of nostalgia. In the folklore of Soviet rocketry, in the foundational myths that laid the narrative basis for rocket engineers' professional culture, even the fear and oppression of the Stalin era were often remembered fondly as productive mechanisms for instilling a strong sense of personal responsibility. For example, Yurii Mozzhorin, director of the Scientific Research Institute No. 88, wrote: "At that time, Joseph Stalin, who did not forgive any mistakes, was still in power, and our branch of industry was supervised by Lavrentii Beria, his henchman. For this reason, the development of technical-tactical specifications for rocket weaponry and its deployment had extraordinary significance and required a responsible approach."[35] Mozzhorin meaningfully pointed out that in the Stalin years no institutions, organizations, or individuals had been allowed to interfere with rocket research and production without special authorization from the Council of Ministers.[36]

Soviet rocket engineers' fond memories about Stalinism as the golden age of rocketry were quite selective. Lavrentii Beria was not, in fact, supervising Soviet rocketry (he was responsible for the atomic bomb), but his prominent presence in rocket engineers' folklore is indicative of their mythology of Stalinism. The perfect order and discipline of the Stalin era were a useful construct that helped the engineers to underscore their critique of the haphazard management of the space program under Khrushchev. In fact, in the late 1940s, top defense industry managers similarly complained of insufficient resources and petitioned to transfer factories from other ministries to their own control to ensure timely supplies.[37] Yet the image of the Stalinist era as the epitome of strong management, strict discipline, and personal responsibility formed the foundation for the professional culture of Soviet rocketry.

TAKING THE INITIATIVE

In May 1964, frustrated by the lack of action on the government resolutions authorizing the lunar program, Korolev decided to appeal to Leonid Brezhnev, then the secretary of the Central Committee for Defense Industry. "There are no firm deadlines, no essential organization, nor sufficient financial or material support," he wrote. "The initial sum of money set aside in 1964 for the Ministry of Defense to build preflight

---

[33] F. 4372, op. 81, d. 1944, l. 43, RGAE.

[34] V. Golovachev, "A Hercules Is Born," trans. Sharon Breit and Wade Holland, *Soviet Cybernetics: Recent News Items*, June 1967, no. 5:70–8, on 72.

[35] Mozzhorin, "Upravlenie raketnogo vooruzheniia GAU," chap. 2 in *Tak eto bylo* (cit. n. 6); http://epizodsspace.testpilot.ru/bibl/mozjorin/tak/02.html (accessed 30 Oct. 2006).

[36] Yurii A. Mozzhorin, "Rol' S.P. Koroleva v razvitii otechestvennoi raketnoi i kosmicheskoi tekhniki za 50 let (1946–1966 gg.)," *Iz istorii aviatsii i kosmonavtiki* 72 (1998); http://epizodsspace.testpilot.ru/bibl/iz-istorii/rol-kor.html (accessed 12 Oct. 2006). The Soviet government decree of May 13, 1946, stipulated, "No institutions, organizations, or individuals shall have the right to interfere with or ask for information concerning the work being conducted on reactive armaments without the special permission of the Council of Ministers." Chertok, *Rockets and People* (cit. n. 20), 2:11.

[37] Bystrova, *Voenno-promyshlennyi kompleks* (cit. n. 25), 244–6.

testing and launch facilities for the N-1, was 11 million rubles, but then the Ministry suddenly reduced this amount to 7 million, most recently to 4 million, and currently the Ministry refused any further financing of the N-1 construction despite the existing Party and Government resolution to this effect."[38] Korolev did not send the letter, perhaps realizing the futility of the effort. The customer—the Ministry of Defense—did not provide funds, subcontractors avoided contracts: Korolev was caught in the middle of a stalled economic and administrative structure.

The overly complicated system of Soviet defense industry management, which was supervised by several party and government agencies with overlapping authority and conflicting interests, relied on the Military-Industrial Commission to coordinate projects across agency lines. The commission, with its limited authority, could hardly manage large and complex space projects, and Korolev regularly complained about its poor performance. He suggested reorganizing the industry along the lines of the Stalin-era nuclear weapons and air defense programs, that is, placing a single central agency in charge.[39] Unable to effect such a radical administrative reform, he decided to facilitate government decision making and enforce the discipline of supply and production by other means. Korolev in effect established an alternative management mechanism, which complemented government structures and helped overcome bureaucratic barriers. He borrowed some of the proven management techniques of the Stalin era and adapted them for the new environment.

First, Korolev vastly expanded and strengthened his personal network. Like all chief designers, he attached special importance to the "vertical" patronage ties with Khrushchev and the chief patron of the defense industry, Dmitrii Ustinov. But his most effective tool was a "horizontal" network, linking top engineers and defense industry and military leaders. The hub of this network was the Council of Chief Designers. He organized this informal body in 1947 to coordinate the efforts of several key institutions involved in the design of the first Soviet ballistic missiles. The six original members of the council were Sergei Korolev (the entire rocket complex), Valentin Glushko (rocket engines), Mikhail Ryazanskii (ground-based guidance systems), Nikolai Pilyugin (onboard guidance systems), Viktor Kuznetsov (the chief designer of gyroscopes), and Vladimir Barmin (launch equipment). The original six were bound together by the ties of personal acquaintance and friendship going back to the 1930s' early rocketry studies or to the 1940s' prolonged joint mission to occupied Germany to collect rocketry artifacts and know-how.[40] The council tackled 90 percent of all engineering problems.[41]

While working on *Sputnik*, the first lunar probes, and the first piloted spacecraft, Korolev realized that a whole host of new problems had arisen—both technical and administrative—that went far beyond the area of expertise and influence of the original "rocketry caste." He invited at least fifteen new members, including leading mathematicians, ballistics specialists, designers of communications systems, new engine designers, ground tracking specialists, physicians, and air force officials.[42]

---

[38] Boris V. Raushenbakh, ed., *S. P. Korolev i ego delo: Svet i teni v istorii kosmonavtiki* (Moscow, 1998), 449–50.
[39] Ibid., 443.
[40] Chertok, *Rockets and People*, (cit. n. 21), 1:5; Chertok, *Rockets and People* (cit. n. 20), 2:6.
[41] Boris E. Chertok, *Fili—Podlipki—Tiuratam*, vol. 2 of *Rakety i liudi*, 3rd ed. (Moscow, 2002), 413.
[42] Ibid., 413–5.

Figure 1. The Council of Chief Designers. From right to left: Viktor Kuznetsov, Vladimir Barmin, Valentin Glushko, Sergei Korolev, Nikolai Pilyugin, Mikhail Ryazanskii, and Aleksei Bogomolov. The first six were the members of the original council, organized by Korolev in 1947. This photo was taken in 1959 during control of a Luna mission to the Moon. (NASA photo no. chiefdesigners1959.)

The sphere of the council's authority spread from pure engineering to organizational issues. Achieving a consensus among the chief designers was crucial not only for resolving internal disputes but also for presenting a joint front to lobby the higher authorities.

Members of the council played a unique role: through personal contacts and alliances with multiple power structures within the party and government apparatus, they lobbied for their projects, obtained official approval, and were able to enforce the execution of government orders, which government officials often failed to do. As Korolev's first deputy, Vasilii Mishin, aptly put it, "Korolev was built into the space program like an engine into a rocket. He fitted into the existing social and economic system so well that he could, in fact, circumvent it."[43] Konstantin Feoktistov, the leading spacecraft designer, confirmed that "strategic decisions were made not by the Party Central Committee or by the Government but by Ustinov and Korolev, and often by Korolev alone. Only later, one way or another, did they manage to obtain an official endorsement of those decisions by the 'competent organs.'"[44]

Instead of hierarchical top-to-bottom decision making, the council practiced consensus-building negotiations. If the chief designers could not reach a decision on a complex issue, the council created a working group to hammer out a compromise.

---

[43] Vasilii Mishin, "My dolzhny spustit'sia s nebes na Zemliu," *Nezavisimaia gazeta*, 12 April 2001; http://www.astronaut.ru/bookcase/article/article22.htm (accessed 30 Oct. 2006).

[44] Konstantin Feoktistov, *Traektoriia zhizni* (Moscow, 2000), 36–7.

As one memoirist recalled, Korolev's working style was to "arbitrate disputes."[45] Korolev was not willing to postpone decisions indefinitely, however. If the council could not eventually reach a consensus, he would make a decision himself.[46]

In the 1960s, the Council of Chief Designers, an informal body whose decisions had no legal binding power, became a de facto steering committee for the Soviet missile and space program. The council often invited to its sessions a large group of defense industry managers, military officials, and academics. At those informal meetings, Korolev and other chief designers could frankly exchange opinions on crucial technical and organizational issues without generating a huge bureaucratic paper trail. For example, in September 1960, the council meeting included eighty-seven participants and discussed the design of the N-1 heavy booster and its potential military applications.[47] In January 1961, the council met again to discuss specifically the choice of fuel for the N-1, taking into account the efficiency, toxicity, and cost of various fuels.[48] At the June 1964 meeting, the council made a crucial decision to use liquid oxygen as the main fuel for the N-1 and to use the N-1 in the lunar landing program.[49] This decision was officially approved in the August 1964 joint decree of the Party Central Committee and the Soviet government, authorizing the Soviet lunar landing program.[50] The deterioration of Korolev's personal network after his death in early 1966 is often pointed to as a key factor in the Soviets' ultimate loss in the lunar race.

Another Stalin-era mechanism that Korolev adapted for the 1960s was the personal responsibility of chief designers for the failure-free operation of their systems. In September 1960, as the Vostok spacecraft for the first human flight was being built, Korolev's design bureau prepared *Basic Guidelines for the Development and Preparation of the Object 3KA* (3KA was the code name for Vostok), which all members of the Council of Chief Designers later signed. They realized that the reliability of Vostok was of paramount importance, but the scale and complexity of the project made efficient quality control exceedingly difficult. The total of 123 organizations from various ministries and agencies, including thirty-six factories subordinated to thirteen different regional economic councils, participated in the construction of the Vostok rocket and spacecraft. The rocket engines had thirty-three chambers, and the spacecraft carried on board 241 vacuum tubes, more than 600 transistors, fifty-six electric motors, nearly 800 relays and switches connected by 880 electric plugs, and almost fifteen kilometers of cables.[51] The *Guidelines* established "personal responsibility of chief designers, factory directors, and heads of services for the quality of technical documentation, for the correctness of design, for the testing and reliability of construction elements, and for the quality of production, assembly, and testing."[52] Korolev believed that to ensure the reliability of the entire system, one had to instill the sense of responsibility not only in the top management but also in every worker involved in the production of Vostok. Every step in the assembly and testing was documented, including the names of the workers responsible for that step. Quality

---

[45] Vladimir Syromiatnikov, interview by author, 25 May 2004, Moscow.
[46] Evgenii Shabanov [Shabarov], quoted in Ishlinskii, *Akademik S. P. Korolev* (cit. n. 27), 259.
[47] Raushenbakh, *S. P. Korolev i ego delo* (cit. n. 38), 305–8.
[48] Ibid., 319–23.
[49] Ibid., 455–60.
[50] Asif A. Siddiqi, "A Secret Uncovered: The Soviet Decision to Land Cosmonauts on the Moon," *Spaceflight* 46 (2004): 205–13.
[51] Iurii V. Biriukov, ed., *Materialy po istorii kosmicheskogo korablia "Vostok"* (Moscow, 2001), 213.
[52] Ibid., 128.

control at major stages of assembly and testing was assigned to military specialists from the Ministry of Defense. All Vostok parts were branded by a special mark and documented as "acceptable for 3KA." Every worker knew that the life of a cosmonaut depended on the quality of that part.[53] It is worth stressing that the *Guidelines* were not imposed on the industry by any official authority. This document was spread around and enforced through the informal network coordinated by the Council of Chief Designers.

For chief designers, greater personal responsibility meant greater personal authority. Korolev "quite consciously sought such authority," remarked the space engineer Georgii Vetrov. "Such authority brought him power, which the Chief Designer needed as much as a military commander."[54] Other memoirists also compared Korolev to a military commander[55] or even to "an absolute dictator."[56] "The huge burden of personal responsibility, which Korolev could not share with any of his subordinates, sometimes made him exigent to the point of despotism, authoritative to the point of arrogance, and intensely focused to the point of alienation and seclusion," recalled Vetrov.[57] Korolev's idea of efficient management was to exercise personal control over every technical and organizational aspect of the space program. "As a leader, he believed he must extend his power over every single element of design," wrote Vetrov.[58] Korolev demanded, for example, "the right of the first information" about any failure during the testing phase.[59]

Korolev often began designing and even building rockets and spacecraft on his own initiative, without any official contract, using only the internal resources of his design bureau. This strategy of "design with anticipation" required "a tough character and strong nerves," for it relied on complete trust in the ultimate success of the pending proposal.[60] This strategy often paid off when Korolev faced a rival: he used to shore up his proposals with a hardware prototype, which, however imperfect, looked to the party and government leaders much more impressive than a stack of draft blueprints. For example, by hastily refurbishing the metal hulls of Soyuz orbital spacecraft for a circumlunar mission, he wrestled this mission's contract away from rival Vladimir Chelomey, whose design, however original and sophisticated, at that time had not progressed much beyond the paper stage.[61]

Although Korolev's formal administrative status in the space industry was not very high—he was just the head of a design bureau, with multiple layers of ministry bureaucracy above him—his fantastic energy and skill in manipulating the system allowed him to take up and win arguments with other chief designers, ministers, and party officials. Instead of focusing on what was possible within the allotted funding and the prescribed time frame, he started new projects, hoping to get funding and adjust the deadlines later. He stubbornly resisted the decisions he disagreed with, even if they came from the very top. For example, he alienated Khrushchev by bluntly

---

[53] Chertok, *Fili—Podlipki—Tiuratam* (cit. n. 41), 425.
[54] Georgii Vetrov, quoted in Ishlinskii, *Akademik S. P. Korolev* (cit. n. 27), 116.
[55] Boris Raushenbakh, quoted in Ishlinskii, *Akademik S. P. Korolev* (cit. n. 27), 375.
[56] Feoktistov, *Traektoriia zhizni* (cit. n. 44), 223.
[57] Vetrov, quoted in Ishlinskii, *Akademik S. P. Korolev* (cit. n. 27), 116.
[58] Ibid., 121.
[59] Georgii Vetrov, "O tvorcheskom stile Koroleva," 1975, f. 1546, op. 1, d. 50, l. 8, Russian Academy of Sciences Archive, Moscow.
[60] Igor' Erlikh, quoted in Ishlinskii, *Akademik S. P. Korolev* (cit. n. 27), 304.
[61] Siddiqi, *Challenge to Apollo* (cit. n. 3), 501.

refusing the Soviet leader's suggestion to use storable rocket fuel. Using this fuel would have shortened launch preparation time and greatly improved combat readiness of ballistic missiles, but Korolev's priorities lay elsewhere. He preferred dual-use rockets that could serve not just as military missiles but also as spacecraft boosters. For this reason he insisted on cryogenic fuels, which were non-toxic and more efficient and therefore more suitable for space launches, even though it took a day to prepare such a rocket for launch, rendering it useless as a practical weapon. The head of the Communist Party and the Soviet government could not persuade Korolev to change his position. Korolev designed his flagship lunar rocket, the N-1, for cryogenic fuels, even though he lost the support of the Ministry of Defense, which did not see any military purpose for this rocket.

Korolev's striving for greater authority also had its roots in the Stalin period. In the late 1940s, chief designers such as Korolev and Glushko reported directly to the deputy chairman of the Special Committee for Reactive Technology, Dmitrii Ustinov, while Ustinov personally reported to Stalin. Placing high-priority projects under his personal supervision, Stalin created a management structure that bypassed the multiple levels of bureaucracy separating him from the people directly involved in these projects. This shortcut to the supreme leader gave the chief designers a clear advantage over their immediate ministry superiors and helped them manage the pressures from various parts of the bureaucracy. The rocket engineers' memory of the Stalin era thus featured a phantom, an ideal supreme leader—omniscient, omnipresent, gracious, and infinitely powerful. In their perception, power resided in particular individuals, not in bureaucratic structures.

Korolev became convinced of the efficiency of this shortcut management structure, and he imitated it within his own design bureau. For every individual project, he appointed the so-called lead designer, who oversaw the production of all components and their integration, cutting across any departmental lines. The lead designer reported directly to Korolev and effectively served as his proxy. As one such designer recalled, "[Korolev] strengthened the authority of the lead designer, augmenting it with his own authority. The lead designer was often called 'the eyes and the ears of the Chief Designer' or 'the little Chief Designer.' It was a big trust, and to bear it was not easy. One had to work very hard, to know everything—what is being done, where, in what condition—down to the smallest detail."[62]

As the historian Susanne Schattenberg has shown, after Khrushchev's denunciation of the "cult of personality," the Stalinist authoritarian management style increasingly became a target of criticism from below. Stalin-era managers were often labeled "little Stalins" and denounced to the authorities. One such denunciation targeted Ustinov, then the minister of the armaments industry. His disgruntled subordinates reported:

> The Minister comrade D. F. Ustinov obviously considers despotism to be the best method of leadership. . . . The collective meeting room functions as a place of execution as under Ivan the Terrible. Cooperation is out of the question, because all members of the staff are frightened and used to voting for decisions made by HIMSELF. Everybody who falls out of Ustinov's favor, even the most talented employee, will be destroyed by him.[63]

---

[62] Oleg Ivanovskii, *Rakety i kosmos v SSSR* (Moscow, 2005), 51.
[63] Quoted in Susanne Schattenberg, " 'Democracy' or 'Despotism'? How the Secret Speech Was Translated into Everyday Life," in Jones, *Dilemmas of De-Stalinization* (cit. n. 8), 64–79, on 73.

However exaggerated, this characteristic captured Ustinov's direct and forceful style, aimed at cutting through the red tape and getting things done. Suspecting that midlevel managers might hide the truth from him, he used to enter factories from the back door and examine the shop floor by himself.[64] Korolev similarly often visited workshops with sudden inspections, swiftly assigning blame, dispatching reprimands, and issuing orders. His lead designers stayed on the shop floor practically around the clock, implementing his ideal of constant supervision. Like the notorious "little Stalins," Korolev's "little Chief Designers" embodied the idea of omniscient, omnipresent authority, which rocket engineers associated with the Stalin era.

In the words of Polly Jones, during the Khrushchev period "the very ideas of stability, control and authority were thrown into question."[65] It was precisely the notions of control and authority that had shaped the professional culture of rocket engineers in the Stalin era. They found these values challenged by the instability and haphazard reforms under Khrushchev, and they resorted to the proven management techniques of the past: professional networking, personal responsibility, and shortcut direct control.

In preparation for launching the first piloted Vostok, however, there was one element in the entire system that lay beyond the effective control of space engineers. This element was the cosmonaut himself.

A TINY SCREW IN A GIANT MECHANISM

A key decision shaping the future of the Vostok spacecraft was made by the Council of Chief Designers in November 1958. The council discussed three alternative proposals for a new spacecraft: an automatic reconnaissance satellite, a piloted spacecraft for a ballistic flight, and a piloted spacecraft for an orbital flight. All three proposals emerged from Korolev's Experimental Design Bureau No. 1; he always preferred to hedge his bets by developing variants for future projects. The spy satellite designers promoted their proposal, stressing its primary importance for defense. This clearly had an appeal to the military, the main customers of Korolev's bureau. The proponents of a piloted ballistic flight offered a quick result and a guaranteed win in the race to get a human into space. The group that designed a more complex orbital piloted mission, led by the integration designer Konstantin Feoktistov, decided to strengthen their proposal by performing what he called a "tactical maneuver." They claimed that their piloted spaceship could easily be converted into a fully automatic spacecraft and used as a reconnaissance satellite, which would be able to return to Earth not just a small container with film but also a large capsule with the entire camera set. This dual-purpose design promised great savings of time and money. The council supported Feoktistov's scheme, and he drafted a formal proposal to the Military-Industrial Commission for an automatic spy satellite, disguising its second function as a piloted spacecraft. Some officials became suspicious when they noticed, for example, that the presumably automatic satellite was equipped with a set of communication devices. "Who is going to talk over this radio?" they inquired. "The photo cameras?"[66] In May 1959, the Soviet government adopted a secret decree au-

---

[64] Ivanovskii, *Rakety i kosmos* (cit. n. 62), 197–8.
[65] Jones, introduction to *Dilemmas of De-Stalinization* (cit. n. 8), 4.
[66] Feoktistov, *Traektoriia zhizni* (cit. n. 44), 62.

thorizing the construction of an automatic reconnaissance satellite. Through personal connections, Korolev was able to add seven words to the text of the decree: "... and also a sputnik for human flight."[67]

The forced unification of two totally different projects—automatic reconnaissance and human flight—in one decree resulted in a peculiar design for the final spacecraft. The piloted Vostok had to be constructed in such a way as to be easily convertible into a spy satellite named Zenit by simply replacing the cosmonaut's couch with a set of photo cameras. Because the Zenit had to be fully automatic, the Vostok turned out to be fully automatic as well. The entire Vostok piloted mission could be flown without the cosmonaut's touching any controls on board the spaceship.

Having a fully automatic spacecraft at hand, the spacecraft designers began carving out a role for the cosmonaut to play. By early 1960, the Experimental Design Bureau No. 1 had completed the design of an automatic control system; only then did they begin working on manual control. Unlike with classical automation, which presumes a transfer of certain functions from a human to a machine, Vostok designers had to work out how to transfer functions from an existing automatic system to a human pilot. What needs to be explained here is not why the Vostok was automated but why it had a manual control system at all. The purposes of the manual control were to back up the automatic system in case of malfunction, to expand the window for controlled descent, and most important, to provide psychological support to the cosmonaut. As the designer of the manual control system, Boris Raushenbakh, put it, "The cosmonaut must be convinced that even if ground control equipment and the onboard automatic system fail, he would be able to ensure his own safety."[68] The engineers assigned the cosmonaut only two manual control functions—attitude control and retrorocket firing—to be used in an emergency as a backup for failed automatics.[69]

The first spacecraft—the Soviet Vostok and the American Mercury—were both fully automated and were flight-tested first in the unpiloted mode. Yet there was one important difference: the range of manual control functions available to and actually performed by Mercury astronauts was much wider than those on Vostok. Astronauts could manually perform such essential tasks as separating the spacecraft from the booster, activating the emergency rescue system, releasing the parachute, dropping the main parachute in case of failure and activating the second parachute, and correcting the onboard control system. Those and many other functions were not available to Soviet cosmonauts.[70] This can be illustrated by a simple comparison of the control panels of Vostok and Mercury. The Vostok instrument panel had only four switches and thirty-five indicators, while the Mercury panel had fifty-six switches and seventy-six indicators.[71]

---

[67] Chertok, *Fili—Podlipki—Tiuratam* (cit. n. 41), 423.
[68] Quoted in Aleksei Eliseev, *Zhizn'—kaplia v more* (Moscow, 1998), 15.
[69] Valentina Ponomareva, "Osobennosti razvitiia pilotiruemoi kosmonavtiki na nachal'nom etape," in *Iz istorii raketno-kosmicheskoi nauki i tekhniki*, no. 3, ed. V. S. Avduevskii et al. (Moscow, 1999), 132–67; Siddiqi, *Challenge to Apollo* (cit. n. 3), 196.
[70] Loyd S. Swenson Jr., James M. Grimwood, and Charles C. Alexander, *This New Ocean: A History of Project Mercury* (Washington, D.C., 1989); Robert B. Voas, "A Description of the Astronaut's Task in Project Mercury," *Human Factors* 3 (Sept. 1961): 149–65.
[71] For a comparison of the technical parameters of manual control panels on American and Soviet spacecraft, see Georgii T. Beregovoi et al., *Eksperimental'no-psikhologicheskie issledovaniia v aviatsii i kosmonavtike* (Moscow, 1978), 62–3.

One reason the American astronauts had a greater role than their Soviet counterparts: while secrecy kept Soviet cosmonaut trainees out of the spotlight, American astronauts enjoyed celebrity status even before their flights. Capitalizing on their prominent public image, they were able to influence the design of control systems. They renegotiated the role of passenger/observer initially assigned to them and gained a greater share in spacecraft control.

On the Soviet side, however, the engineers and designers saw the potential expansion of the number of manual control functions as a threat to the well-organized scheme of ensuring the quality and reliability of every system component. Space engineers plainly did not trust the cosmonaut's untested ability to operate onboard equipment while in orbit, in the unusual conditions of zero gravity and psychological stress. Feoktistov openly told the cosmonauts that "in principle, all work will be done by automatic systems in order to avoid any accidental human errors."[72] Korolev strongly believed that automation produced much more reliable results, and he pressured his subordinates and subcontractors to automate every possible step in the production and operation of space equipment.[73] He trusted a ground operator no more than he trusted a cosmonaut. His discovery that employees at an automatic ground control station performed one of the procedures manually enraged him.[74]

Manual control also undermined the space engineers' efforts to maintain maximum control over every element of the space system. Boris Chertok, who was responsible for the Vostok's entire control system, has formulated Korolev's approach to system engineering as follows: "The properties of every element, every device, every unit, even the human being and his activity must be subordinated to the common interest of system synthesis."[75] As the historian Asif Siddiqi has argued, Vostok designers "not only did not trust a pilot's capability to function adequately, but they also wanted to design the craft, fly it, and land it all on their own."[76]

The division of control functions between the human and the machine on board also reflected, as well as affected, the relative influence of the institutions responsible, respectively, for the training of cosmonauts and for the design of automatics. Under Korolev's leadership, Experimental Design Bureau No. 1 acquired unprecedented control over multiple aspects of the space program. The air force, which supervised cosmonaut selection and training, tried to acquire greater say in mission programming, but Korolev and his bureau did not yield decision-making power to any outside agency. Korolev, in particular, played a central role in decision making on a whole range of issues going far beyond engineering, such as spacecraft procurement, crew selection, cosmonaut training curriculum, mission planning, and ground flight control. His own role as the unquestionable leader within the Experimental Design Bureau No. 1 mirrored the central role of his bureau in the entire space industry.[77]

The automation of Vostok set a trend that dominated the Soviet piloted space program for decades. Despite the expansion of manual control functions on the Soyuz ships, the role of the cosmonaut as an emergency backup for automatics did not fun-

---

[72] Vladimir Komarov, workbook no. 39, 1961, Gagarin Memorial Museum Archive, Gagarin, Smolensk, Russia; http://web.mit.edu/slava/space/documents.htm (accessed 28 Aug. 2006).
[73] Ishlinskii, *Akademik S. P. Korolev* (cit. n. 27), 449.
[74] Feliks Meshchanskii, *Obratnaia storona* (Boston, 2001), 69.
[75] Boris Chertok, quoted in Ishlinskii, *Akademik S. P. Korolev* (cit. n. 27), 462.
[76] Siddiqi, *Challenge to Apollo* (cit. n. 3), 198.
[77] Ishlinskii, *Akademik S. P. Korolev* (cit. n. 27), 292.

Figure 2. *Three female cosmonauts at the cosmodrome prior to the launch of* Vostok 6 *on June 16, 1963. Left to right are second backup Valentina Ponomareva, first backup Irina Solov'eva, and prime crewmember Valentina Tereshkova. Behind the women are (from left) Sergei Korolev, the State Commission chairman Georgii Tyulin, and the Strategic Missile Forces commander in chief Sergei Biryuzov. Tereshkova was the first woman in space, spending three days aboard* Vostok 6. *(NASA photo no. womencosmonauts.)*

damentally change. Although Gemini and Apollo astronauts routinely performed manual rendezvous, Soyuz cosmonauts only occasionally had an opportunity to try a manual procedure. Such efforts often failed because engineers' main efforts were aimed at the perfection of automatics and the cosmonauts were not fully equipped, properly trained, or authorized to dock manually in difficult conditions.[78]

The cosmonauts resented this general trend toward automation. Some of them saw its origins in the ideological foundations of the Soviet system. For example, the former cosmonaut candidate Valentina Ponomareva, who served as a backup for Valentina Tereshkova, wrote in her memoirs:

> "The emphasis on automation" is the result and inherent part of the total mistrust of the individual, the mistrust peculiar to our ideology. . . . Propaganda tried to impose on people's minds the idea that technology decided everything. From this it directly followed that the individual was small and insignificant, and that he was only a tiny "screw" in a giant mechanism.[79]

In my view, this sentiment against Soviet ideology is misplaced. It was not the party or the government that encouraged automation; it was the space engineers' mindset of control, which they developed in an attempt to reduce uncertainty and risk while they dealt with the inadequacies of the overall organization of the space program.

[78] Slava Gerovitch, "Human-Machine Issues in the Soviet Space Program," in *Critical Issues in the History of Spaceflight*, ed. Steven J. Dick and Roger D. Launius (Washington, D.C., 2006), 107–40.
[79] Valentina Ponomareva, *Zhenskoe litso kosmosa* (Moscow, 2002), 207.

There is historical irony here. Spacecraft designers—some of the most talented, innovative engineers in the country, the cream of the crop of the Soviet technical intelligentsia—at the height of the cultural thaw built a spacecraft that embodied the notions of control and authority derived from their idealization of the Stalin era. The Vostok spacecraft became the technological analog of the totalitarian myth, an omniscient panopticon that monitored the cosmonaut's every move. Like any other technological artifact, the Vostok spacecraft reflected the professional culture of its designers. To the extent that this culture bore an imprint of the Stalin era, one could argue that the Vostok, the most celebrated artifact of the thaw, was a flying example of mythologized Stalinism.

DUAL USE, DUAL IDENTITY

The dual controls of the Vostok spaceship—automatic and manual—reflected the dual use of this spacecraft for military and civilian purposes. As a highly publicized project carried out in closed defense institutions, the space program represented an anomaly, a centaurlike creature. This fostered an unusual split identity of space engineers working at organizations such as Korolev's Experimental Design Bureau No. 1, which developed new types of military missiles alongside space rockets and spaceships. Their original professional identity as secret rocket designers, working in isolation from the rest of the world, clashed with their newly found sense of being (though anonymously) in the spotlight.

Space engineers were excited to see huge popular enthusiasm over *Sputnik* and the first human space flights. As Boris Chertok has recalled:

> [T]he effect produced by Sputnik proved totally unexpected. Workers, engineers, and scientific researchers from numerous institutes, design bureaus, and the cosmodrome had believed that they had been doing very important but ordinary work. Suddenly they realized that it was quite extraordinary. Every participant in the design, production, preparation, and launch [of Sputnik] felt connection with a scientific feat, with a lustrous day in the history of humankind.[80]

Yet the Soviet leadership decided not to reveal the identity of Korolev or any other leading space engineer, on the grounds that all of them were involved in top secret missile work. The spotlight focused squarely on the young, photogenic, smiling cosmonauts, while the chief designers were prominently absent from public ceremonies. Other individuals, often not involved in the space program at all, traveled abroad, gave speeches, and received honors. Korolev was designated in the press only as "the Chief Designer" and remained anonymous until his death in 1966. In September 1963, well after *Sputnik* and Gagarin's flight, Korolev was vacationing on the Black Sea and decided to attend a public lecture about Soviet triumphs in space. Nobody in the audience, including the lecturer, had any idea who he was.[81] Even the prospect of receiving the Nobel Prize for *Sputnik* and later for Gagarin's flight did not move the Soviet leadership to reveal Korolev's identity. In response to an inquiry from the Nobel Committee, Khrushchev reportedly said that "the creator of Sputnik is the Soviet

---

[80] Chertok, quoted in Ishlinskii, *Akademik S. P. Korolev* (cit. n. 27), 461.
[81] Ivanovskii, *Rakety i kosmos* (cit. n. 62), 140–1.

people."⁸² Once Korolev bitterly remarked to an old friend, "I have no public identity. And will probably never have one."⁸³

The cultural thaw opened new exciting opportunities for academic scientists to establish contacts with foreign colleagues, travel to international conferences, and publish abroad. But rocket engineers, despite their direct contribution to a research enterprise of huge international significance, remained isolated from the West. Chertok has recalled that in the wake of World War II Korolev and his team

> dreamed that instead of the confrontation that had begun to emerge, the interaction of the scientists from the victorious countries would be a natural continuation of the military alliance. In late 1946, Korolev, who had returned from some meeting in Berlin, [told] me, "Get ready to fly across the ocean." Alas! Until the very day he died, neither Korolev, nor any one of his closest associates was ever "across the ocean."⁸⁴

Joint work on classified projects, which had to be kept secret even from family members, the specific lifestyle during prolonged stays in the harsh climate and primitive conditions of the cosmodrome, the sense of pride for internationally recognized achievements, and the bitterness about the lack of public recognition—all this strongly reinforced the group identity of rocket engineers. Shared lives bred identical values and interests. One engineer, returning to Moscow from the cosmodrome, picked up the wrong suitcase at the airport and discovered, to his astonishment, that the contents of the suitcase were nearly identical to those of his: "A shaving set, just like mine, lay in a box; next to it, a few issues of a literary magazine, the same ones I had; the same gloves, helmet, pilot's pants, underwear, and toiletry. Several recently purchased books were almost the same as the ones I had bought while staying at the cosmodrome."⁸⁵ The owner of the suitcase was soon found; naturally, he turned out to be another space engineer.

Boris Chertok stressed that rocket engineers often lacked the cultural sophistication usually associated with the intelligentsia: "To act, not to chat; to take risks; to make a decisive impact on the course of events—such was our work style. Those who did not care left very quickly. Many in our group lacked such intelligentsia traits as cultured conversation, tactfulness, or politeness. But we appreciated the sense of humor and were attentive to each other's work, trying to help if necessary."⁸⁶ Indeed, as another memoirist remarked, the manners of original Council of Chief Designers member Mikhail Ryazanskii, who was "overtly intelligentsia-like, invariably polite, courteous, and friendly," were an exception to the prevailing pattern among the council members.⁸⁷ Chief designers—tough negotiators and strict administrators—often regarded refined manners as a sign of weakness. For example, Korolev at first distrusted the communication systems designer Yurii Bykov. Korolev was "suspicious of Bykov's overt politeness, of his inner and outer intelligentsia-like traits," recalled Chertok. "Would Bykov falter at a difficult, decisive moment, when a cosmonaut's

---

⁸² Golovanov, *Korolev* (cit. n. 3), 585–6.
⁸³ Mark Gallai, quoted in Ishlinskii, *Akademik S. P. Korolev* (cit. n. 27), 64.
⁸⁴ Chertok, *Rockets and People* (cit. n. 20), 2:27.
⁸⁵ Meshchanskii, *Obratnaia storona* (cit. n. 74), 75.
⁸⁶ Boris E. Chertok, *Lunnaia gonka*, vol. 4 of *Rakety i liudi* (Moscow, 2002), 356.
⁸⁷ Meshchanskii, *Obratnaia storona* (cit. n. 74), 61.

life and the nation's prestige are at stake?"[88] Eventually, Korolev overcame his suspicions, but his initial doubts are telling.

Civilian engineers were always surrounded by military personnel: by detachments servicing the launch facility, by military specialists testing rocket and spacecraft equipment, and by the military's top brass supervising launches. In the intermingling of the military and civilians in the rocket and space industry, a curious transitional category of "civilian military personnel" emerged: such was the nickname for military engineers assigned to civilian engineering groups to facilitate the development of rocket technology.[89] Several key positions in the leadership of the rocketry sector of the defense industry were occupied by military officers, for whom an exception was made that allowed them to work at civilian institutions while remaining on active duty.[90] The professional culture of rocket engineers became permeated with the spirit and values of military service. Yet their identities remained split: the commitment to the construction of missiles for the defense of the socialist fatherland clashed with the aspiration to explore space. And they often viewed the former as merely the means for the latter.

CONCLUSION

Writing his memoirs in the post-Soviet era, Boris Chertok has formulated several traits, which, in his view, described typical Soviet rocket engineers: they found the meaning of life in creative engineering work; combined technical work with organizational activity; bore personal responsibility for project results; worked in isolation from their Western counterparts and relied exclusively on domestic technologies; worked in cooperation with researchers from other fields; and identified themselves as members of a "gigantic technocratic system closely associated with the state and with the ideology of a socialist society."[91]

The engineers' belief in a technological utopia fitted well with the Marxist view of scientific and technological progress as a foundation for building a better society. Unlike many writers, artists, and scientists, space engineers kept a safe distance from any sensitive political issues. As one spacecraft designer admitted, "There were no dissidents among us."[92] Space engineers needed the regime to implement their ambitious space plans, while the regime needed their help in strengthening the defense and raising the nation's prestige. The top engineers were gradually integrated into the country's political elite. The chief designers Korolev, Glushko, Yangel, and Chelomey became delegates at party congresses, and Glushko and Yangel joined the ruling Areopagus, the Central Committee of the Communist Party.

In his landmark study, *Technology and Society under Lenin and Stalin*, Kendall Bailes remarked on the following "paradoxical relationship" between the technical intelligentsia and the Soviet state:

---

[88] Chertok, *Fili—Podlipki—Tiuratam* (cit. n. 41), 420.
[89] Meshchanskii, *Obratnaia storona* (cit. n. 74), 8.
[90] Bystrova, *Voenno-promyshlennyi kompleks* (cit. n. 25), 214–28.
[91] Chertok, *Rockets and People* (cit. n. 21), 1:8.
[92] Vladimir S. Syromiatnikov, *100 rasskazov o stykovke i o drugikh prikliucheniiakh v kosmose i na Zemle*, vol. 1, *20 let nazad* (Moscow, 2003), 423.

> Just as the Russian nobility staffed the upper levels of the Tsarist bureaucracy before 1917, and provided the core of the "critically-thinking" intelligentsia during the nineteenth century, since Stalin's death, the Soviet technical intelligentsia has emerged as the single largest element from which the ruling elite has been recruited, and also has been a large segment of the new, critically-minded intelligentsia.[93]

This statement appears as a paradox only if one assumes that the ruling elite was monolithic and lacked a critical attitude. The case of the space engineers suggests, however, that this most privileged group of the Soviet technical intelligentsia had a torn identity: the secretive world of postwar rocketry reinforced their affinity with the military, while working on cutting-edge technologies nurtured their sense of belonging to the international technoscientific elite. Continuous disputes over military and academic priorities of space missions reflected this deep-seated tension in the space engineers' identity. The chief designers constantly argued over space policy issues both with party and government leaders and among themselves.

Korolev and other chief designers did not simply use their personal networks to execute government orders more efficiently. They used these networks to recruit allies in the pursuit of their own visions of space exploration. Korolev designed and build rockets that nominally had a dual use but in fact were much better suited for space exploration than for combat. Chelomey, by contrast, built rockets on storable fuels, won support of the military, and tried to leverage this support to advance his own space projects.

The rocket engineers manipulated the system at least as much as the system manipulated them. In 1966, trying to curb independent activity on the part of the chief designers, the Military-Industrial Commission set up a formal procedure by which the chief designers could no longer bypass the commission in their lobbying efforts and had to clear their proposals with the commission and the Ministry of Defense before appealing to the political leadership. As a former deputy chairman of the commission admitted, "Naturally, this procedure was not always thoroughly followed."[94]

The professional networks of rocket engineers did not merely facilitate the work of Soviet defense industry. They became channels through which Soviet space policy was formulated, debated, and reshaped. By working often in parallel with, and sometimes in opposition to, the established administrative hierarchies, the chief designers were able to develop and promote their own policy initiatives. It was their proposals, reluctantly approved by the Soviet government, that produced *Sputnik*, Gagarin's flight, and the ambitious interplanetary and lunar programs. It was their technocratic vision of the future as a technological utopia that captured the public imagination in the early 1960s. Ironically, just as Soviet society tried to shed the political legacy of Stalinism, it was inspired by products of the engineering culture of the Stalin era.

---

[93] Bailes, *Technology and Society* (cit. n. 11), 3.
[94] Stroev, "Voennaia aviatsia" (cit. n. 23), 280.

INTELLIGENTSIA AS UTOPIA

Taming the Primitive:
Elie Metchnikov and His Discovery of Immune Cells

By Kirill Rossiianov*

ABSTRACT

This essay argues that the idea of race had a profound, yet unacknowledged, impact on Elie Metchnikov's research on immunity. Making the phagocytes, the most primitive cells of the body, responsible for the organism's integrity, Metchnikov attempted to reconcile the cultural idea of primitivity as a threat with the notion of the body as capable of self-defense and self-organization. The vision of phagocytes as being both the agents of an organism's "harmony" and its potential enemies reflected the complications that the ideas of race and racial primitivity met in Russian intellectual contexts.

INTRODUCTION

Elie (Il'ya Il'ich) Metchnikov was one of the most famous Russian scientists of his time. Born in 1845, he did important research in comparative and evolutionary embryology. In 1882–1883, Metchnikov made his most significant discoveries, bringing him fame as the founding father of the new science of immunology. Observing how the white blood cells, which he later called phagocytes, engulfed and digested pathogenic bacteria, Metchnikov identified this behavior as a reaction of immune defense and formulated the first theory of immunity. Following the discovery of phagocytes, Metchnikov left embryology and devoted himself to research on immunity, human pathology, and bacteriology. In 1888, he moved to Paris, joining the staff of the newly established Pasteur Institute, and worked in France until his death in 1916. In 1908, Metchnikov, together with the German immunologist Paul Ehrlich, was awarded the Nobel Prize in Physiology and Medicine.[1]

---

* Institute of the History of Natural Sciences and Technology, Staropansky per. 1/5, Moscow, Russian Federation; rossiianov@yahoo.com.

I am grateful to Alexei Kojevnikov and Michael Gordin for comments and editorial advice. I am also indebted to Dr. Kornelia Grundmann, who helped me locate important archival materials at the Emil von Behring Archive in Marburg. My research was made possible by aid from the Gerda Henkel Stiftung, Germany, Fellowship no. 01/SR/03.

[1] Alfred I. Tauber and Leon Chernyak, *Metchnikoff and the Origins of Immunology: From Metaphor to Theory* (Oxford, 1991); see also Daniel P. Todes, *Darwin without Malthus: The Struggle for Existence in Russian Evolutionary Thought* (New York, 1989), chap. 5, 82–103; A. E. Gaissinovitch, "Epistoliarnoe nasledie I. I. Metchnikova," in I. I. Metchnikov, *Pis'ma, 1863–1916* [Letters, 1863–1916] (Moscow, 1974), 3–37; Gaissinovitch, "Khronika zhizni i tvorchestva I. I. Metchnikova," in I. I. Metchnikov, *Pis'ma O. N. Metchnikovoi, 1876–1899* [Letters to Ol'ga Metchnikov, 1876–1899] (Moscow, 1978), 3–21; Gaissinovitch, "Trudy i dni I. I. Metchnikova," in I. I. Metchnikov, *Pis'ma*

Metchnikov was the first scientist to describe immunity as the active response to infection, claiming that phagocytes form the organism's "system" of defense. According to Alfred Tauber and Leon Chernyak, Metchnikov's novel approach to the problem of bodily integrity arose both from his scientific observations and from a new conception of the organism as a principally disharmonious structure that had to be harmonized. Metchnikov was nearly obsessed with the notion of painful, almost irreconcilable, contradictions between the evolutionarily advanced and primitive elements coexisting within the human organism. However, emphasizing the significance of disharmonies, Chernyak and Tauber say little about the racial dimension of Metchnikov's ideas. Writing about the primitive elements of the human body, Metchnikov used the racial imagery of the atavistic self. In the 1870s, he did important field research on the Kalmucks, the Mongoloid inhabitants of the steppe region near the Caspian Sea, and wrote extensively about "inferior" human races in a number of popular articles about anthropology and Darwinism.

I argue in this essay that immunity, as Metchnikov described it, was the culturally transformed activity of the atavistic animal. In one of his first publications on immunity, Metchnikov compared the phagocytes to the armies that the European powers conscripted in their African colonies: "Like a highly organized state, that, fighting against the savage tribes, conscripts an army whose members are in every respect close to savages . . . , the organism that fights against the lower plants [i.e., bacteria] deploys such elements that most of all remind the lower animals—against the bacteria it sends the army of amoeboid cells."[2] The link between immunity and primitivity remained remarkably persistent in Metchnikov's later works. The primitive nature of phagocytes explained, in fact, why they possess the capacity for independent movement—the capacity that enables them to find and attack parasitic microorganisms—acting, in Metchnikov's own words, with all "primeval ferociousness." It is remarkable that under certain conditions phagocytes might return to the state of "wildness," attacking the "higher," or "noble," cells of an organism. In this respect, taming—as I call the cultural transformation of an aggressive beast into the organism's primitive soldier—was incomplete. But why was the hostile, internal animal tamed rather than suppressed?

This question becomes central to our understanding of Metchnikov's ideas about immunity in the human body. I show that in making the primitive cells the agents of the organism's integrity, Metchnikov was responding to larger cultural challenges. Racial and cultural primitivity had strong positive meanings in the Russian context. In the 1860–1870s, when Metchnikov developed his views about the human body and race, the old myth of the noble savage was reinforced by egalitarian ideas upheld by

---

O. N. Metchnikovoi, 1900–1914 [Letters to Ol'ga Metchnikov, 1900–1914] (Moscow, 1980), 3–42; Semion Reznik, *Metchnikov* (Moscow, 1973); Alexander Besredka, *The Story of an Idea: E. Metchnikoff's Work*, trans. A. Rivenson and R. Gestreicher (orig. in French, 1921; English trans., Bend, Ore., 1979); Heinz Zeiss, *Elias Metschnikow: Leben und Werk* (Jena, 1932); Olga Metchnikoff [Metchnikov], *Zhizn' Il'i Il'icha Metchnikova* (Moscow, 1926); English translation: Metchnikoff, *The Life of Elie Metchnikoff*, trans. E. Ray Lankester (London, 1921); the full list of Metchnikov's works is given by V. V. Khizhniakov, G. M. Vaindrakh, and N. V. Khizhniakova, *Tvorchestvo Metchnikova i literatura o nem (Bibliograficheskii ukazatel')* [The works of Metchnikov and the publications about him (Bibliographical index)] (Moscow, 1951).

[2] I. I. Metchnikov "O tselebnykh silakh organizma" [On the curative forces of the organism], in *Protokoly sed'mogo s"ezda estestvoispytatelei i vrachei* [Proceedings of the seventh meeting of naturalists and physicians] (Odessa, 1883), 1–14, on 13.

the emerging Russian intelligentsia. At the same time, racist anthropology appeared to Russian intellectuals as an important part of modern, Western science.[3] The contradictory visions of primitive peoples caused tensions when Metchnikov addressed the problem of the primitive elements of the human body. Explaining why the savage is more harmonious than civilized man, he made important biological assumptions that radically undermined the power of the "higher" elements of the human body and their capability to maintain harmony by suppressing the "lower," primitive elements. This dilemma informed Metchnikov's theory of immunity, which envisioned a new and rather paradoxical type of relationship between the high and the low. Metchnikov redirected the destructive activity of the internal "animal," with microorganisms as the new target. The morally painful conflict was reshaped, emerging in ethically acceptable form as a struggle between primitive phagocytes and primitive microbes.

The first section of this essay describes Metchnikov's early ideas about the human body as a harmonic whole and explores the impact of his anthropological research on these ideas. The second section examines the question of why suppression—a standard cultural model for coping with the primitive—did not work well in Russia. The third section tells the story of Metchnikov's discovery of phagocytes and analyzes the rhetorical role of primitivity in the discussions between Metchnikov and the adherents of humoralist theories of immunity. The fourth section describes the new intellectual possibilities, as well as the limitations, of Metchnikov's approach to immunity and organismic integrity. In the conclusion, I discuss the significance of Metchnikov's ideas for our historical understanding of the relationship between moral thought and evolutionary biology.

THE DISCOVERY OF PRIMITIVITY (C. 1868–C. 1874)

Born to the family of a retired guards officer, Metchnikov became interested in natural sciences at a very early age. After his graduation from the University of Khar'kov in 1864, Metchnikov went to Europe and worked for a number of years in different laboratories in Germany, Switzerland, and Italy. He devoted himself to embryology, concentrating, with his friend Aleksandr Kovalevskii (1840–1901), on the study of the fine structure of development of invertebrates. Later, in an essay remembering his youth, Metchnikov wrote that the young people of his generation believed that natural sciences would solve the "main problems tormenting mankind."[4] Educated Russians combined interest in science with egalitarian values. What would happen, many wondered, if Westernizing reforms increased, rather than diminished, social and cultural inequality? In the late 1860s, the Russian intelligenty began asking this question with increasing frequency.

In 1869, Nikolai Mikhailovskii—who several years later would become a famous

---

[3] Historians have not yet paid sufficient attention to the Russian discourse about race. Some important information can be found in Yuri Slezkine, *Arctic Mirrors: Russia and the Small Peoples of the North* (Ithaca, N.Y., 1994). See also Willard Sunderland, "Russians into Iakuts? 'Going Native' and Problems of Russian National Identity in the Siberian North, 1870s–1914," *Slavic Review* 55 (1996): 806–17; and Allison Blakeley, *Russia and the Negro: Blacks in Russian History and Thought* (Washington, D.C., 1986).

[4] I. I. Metchnikov, "A. O. Kovalevckii (Ocherk iz istorii nauki v Rossii) [A. O. Kovalevskii (An essay from the history of Russian science)]," in *Stranitsy vospominanii: Sbornik avtobiograficheskikh statei* [The pages of memory: A collection of autobiographical articles], ed. A. E. Gaissinovitch (Moscow, 1946), 14–44, on 44.

author and an almost iconic figure for the radical intelligentsia—published the article titled "What Is Progress?" which was a critique of Herbert Spencer's theories of the "social organism" and social progress. Mikhailovskii claimed that the increasing division of labor in modern society was immoral and was the main reason for social inequality. Commenting later on Emile Durkheim, he reversed the latter's argument: "mechanical solidarity" was superior to the "organic." In Mikhailovskii's view, the division of labor should be minimized in society; as a result, self-sufficient and equal individuals would be able to freely cooperate with each other. A strong division of labor was against the "nature" of man, who should be able to exercise all his natural capabilities. Mikhailovskii based this conception of man on what he regarded as obvious facts of physiology, postulating some inborn harmonic organization of the human body. Real progress should lead, in his view, to a minimal division of labor between people and to the maximal division of labor between the different organs of the harmonic body. Proposing this theory of progress, Mikhailovskii believed that, taken as model of social development, it would save the Russian people from the horrors of capitalism.[5]

These and similar theories that made a connection between the notion of the authentic, unalienated self and the idea of the harmonic human body had strong moral appeal. Educated people found unacceptable and intolerable the division, the real abyss that existed because of the unfair division of labor, between the "thinking" part of society and the simple people doomed to hard manual labour. Leo Tolstoy, for example, who, in contrast to Mikhailovskii and other Populists, was deeply distrustful towards science, upheld a similar position: Defending the moral idea of simple life (*oproshchenie*), he referred to the natural division of labor between different organs, which he regarded as a self-evident fact.

It was astonishing to discover from Metchnikov's diaries that in the late 1860s, he also believed in the idea of the harmonic body, agreeing fully with the proponents of these views, whom he called "harmonists." Criticizing Spencer, Metchnikov cited Mikhailovskii and claimed that the "normal functioning" and the "proportional, even development" of different capabilities and organs of the human body "should constitute a harmonic whole." "This harmonic work of different organs of the human body should fill," he wrote in the diary, "the life of man, and from this we should deduce the rights and the duties of man." Metchnikov later remembered that in the late 1860s—when he was a docent at St. Petersburg University—he made an attempt to organize his life following the idea of the minimal division of labor between people; refusing for principled reasons to hire a servant or servants, he did all the housework himself.[6]

---

[5] N. K. Mikhailovskii, "Chto takoe progress?" [What is progress?], in *Sochineniia* [Works], 4th ed., vol. 1 (St. Petersburg, 1906). See James H. Billington, *Mikhailovsky and Russian Populism* (Oxford, 1958); Andrzej Walicki, *The Controversy over Capitalism: Studies in the Social Philosophy of the Russian Populists* (Oxford, 1969), 26–7; Alexander Vucinich, *Social Thought in Tsarist Russia: The Quest for a General Science of Society, 1861–1917* (Chicago, 1976).

[6] I. I. Metchnikov, "Vospitanie s antropologicheskoi tochki zreniia" [Education from the anthropological point of view] (1871), in I. I. Metchnikov, *Sorok let iskaniia ratsional'nogo mirovozzreniia*, 3rd ed. (Moscow, 1925), 37–58, on 58; Metchnikov's notebooks, 1866–1869, 1871, entry 10 Aug. 1868, f. 584, op. 1, d. 4, 1. 113, Archives of the Russian Academy of Sciences (Arkhiv Rossiiskoi Akademii Nauk; hereafter cited as ARAN); Metchnikov's notebooks, 1873–1874, undated entry, f. 584, op. 1, d. 271, ARAN; Abba Evseevich Gaissinivitch, "Primechaniia" [Notes], in Metchnikov, *Stranitsy vospominanii* (cit. n. 4), 246–7.

However, in 1873 or 1874 (there is no exact date next to this entry), a remarkable confession appears in Metchnikov's diary: his old ideas could no longer be supported by facts. In the past, Metchnikov wrote, he postulated a "harmonic organization of the body that does not exist in reality." At this time, he was going through a deep personal crisis, and we know that he even attempted suicide following the premature death of his wife on April 20, 1873. Due to developing eye illness, he was also required to stop his embryological research; he decided to go to the steppes to do anthropological research on Kalmucks.[7]

As a result of the racial discoveries that Metchnikov made during his expedition, he came to the idea of "developmental arrests" (*das Stehenbleiben in der Entwicklung*; *ostanovka razvitiia*), and this idea undermined his earlier views about the harmonic organization of the human body. Examining the Kalmucks, he discovered that certain anthropological traits in adult Mongoloids correspond with the characters that Europeans demonstrated as infants. His further reasoning was based on Ernst Haeckel's "bio-genetic law." Metchnikov applied what he "observed" in the individual, postembryonic development of Kalmucks to the phylogenetic development of races, claiming that not only the specific organs or characteristics but also the "inferior" races as such demonstrated an arrest in the evolutionary development.[8] He wrote: "It looks as if the lower races retard on a low stage of [individual] development." He compared this retardation to the development of the triton, which, like the frog, belongs to the class of Amphibia but, in contrast to the frog, does not go through a full metamorphosis, preserving a long tail as an adult animal. The existence of the arrested, or primitive, traits and organs became a "factual" basis for Metchnikov's conclusions about "inferior" races. The emphasis on the significance of primitive traits for the scientific understanding of human races made the primitive organs also important for the analysis of the human body, delivering a fatal blow to the moral theory of the harmonic organism.[9]

In a draft titled "Anthropological Essays about Morality," Metchnikov claimed that many naturalists believed that the animal organism was a machine. But "[t]he animal organism is a machine in which the excessive screws have been preserved, rather than removed."[10] "If we had decided to exercise all the rudiments that are capable of functioning, the human organism would have regressed to an incomparably lower, animal-like state.... This will happen if we decide to follow the principle of 'the harmonic development of all human capabilities.'" Claiming in his later article that "a lower animal is included in our body," he reassessed the supposed link between morality and the free, authentic self: "Let him [the lower animal within our body] . . .

---

[7] Metchnikov's notebooks, 1873–1874, undated entry, f. 584, op. 1, d. 271, l. 1, ARAN; Metchnikoff, *Zhizn' Il'i Il'icha Metchnikova* (cit. n. 1), 169–77.

[8] Elias Metschnikoff, "Über die Beschaffenheit der Augenlider bei den Mongolen und Kaukasiern," *Zeitschrift für Ethnologie* 6 (1874): 153–60; for Metchnikov's expedition and his anthropological works, see also R. I. Belkin and M. A. Gremiatskii, "Kommentarii k stat'iam po antropologii i etnografii," in I. I. Metchnikov, *Academic Collection of Works*, vol. 16 (Moscow, 1964), 365–86; for the historical analysis of biogenetic law, see Robert J. Richards, *The Meaning of Evolution: The Morphological Construction and Ideological Reconstruction of Darwin's Theory* (Chicago, 1992).

[9] Metchnikov, "Vozrast vstupleniia v brak" [The age of marriage] (1874), in *Sorok let iskaniia* (cit. n. 6), 59–101, 62.

[10] Metchnikov, "Antropologicheskie etudy o nravstvennosti" [Anthropological essays about morality], f. 584, op. 1, d. 266, l. 10, ARAN.

develop freely, assuming that such a development is natural to our organism, and he will immediately break out."[11]

So, Metchnikov's old faith was lost and, what was perhaps worse, not simply lost, but destroyed by "facts." As a result of his racial findings, morality was relocated: it was now associated with a part of the body—its higher elements—rather than with the body as a whole. To what extent could the new evolutionary hierarchy of the high and the low compensate for the loss of his old faith in the naturally free, authentic self? Was he ready to abandon the egalitarian beliefs of his youth? To answer these questions, we need to examine Metchnikov's racial ideas more closely.

THE AMBIGUITY OF RACE (C. 1874–C. 1878)

In 1864, Varfolomei Zaitsev—a well-known author and a radical social critic—published an article defending the view that the Negroes represented a biologically inferior human type and that the liberation of slaves in America might only lead to their extinction because they were not capable of living in civilized society. Zaitsev portrayed his position as a logical and inevitable conclusion from the Darwinian theory of evolution and modern racial theory. His article caused a heated debate, in which most participants rejected his conclusions. In their view, even the supposed fact of racial inferiority did not justify slavery. Most historians have considered Zaitsev's article and the discussion that followed it an isolated appearance of scientific racism on the Russian soil.[12] Metchnikov's works, however, provide evidence of the much wider impact of racist ideas.

Following his articles about the anthropology of Kalmucks, Metchnikov discussed the racial theme in a number of popular publications that appeared in the liberal journal *Vestnik Evropy* (Herald of Europe). He obviously believed in the existence of inferior races: "Without any doubt, there are the outstanding personalities among the Negroes, but it cannot change our general conclusion because it is equally correct that you can sometimes find European faces among the Negroes . . . I am not at all surprised by the growing evidence that in Liberia, on Haiti, and in America the Negroes return to wilderness and paganism and even to introduce cannibal customs."[13]

Yet it was hard to use the same language when writing on primitive peoples who lived in the Russian empire. The Kalmucks whom Metchnikov studied during his expedition were Mongoloids who appeared to him as the most ancient and the most primitive of all human races, not excluding Negroes. But, in contrast to what he wrote about Negroes (of whom he knew only from books), he never used derogatory moral language when he wrote about the Kalmucks. Some time after his expedition had taken place, Metchnikov remembered that every male Kalmuck whom he had met in the

---

[11] Metchnikov "Zakon zhizni: Po povodu nekotorykh proizvedenii gr. Tolstogo" [The law of life: Concerning some works of Count Tolstoy] (1891), in *Sorok let iskaniia* (cit. n. 6), 200–26, on 208.

[12] V. A. Zaitsev, "Retsenziia na knigu Katrfazha 'Edinstvo roda chelovecheskogo,'" *Russkoe slovo*, 1864, no. 6:93–100; M. A. Antonovich, "Russkomu slovu (Predvaritel'nye ob'iasneniia)," *Sovremennik*, 1864, nos. 11–12; V. A. Zaitsev, "Otvet moim obviniteliam po povodu moego sochineniia o tsvetnykh plemenakh," *Russkoe slovo*, 1864, no. 12: bibliograficheskii listok, 20–4, 81–2; N. D. Nozhin, "Po povodu statei 'Russkogo slova' o nevol'nichestve," *Iskra*, 1865, no. 8; D. I. Pisarev, "Posmotrim!" *Russkoe slovo*, 1865, no. 9:10–3. See also Loren R. Graham, *Science in Russia and the Soviet Union: A Short History* (Cambridge, UK, 1993), 60–4; Slezkine, *Arctic Mirrors* (cit. n. 3), 118.

[13] Metchnikov, "Vozrast vstupleniia v brak" (cit. n. 9), 61.

steppe appeared to him as the very image of Professor I. M. Sechenov, his colleague and close friend at the University of Odessa.[14] It was a joke, but everybody who saw photographs of Sechenov had to admit that Sechenov had distinct Mongoloid facial features, a clear sign of his Mongoloid ancestry. Racial boundaries were never so clearly defined in the Russian empire as in Europe and the European colonies overseas. And it would hardly be possible to establish strict moral borders between the races.[15]

The Jews who had lived in the Russian empire since the eighteenth century remained a closed ethnic group, largely due to the discriminatory policies of the Russian government. It is interesting that Metchnikov cites Jewish people as one of his examples in the article "The Struggle for Existence in a General Sense," describing the relationship between the more "civilized" races and nations, on the one hand, and the less civilized, on the other, in terms of the Darwinian struggle for existence. He mentions the Europeans and the aboriginal people of Australia, the Chinese and the Mongols, and then he writes about the Jewish people exploiting Ukrainians and Poles. Apparently, he meant the Jewish tenant farmers and managers of estates who were often in real confrontation with Slavic peasants over land and other economic issues.[16] Metchnikov did not develop his argument, but it is clear from the general context of his reasoning that he considered the success of the Jews to be due to biological superiority.[17] But this example is curious. First, unlike other Russian authors who wrote about the supposed exploitation of Slavs by the Jews, Metchnikov was never anti-Semitic, and therefore he never considered the superiority of Jews over Slavs as a threat. After he had obtained a laboratory at the Pasteur Institute in Paris, he invited a number of Jewish scientists who could not find a job or receive proper education in Russia to work with him there. Second, Metchnikov's mother, Liudmila Nevakhovich, was born to the family of a baptized, assimilated Jew, and it remains an open and interesting question to what extent Metchnikov tended to think of himself, his family, and his ancestry in racial terms.[18]

Metchnikov believed that the inequality of races that resulted from the Darwinian struggle for existence was real and, therefore, normal. But the article "The Struggle for Existence in a General Sense," in which Metchnikov describes the competition between races, nations, and individuals in Darwinian terms, was perhaps the most pessimistic of all the articles he wrote during the 1870s. As Daniel Todes has pointed out, "[A] distaste for the competitive individualism" was one of the reasons why Russians looked with such suspicion upon Darwin's views on overpopulation and the role of intraspecific competition in biological evolution and why many prominent Russian scientists rejected those views in their analysis of biological evolution and its mechanisms.[19] Metchnikov's case, however, demonstrates a different and interesting

[14] I. M. Sechenov, *Avtobiograficheskie zapiski* [Autobiographical notes], 4th ed. (Nizhni Novgorod, 1998), 14.
[15] See Sunderland, "Russians into Iakuts?" (cit. n. 3).
[16] A vivid description of one such conflict that took place on Metchnikov's family estate comes from Metchnikov's second wife, Olga. See Metchnikoff, *Zhizn' Il'i Il'icha Metchnikova* (cit. n. 1), 91–3.
[17] I. I. Metchnikov, "Bor'ba za sushchestvovanie v obshirnom smysle" [The struggle for existence in a general sense] (1878), in *Sorok let iskaniia* (cit. n. 6), 121–86, 157.
[18] S. Shtraikh, "Evreiskie otnosheniia Metchnikova" [The Jewish relations of Metchnikov], *Novyi put'*, no. 26 (1916): 7–9.
[19] Todes, *Darwin without Malthus* (cit. n. 1), 168.

pattern: The evolutionary inequality of different races and people appeared to him as an undeniable fact, but to what extent was this evolutionary inequality moral?

"It is a fact," Metchnikov noted in his diary as far back as 1871, "that, with respect to their morality, many savages stand much higher than Gauss, Bacon, Rousseau and others who represent a much more perfect human type." In the article "The Age of Marriage" (1873), he also emphasized the moral harmony of primitive people untouched by civilization and culture and made this statement in a typically Rousseauistic manner.[20] But, in contrast to Rousseau's view, the difference between civilization and primitivity appeared to Metchnikov as evolutionary and biological, rather than cultural. He did not think of human nature as plastic, believing, in full accord with racial anthropology of his time, in the relative stability of different racial types.

There are practically no historical studies on the biological conception of race in Russian science and culture, which in itself is an interesting historiographical phenomenon. It is likely that the "syncretic" understanding of racial inferiority that appeared as being both "bad" and "good" was not uncommon for the educated Russians of that time. For example, Afanasii Shchapov, a Russian historian influenced by socialist ideas, complained about the insufficient development of the "nerve and brain capabilities" in Buryats, a Siberian ethnicity of Mongoloid descent. In his view, this insufficiency made their prospects of cultural progress profoundly dubious. However, Shchapov also believed that *ulus*, the Buryat peasant commune, was much closer to the collectivist ideal than was the Russian commune. Following Mikhailovskii and others, many Russian ethnographers viewed primitivity in a positive way.[21] It is likely that this positive assessment existed side by side with scientific claims about the evolutionary hierarchy of races. Apparently, there was no contradiction because the nonwhite minorities appeared as neither a moral nor a significant political threat to Russian society at that time.[22]

It was harder to ignore the threat when it was located within the organism. In Metchnikov's view, not only the primitive, atavistic elements of the body made modern man suffer. He argued that evolution always led to the progressive development of few organs, causing disharmonies, that is, contradictions between "younger" and "older" organs. Only later, when the period of progressive evolution was over, could natural selection make the constitution more balanced.[23] A younger species, such as man, or younger races, such as the white race, were always more disharmonious than more ancient species or races. Development is disharmonious for two principal reasons: first, the atavistic organs and instincts pose much greater threats to a civilized man than to a "savage"; second, the advanced organs can also destabilize the harmony or balance.

In this respect, Metchnikov's theory of the human body reflected his conflicting racial values, explaining why white man is superior to the primitive and why primitive man is more harmonic than civilized man. On the one hand, we can read Metchnikov's reasoning about disharmonies as a particular version of the "degeneration" theories that portrayed degeneration, that is, the reversion to the lower, atavistic state,

---

[20] Metchnikov's notebooks, 23 Jan. 1871 entry, f. 584, op. 1, d. 4, l. 123, ARAN; Metchnikov, "Vozrast vstupleniia v brak" (cit. n. 9), 100–1.

[21] See Slezkine, *Arctic Mirrors* (cit. n. 3), 118.

[22] Sunderland, "Russians into Iakuts?" (cit. n. 3).

[23] See I. I. Metchnikov, "Ocherk vozzrenii na chelovecheskuiu prirodu" (1877), in *Sorok let iskaniia* (cit. n. 6), 102–20.

as a central threat to civilization and the moral self.[24] On the other hand, "disharmony" was obviously a broader concept.

And as a moral concept it was profoundly contradictory. Was the suppression of the primitive forces of human nature the moral duty of a civilized man? Yes, it was the moral mission of civilized man not to live a natural and harmonic life. Otherwise, man would regress, and the primitive animal would get out. However, it was simply not sufficient to suppress the low because the high, the part of the self responsible for moral control and order, was also a disharmonizing and destabilizing factor. If some elements of the human body got ahead while others lagged behind, one could not restore the harmony by suppressing the primitive because the problem had an almost symmetric profile: the higher elements also destabilized the harmony. And Metchnikov, who always ascribed great significance to science and scientific creativity, would never agree with Dostoevsky's character Shigalev, who was enthusiastic about the idea of equality and an egalitarian social order but hoped to achieve them by suppressing geniality and outstanding human capabilities.[25]

This raised the question of whether scientific, rational morality was at all possible. As mentioned, Metchnikov abandoned his old beliefs that linked morality to the notion of the organism as a whole in favor of an evolutionary model, in which the atavistic "animal within" was a major threat to the organism's higher elements that were to suppress the animal in the interests of morality. In fact, however, civilization and progressive development were possible only at the cost of harmony and wholeness. As a result, morality coud be associated neither with the evolutionarily advanced nor with the primitive part of the human self. At the height of the disappointment, Metchnikov attempted suicide again.

The solution came when he moved from studying primitive organs to studying primitive cells and discovered a new type of relationship between the high and the low, which became his main contribution to the evolutionary understanding of wholeness. Rather than causing disharmonies, the "primitive" phagocytes enabled the functioning of the organism as a whole.

THE CELL STATE AND ITS CITIZENS (C. 1878–C. 1894)

Metchnikov based his scheme of immune response on the activity of primitive, evolutionarily ancient cells that were normal constituents of the organism. The existence of atavistic organs was common knowledge at that time, but it is likely that Metchnikov was the first scientist to apply this concept to cells. We know that the so-called cell state theories strongly emphasized the inequality of different cells in the organism. Rudolf Virchow wrote about different, higher and lower, dignity (*Dignität*) of different

---

[24] It is likely that Metchnikov was the only Russian scientist who wrote at length about atavism and degeneration as an evolutionary problem. Anton Dohrn and Ray Lankester explored this problem at about the same time Metchnikov did. See Anton Dohrn, *Der Ursprung der Wirbelthiere und das Princip des Functionswechsels: Genealogische Skizzen* (Leipzig, 1875); E. Ray Lankester, *Degeneration: A Chapter in Darwinism* (London, 1880); see also Daniel Pick, *Faces of Degeneration: A European Disorder, c. 1848–c. 1918* (Cambridge, UK, 1989). For the interest in the problem of atavism and degeneration among Russian psychiatrists, see Mark B. Adams, "Eugenics in Russia, 1900–1940," in *The Wellborn Science: Eugenics in Germany, France, Brazil, and Russia*, ed. Mark B. Adams (Oxford, 1989), 153–216; Irina Sirotkina, *Diagnosing Literary Genius: A Cultural History of Psychiatry in Russia, 1880–1930* (Baltimore, 2002).

[25] Fyodor Dostoevsky, *Demons*, trans. Richard Pevear and Larissa Volokhonsky (New York, 1995).

cells, such as the brain cells and the epithelial cells of the skin. Ernst Haeckel linked the inequality of cells in the cell state to the notion of inequality between people; and for him it was more than a metaphor because the inequality of cells and the inequality of people appeared to him as particular manifestations of "the great law of the division of labour" that was valid for biological organisms, for colonies of social insects, and for human societies.[26]

But it is interesting that there was no specific mechanism—except for the moral sense, the interiorized discipline of cells-citizens—that provided the subordination of the inferior to the superior within the grand scheme of the cell state. To Haeckel it was, indeed, evolution that shaped the cell state, with heredity as the conservative force that maintained the structural and functional inequality of different cells. But, as we know, the evolutionary adjustment is never absolute. What happens if some cells refuse to obey? In Virchow's view, this would lead to disease. In this respect, there is no contradiction between the "disobedience" and the moral harmony because the disobedience is understood as a pathological phenomenon.

However, in Metchnikov's theory of immune response, the normal, healthy organism includes the primitive cells whose "interests" do not fully coincide with the interests of the higher elements of the organism. The concept of primitive cells arose from Metchnikov's embryological research. Since the mid-1870s, he had been in deep disagreement with Ernst Haeckel over the so-called Gastraea theory, which Haeckel had formulated to explain the evolution of the first multicellular organisms. According to this dominant theory of the time, the "introgression," or "invagination," was the most ancient mode of formation of a primeval two-layer organism from a one-layer colony of unicellular animals, while other possible mechanisms were regarded as falsifications, that is, secondary and much more recent alterations. In his works on the embryology of sponges, as well as medusae and other Hydrozoa, Metchnikov questioned the universal role of invagination in the phylogenetic and ontogenetic development of multicellular organisms.[27]

Ernst Haeckel wrote about the two layers as the two physiological organs of the primeval multicellular organism, ascribing locomotive function to the external layer and digestive function to the internal one. However, in Metchnikov's view, this model had serious defects. To understand how the first two-layer organism emerged, one needed to explain how this organism functioned. What evolutionary advantage could that organism gain from the central gastral cavity Haeckel postulated in his model? The central cavity could provide the inflow of water and better contact of the internal, digestive layer with particles of food. But this was true only after the cavity formation had been completed. Like many evolutionists of his time, Metchnikov believed in the gradualist character of evolution. And, in this respect, Haeckel's model was logically inconsistent, postulating, in fact, a leap, a real abyss between the two stages: the colony and the first multicellular organism with the fully developed gastral cavity.

According to Metchnikov's model, the transformation of a one-layer colony into

---

[26] Rudolf Virchow, *Die Cellularpathologie in ihrer Begründung auf physiologische und pathologische Gewebelehre* (Berlin, 1858); Virchow, "Atome und Individuen" (1859), in *Reden über Leben und Kranksein*, ed. F. Krafft (Munich, 1971), 33–67; Ernst Haeckel, *Über Arbeitstheilung in Natur- und Menschenleben* (Berlin, 1869).

[27] See I. I. Metchnikov, *Embryologische Studien an Medusen: Ein Beitrag zur Genealogie der Primitivorgane* (Vienna, 1886); Russian translation: "Embriologicheskie issledovaniia nad medu-zami," in I. I. Metchnikov, *Izbrannye biologicheskie proizvedenia* (Moscow, 1950), 271–472.

a two-layer multicellular organism proceeded continously. Engulfing the food particles, the ciliated cells of the primeval colony migrated inside, which led to a division of labor (the terms "division of labor" and "inequality" were routinely used by Metchnikov, Haeckel, and other researchers in works on embryology), as well as to morphological differentiation. However, the internal cells that had lost their cilia could still move, approaching the surface and engulfing food. The internal embryonic layer emerged, then, as a central parenchyma, as a mass of amoeboid movable cells, rather than as the epithelial wall of the gastral cavity. The digestive parenchyma appeared to Metchnikov as the ancestral tissue for both entoderm and mesoderm, while the gastral cavity emerged only much later in evolutionary development.

This first division of labor led to the emergence of a large family of the primitive amoeboid cells that existed almost unchanged in all groups of multicellular animals. At a later time, Metchnikov established a link between the amoeboid cells of invertebrates and similar amoeboid cells in higher animals. In fact, he had already known about the findings of German pathologists who had discovered the so-called *Wanderzellen*, migrating cells that are present in blood and in bodily fluids as well as in pus.[28]

But, then, the key question was: If these cells were direct descendants of the cells of primitive digestive parenchyma, why did they still exist? At this point, Metchnikov radically reshaped the notion of primitivity: primitive cells were preserved by evolution because their primitive intracellular digestion served the purpose of the organism's protection. In December of 1882, Metchnikov observed in an experiment with starfish how the amoeboid, phagocytic cells surrounded alien bodies, attempting to engulf them.[29] This observation was followed by experiments with pathogenic bacteria. On the suggestion of Virchow, whom he met in Messina, Metchnikov began to work with Daphnia, which, like other transparent sea organisms, appeared as particularly suitable for such research. Later he moved to experiments on higher animals. In August 1883, just months after his first experiment, Metchnikov claimed that there existed a "whole system" of "therapeutic (and perhaps prophylactic) digestion."[30]

The notion of phagocytes as independently moving and almost autonomous entities became the main target for numerous critics of his theory. Metchnikov's publications that appeared in *Virchows Archiv* and elsewhere led to a real battle with the younger generation of German pathologists who believed that the amoeboid cells appeared in bodily fluids and in tissues more or less passively, as a result of molecular lesions in the walls of blood vessels. In the late 1880s and early 1890s, the proponents of the new, so-called humoralist understanding of immunity also criticized the idea of the active and seemingly independent, purposeful, movements of the phagocytes. Believing that immunity was due to chemical factors circulating in blood, rather than to the cells, the humoralists denied the protective role of phagocytes. Many scientists

---

[28] See Todes, *Darwin without Malthus* (cit. n. 1), 95–8; L. J. Rather, *Addison and the White Corpuscles: An Aspect of Nineteenth-Century Biology* (Berkeley, Calif., 1972).

[29] See I. I. Metchnikov, "Untersuchungen über die intracelluläre Verdauung bei wirbellosen Thieren," *Arbeiten aus dem Zoologischen Institut zu Wien* 5 (1883): 141–68; Metchnikov,"O tselebnykh silakh organizma" (cit. n. 2); Metchnikov, "Ueber eine Sprosspilzkrankheit der Daphnien, Beitrag zur Lehre über den Kampf der Phagocyten gegen Krankheitserreger," *Virchows Archiv* 96 (1884): 177–95; Metchnikov, "Ueber die Beziehung der Phagocyten zu Milzbrandbazillen," *Virchows Archiv* 97 (1884): 502–26; Metchnikov, "Moe prebyvanie v Messine (Iz vospominanii proshlogo)," in Metchnikov, *Stranitsy vospominanii* (cit. n. 4), 70–6, on 74–5.

[30] Metchnikov, "O tselebnykh silakh organizma" (cit. n. 2), 197, 199.

perceived Metchnikov's ideas as remnants of the old teleological vitalism. According to Paul Baumgarten, Metchnikov was resurrecting the idea of ancient Greek gods and goddesses in his theory of phagocytes.[31]

Neither Metchnikov nor Virchow (who continued to support him) agreed with this interpretation. In Metchnikov's view, evolution had shaped the purposeful nature of phagocytic attacks. The much-disputed ability of phagocytes to move independently, rather than flow passively with blood, and to act on their own, finding and engulfing the prey, was a legacy of their evolutionary past. Phagocytes have retained the destructive potential of their primeval, aggressive, and ferocious ancestor cells. In contrast to cell state theories that postulated the natural and self-evident hierarchy of superior and inferior cells, Metchnikov wrote about immunity and integrity in language of moral drama, rather than of moral harmony or evolutionary theodicy.[32]

Formulating his model of immunity, Metchnikov moved away from the idea of irreconcilable conflict between evolutionarily advanced and primitive elements of the body—the idea that had tormented him in the 1870s. It was a struggle, or war, but it was a war between the organism's own primitive cells and primitive microbes. This opened a path to the moral integration of the body. Channeling the activity of the atavistic animal away from the higher self, Metchnikov significantly broadened the repertoire of possible interactions between the primitive and evolutionarily advanced elements of the organism. They could now coevolve to greater harmony and could coexist within the same organism, rather than inevitably clash. The primitivity of the phagocytes appears both as the legacy of the past and as an evolutionary acquisition that was useful and important for the organism as a whole. This meant, however, that the connections between the high and the low elements of the body were rather loose. To those who thought of organismic integrity in terms of hierarchy and discipline, such loose, symbiotic relationships might appear as undermining the very idea of wholeness. What were the cultural costs of taming the phagocytes, as well as the immediate, biological consequences of hosting the primitive cells for the organism as a whole?

THE CELL WARS AND THE PROBLEM OF PEACE (C. 1894–C. 1916)

In the 1880s, Metchnikov's idea of primitive and independently moving immune cells led to severe disagreements with a number of influential pathologists and microbiologists. In the mid-1890s, however, he achieved a compromise with the proponents of a humoralist understanding of immunity. He accepted the real existence of the chemical factors of immune defense (or antibodies, as Paul Ehrlich called them), but he continued to think about phagocytic immunity as the more primitive, more ancient, and more general mechanism of immunity. It is likely that the growing emphasis on

---

[31] Paul Baumgarten, "Zur Kritik der Metschnikoff'schen Phagocytentheorie," *Zeitschrift für klinische Medizin*, 15 (1888): 1–41, on 15; on the discussions between Metchnikov and his opponents, see Tauber and Chernyak, *Metchnikoff and the Origins of Immunology* (cit. n. 1), 135–74; Arthur M. Silverstein, *A History of Immunology* (San Diego, 1989), 38–58; Gaissinovitch, "Trudy i dni I. I. Metchnikova" (cit. n. 1), 14–42.

[32] For Virchow's support of Metchnikov and his ideas, see *Berliner klinische Wochenschrift* 34 (1883): 526; Rudolf Virchow, "Der Kampf der Zellen und der Bakterien," *Virchows Archiv* 101 (1885): 1–13, 12–3; Christiane Groeben und Klaus Wenig, eds., *Anton Dohrn und Rudolf Virchow: Briefwechsel, 1864–1902* (Berlin, 1992), 61–3.

the primitive nature of phagocytes led Metchnikov to the conclusion that under certain conditions the phagocytes might represent a threat to the organism as a whole.

In 1890, the German microbiologist Emil von Behring discovered the diphtheria antitoxin and showed that diphtheria could be cured by injections of the antitoxin. However, the use of this treatment had remained limited until Émile Roux immunized horses against diphtheria in 1894 and obtained the antiserum in larger quantities. The Pasteur Institute in Paris became a major center of the production of the antiserum, which led to the institute's expansion and strengthened its reputation in France and worldwide. Other centers emerged in Germany under Emil von Behring and Paul Ehrlich, both prominent humoralists. Roux, Metchnikov, and other scientists at the Pasteur Institute were able to develop a working relationship with Behring, exchanging samples of sera and working together on the problem of a "standard" serum against diphtheria. The correspondence between Behring and Metchnikov provides evidence about the depth and intensity of their collaboration, as well as about the role of Metchnikov as a liaison between Behring and Roux. It is also important that, following the discovery of the diphtheria antitoxin, scientists in Metchnikov's own laboratory in Paris started to experiment with different antitoxins and antibodies. One of his pupils, future Nobel laureate Jules Bordet, who worked in Metchnikov's laboratory in Paris during the 1890s, made several important discoveries that strengthened the knowledge base of the humoral, rather than of the cellular, theory of immunity.[33]

The theories of humoralists, such as Behring and Ehrlich, corresponded with their experimental practices, which were based on chemistry, rather than on the microscopic observation of immune cells. But Metchnikov's situation was different: remaining a staunch adherent of phagocytosis, he negotiated the material exchange and experimental practices that dealt with antitoxins, antibodies, and other chemical factors, rather than with cells. It is also striking that his faith in the role of phagocytes as the main agents of immune response grew stronger, rather than diminished, during the second half of the 1890s, when so many discoveries about the chemistry of immunity were made.

According to Metchnikov, phagocytic immunity was more ancient than humoral immunity. Speaking at the Eighth International Congress for Hygiene and Demography in 1894, Metchnikov compared cellular and humoral immunity with intracellular and extracellular digestion, at different levels of the evolutionary ladder: "In lower animals, digestion takes place exclusively within the cells, but, in the course of progressive evolution, digestion acquires a mixed character: remaining intracellular, it can also proceed extracellularly. From a theoretical point of view, it may appear reasonable that . . . the intracellular destruction of microbes will become extracellular. Yet, in reality, we are still far from this ideal." The word "ideal" is remarkable: humoral immunity appeared to him as a more perfect form of defense than phagocytic immunity. But it also meant that, being a more ancient mechanism of immunity, phagocytosis had a more general significance as a defense mechanism throughout the animal kingdom. In 1897, Metchnikov showed, in fact, that invertebrates possessed a

---

[33] Most of Metchnikov's letters to Behring have been published in Metchnikov, *Pis'ma, 1863–1916* (cit. n. 1); the originals are in the Emil von Behring Archiv, Univ. of Marburg, Marburg, Germany; Behring's letters are in the Archives of Russian Academy of Sciences.

phagocytic system of defense but lacked a humoral one.³⁴ In other words, immune response was possible in the absence of antibodies, and even if humoral factors played certain roles in immunity, these roles were secondary to the more ancient form of defense.

In this way, responding to new discoveries in immunochemistry, Metchnikov emphasized the primitive (and the more basic) nature of phagocytic immunity. The stress that Metchnikov laid on the primitivity of phagocytes also made them potentially dangerous to the organism as a whole. Studying the tissues of old animals and people beginning in the mid-1890s, Metchnikov discovered that phagocytes could attack and obliterate the "noble" cells of muscles, nerves, the brain, and other organs. Apparently, these were the senescent or malignant cells that, as we now know, are normally destroyed by phagocytes. But Metchnikov interpreted his observations as evidence that the "primitive" and "aggressive" phagocytes might get out of control. Attacking the healthy and normal cells of muscles, nerves, or brain, the phagocytes caused, in his view, the atrophy of these noble tissues, which led to senility. In his view, the protective layer that existed around the highly specialized, noble cells thinned with age and, finally, disappeared.³⁵ As a result, he broadened the scope of his theory, describing the process of aging in terms of cell wars: "In other words, a conflict takes place in old age between the higher elements and the simpler or primitive elements of the organism, and the conflict ends in the victory of the latter. This victory is signaled by a weakening of the intellect, by digestive troubles, and by lack of sufficient oxygen in the blood. The word "conflict" is not used metaphorically in this case. It is a veritable battle that rages in the innermost recesses of our beings."³⁶

He ascribed a certain role in this process to microbes that inhabit the large intestine of man and increasingly poison him with their toxins. The toxins destroy the protective layer around the noble cells, opening them to the attacks of the phagocytes. Attempting to prevent "precocious" aging, Metchnikov worked along several lines. First, he recommended the use of lactobacilli to suppress the poisonous bacteria in the large intestine. Second, he worked with the specific toxic substances that, he believed, would revitalize and strengthen the noble cells when applied in small quantities. Third, he encouraged the British surgeon William Lane to remove large fragments of the large intestine. Finally, and most strikingly, Metchnikov attempted to develop a specific serum that would act selectively against the phagocytes, "switching them off" in old age. (The patients, then, would be placed in an aseptic environment.)³⁷

In this way, Metchnikov came to the conclusion that in old age the primitive cells should be suppressed. But it is likely that his support of the evolutionarily advanced cells against the primitive, of "civilization" against "wildness," was not as absolute

---

³⁴ Elie Metchnikoff, "L'état actuel de la question de l'immunité," *Annales de l'Institut Pasteur* 8 (1894): 706–21, 716–7; Metchnikov, "Issledovanniia nad proiskhozhdeniem antitoksinov: O vliianii organizma na toksiny," *Russkii arkhiv patologii* 4 (1897): 379–86; Gaissinovitch, "Trudy i dni I. I. Metchnikova" (cit. n. 1), 26.

³⁵ Gaissinovitch, "Trudy i dni I. I. Metchnikova" (cit. n. 1), 30–40.

³⁶ Elie Metchnikoff [Metchnikov], *The Nature of Man: Optimistic Studies*, trans. P. Chalmers Mitchell (New York, 1903), 239, quoted in Todes, *Darwin without Malthus* (cit. n. 1), 101.

³⁷ Gaissinovitch, "Trudy i dni I. I. Metchnikova" (cit. n. 1), 30–40; see also Metchnikov's letter to D. N. Anuchin, 7 Jan. 1900, in Metchnikov, *Pis'ma, 1863–1916* (cit. n. 1), 162–3, 265–6; Ann Dally, *Fantasy Surgery, 1880–1930: With Special Reference to Sir William Arbuthnot Lane* (Amsterdam, 1996); Todes, *Darwin without Malthus* (cit. n. 1).

as it may appear. He continued to believe in the ideals of harmony, rather than of war and suppression. This is quite evident from his theory of "natural death"—a theory that he was developing since the 1890s.

It is likely that August Weismann was the first scientist to write about death as an adaptation in the Darwinian sense. Assuming that the first organisms did not die, he came to a conclusion that the absence of death made further evolutionary perfection impossible, leading to the extinction of the primeval "immortal" organisms.[38] Like Weismann, Metchnikov thought about death in evolutionary terms, believing that natural limits of individual life exist in each species of animal. In his view, man died before reaching the age at which evolution destined him to die. At the same time, the experience of natural death should be associated with the feeling of deep happiness: "The approach of natural death is one of the most pleasant sensations that can exist . . . Old men . . . die in the fear of death, without having known the instinct of death . . . The normal end, coming after the appearance of the instinct of death, may truly be regarded as the ultimate goal of human existence."[39] It appeared to Metchnikov that natural death might be the case in mayflies (Ephemeridae).

Kirillov, the Dostoevsky character, believed that on the verge of disappearance a suicide could experience a feeling of deep satisfaction and happiness because, choosing death freely, he gains the power over life and death.[40] But for Metchnikov, who had twice attempted suicide, natural death was a kind of reward in the direct, physiological sense—a reward for a harmonious, completed life cycle. He also emphasized that metaphysics, religious faith in particular, offered spiritual consolation to the very few—those capable of spiritual growth—while in others, in the majority, it aroused a deep fear of death. His ideas about the prolongation of human life, up to its natural limits, inspired a grand vision of science and hygiene, acting on a global scale and fighting not only against the disease but also against the very fear of death.

There is a clear analogy between the theory of natural death and the moral beliefs of Metchnikov's youth: the notion of the completed life cycle came to replace the ideal of the harmonious body as something static and given and, therefore, inconsistent from the evolutionary point of view. The notion of the biological harmony—understood as a fully completed life cycle—continued to be linked to the notion of an authentic, unalienated self. One could also describe Metchnikov's intellectual evolution in terms of "price" that he was to pay to reconcile the moral ideas of his youth with the harsh realities of the existing world, in which inequality and the relentless struggle for existence were normal and natural. The price was high: the reconciliation between morality and scientific truth required the acceptance of death as a positive moral value.

CONCLUSION

One of Metchnikov's pupils called him an "outstanding European," emphasizing his cosmopolitanism and his deep connections with European culture and scientific thought.[41] Metchnikov shared important moral values and expectations with his

---

[38] August Weismann, *Aufsätze über Vererbung und verwandte biologische Fragen* (Jena, 1892).
[39] Elie Metchnikov, *The Prolongation of Life: Optimistic Studies*, trans. P. Chalmers Mitchell (New York, 1908), 288.
[40] Dostoevsky, *Demons* (cit. n. 25).
[41] Etienne Burnet, "Un européen: Elie Metchnikoff," *Europe* 45 (1926): 61–77.

generation of educated Russians. But some of his ideas were meaningful only within the broader context of European science and culture.

Darwin's theory of biological evolution produced a feeling of cultural and intellectual shock in many scientists, philosophers, and moralists in the West. Thomas Huxley, one of the most ardent supporters and advocates of Darwin's theory, came to a deeply pessimistic view of human nature at the end of his life. Unlike Darwin and Spencer, Huxley did not believe that natural selection was a force that might bring about moral progress in human society. Darwin's theory of "struggle for survival" also led to a new understanding of nature as profoundly and "naturally" ruthless. This understanding was greeted by some, such as Friedrich Nietzsche, as a liberation from illusions, but in others it produced a feeling of painful conflict between science and values.[42] Metchnikov's vision of the human psyche at war with itself fit within this intellectual framework and was one of the early examples of a disillusionment that became really widespread closer to the end of the nineteenth century.

Examining the Russian context of Metchnikov's discoveries is another way to understand his intellectual biography, answering the question of why the idea of wholeness was of paramount and lasting importance for the evolution of his ideas. Metchnikov believed in wholeness and organismic integrity as a moral concept, but in a specifically Russian, egalitarian, and antiorganicist form. Many Russians—from Mikhailovskii to Tolstoy—ascribed a "natural" wholeness to the human organism, making the "harmonic" body a metaphor for the natural, unalienated self, but, at the same time, they denied wholeness to society. The organicist, holistic approach to society meant inequality, or social hierarchy, an idea that the Russians wanted to avoid both in its "pyramidal form"—as Prince Kropotkin defined the way in which the autocratic, pre-reform Russian society had been organized—and in its new, capitalist appearance.

But Metchnikov was perhaps the only one among his generation who decided to bring the belief about the harmonious organization of the human body to the test. His evolutionary and racial "discoveries" smashed his old beliefs; as a result, he came to a vision of the organism as an evolving, rather than a static and given, structure. In his view, the evolutionary asynchronies in the development of the organism's different parts caused basic and fatal disharmonies of human nature. Discovering immunity as a "useful" activity of primitive and aggressive cells, Metchnikov was able—to a greater or lesser extent—to reconcile the integrity of the body as a moral imperative with the notion of the "obvious" evolutionary asynchronies that undermined that integrity. Immunity became a biological mechanism for the moral integration of the body, a mechanism that was different from suppression.

Metchnikov's military metaphor of immunity had a profound impact on other researchers, including the humoralists; disagreeing with Metchnikov about the nature of immunity, they accepted his vision of immunity as a war and even described the

---

[42] See Kurt Bayertz, "Biology and Beauty: Science and Aesthetics in Fin-de-Siècle Germany," in *Fin De Siècle and Its Legacy*, ed. Mikulas Teich and Roy Porter (Cambridge, UK, 2001), 278–95; James Paradis, "*Evolution and Ethics* in Its Victorian Context," in *James T. H. Huxley's Evolution and Ethics with New Essays on Its Victorian and Sociobiological Context*, ed. James Paradis and George C. Williams (Princeton, N.J., 1989), 3–52; Paul J. Weindling, *Darwinism and Social Darwinism in Imperial Germany: The Contribution of the Cell Biologist Oscar Hertwig, 1849–1922*, vol. 3 of *Forschungen zur neueren Medizin- und Biologiegeschichte*, ed. Gunter Mann and Werner F. Kümmel (Stuttgart, 1991).

antibodies as the organism's specific weapons.[43] Metchnikov understood this war in terms of health and disease. Described by Metchnikov as a biological and medical concept, the integrity of the human body continued to be a moral idea. However, what was a "fact" in his youth—that bodily and moral harmony was inborn and was only distorted by incorrect education and unsuitable social conditions —became an ideal that could be approximated in the new, dynamic model of the body, or a state of psyche that could be experienced at the very moment of death— but not be achieved literally as something lasting. Starting his moral discourse with the notion of the harmonious human body, Metchnikov ended it by formulating a theory of harmonious death and inventing a new space for war—in his own words, the cell wars "rage in the innermost recesses of our beings."[44]

---

[43] J. Andrew Mendelsohn, "The Model of War and the Problem of Peace in Immunity Research Circa 1900" (talk, Berliner Seminar für Wissenschaftsgeschichte an der Technischen Universität, Berlin, 3 June 1996).

[44] Metchnikov, *The Nature of Man* (cit. n. 36), 239.

The Schooling of Lev Landau:
The European Context of Postrevolutionary Soviet Theoretical Physics

By Karl Hall*

ABSTRACT

Theoretical physicist Lev Landau is internationally renowned for his school, whose social and intellectual coherence rests heavily on his personal agency: his brilliance, the scope of his research, his pedagogical program (textbooks and Theoretical Minimum). This essay considers the broader historical factors that made the Landau school possible and sustainable. The collective values associated with it were already in circulation in revolutionary Russia. Landau's own training took place in an atmosphere of unstable social relations that rewarded behaviors that were neither previously nor subsequently tolerated. Yet the Landau school was not sui generis Soviet. It was also the product of Russia's long-standing ties with German physics and of Landau's own intellectual development in interaction with crucial figures such as Niels Bohr. Although Landau famously claimed Bohr as his only teacher, he adopted the Bohr "style" in only a very restricted sense and actively discouraged his students from embracing Bohr's epistemological concerns. His school was a "purely practical" version of a long intelligentsia tradition.

INTRODUCTION

> Let me remind the reader that although, at the beginning of the twentieth century, the large experimental school of Lebedev in Moscow and the small groups of Joffe and Rozhdestvenskii in Petersburg began to form, theoretical physics as a branch of science did not, in fact, exist anywhere in Russia. It is true that, just before the revolution, theoretical research conducted by individual scientists of high caliber started appearing more frequently. However, they did not have their own schools.
> —I. E. Tamm, "Theoretical Physics" (1967)[1]

With its move from Leningrad to Moscow in August 1934, the Soviet Academy of Sciences found that it had to become ever more practiced at painting itself as a pro-

---

*History Department, Central European University, Nádor u. 9, H-1051 Budapest, Hungary; hallk@ceu.hu.

I thank the editors for their timely criticisms. I am grateful to the Max Planck Institute for History of Science for support in the course of writing this essay. Permission to use manuscript materials in this essay was kindly granted by Jo Hookway and the trustees of the Bodleian Library, Oxford, and the Niels Bohr Archive, Copenhagen. Unless otherwise indicated, all translations are my own.

[1] From an essay for *Oktiabr' i nauchnyi progress*, in B. M. Bolotovskii and V. Ya. Frenkel, eds., *I. E. Tamm: Selected Papers* (Berlin, 1991), 284–90, on 285.

ductive smithy for the tools of socialist construction, while retaining a role for "the basic fundamental questions of modern physics."[2] This task was not eased the following year by the revised academy statute that elevated the Technology Group to division status, ensuring a growing role for engineers—many with party ties—in the upper tiers of the academy.[3] Having approved the new academy charter, the Council of People's Commissars (Sovnarkom) further proposed to hold a special session of the academy devoted to critical reports from its leading figures in all the disciplines: the physiologist I. P. Pavlov (1849–1936), the geologist A. E. Fersman (1889–1972), the physicist A. F. Joffe (1880–1960), and others.[4] It became apparent rather quickly that this exercise in public accountability threatened to expand into a bureaucratic babel of superficial analyses. This situation and the death of Pavlov in February 1936 helped resolve the government to focus on one discipline: physics. Even then the many competing research agendas threatened to make the session superficial and disjointed, and Joffe argued that he alone should shoulder the responsibility for conveying a "complete picture of a unified grand enterprise."[5] The government partly relented, narrowing the exercise to three institutions: the Leningrad Physico-Technical Institute (LFTI), the State Optical Institute (GOI) in Leningrad, and the Lebedev Physics Institute (FIAN) in Moscow. The choice was peculiar as only FIAN was an academy institute at the time and the focus of the session would be on LFTI and GOI. In Soviet practice, however, it made perfect sense, for scientific discipline and social leadership were strongly linked, and each institute was tied to a senior figure in the academy: Joffe (LFTI), D. S. Rozhdestvenskii (GOI), and L. I. Mandelstam (FIAN).[6] The academy vice president emphasized the connection when he opened the session by announcing that the assessments of Soviet physics that the assembled worthies were about to hear were directed at "the most important schools in physics."[7]

Standing in the wings were three younger corresponding members of the academy, each invited to give supplemental reports on recent developments in theoretical physics. As none of the three theorists would attend to applied themes nor would any of their talks generate much debate, the motives for having them share the stage at the March session were mixed. Yet there had to be some acknowledgment of FIAN and of Mandelstam (1879–1944), who was respected equally for his experimental and theoretical work, while his nervous constitution left him in abhorrence of these Soviet rituals. The head of FIAN's small Theory Department, Igor Tamm (1895–1971) spoke publicly for his older colleague and friend, a role he had played for more than a

---

[2] See the academy's report to the Seventh Congress of Soviets, *VII Vsesoiuznomu s"ezdu sovetov—Akademiia nauk SSSR* (Moscow, 1935), 33–6. On the move, see V. D. Esakov, "Puteshestvie iz Peterburga v Moskvu: Pereezd Akademii nauk," *Priroda*, 1997, no. 9:131–8; "Shtab sovetskoi nauki meniaet adres," *Vestnik rossiiskoi akademii nauk* 67 (1997): 840–8.

[3] See *Ustavy Akademii nauk SSSR* (Moscow, 1974), 142–50.

[4] Vladimir Vizgin has written the authoritative study of the "March session" in his "Martovskaia (1936 g.) sessiia AN SSSR: Sovetskaia fizika v fokuse," *Voprosy istorii estestvoznaniia i tekhniki*, 1990, no. 1:63–84, and 1991, no. 3:36–55. Further accounts may be found in V. Ia. Frenkel', "K 50-letiiu martovskoi sessii Akademii nauk SSSR (1936 g.)," *Chteniia pamiati A. F. Joffe 1985* (1987): 63–86; Paul Josephson, *Physics and Politics in Revolutionary Russia* (Berkeley, Calif., 1991), 295–305; David Holloway, *Stalin and the Bomb* (New Haven, Conn., 1994), 16–24.

[5] Joffe quoted in Vizgin, "Martovskaia sessiia" (cit. n. 4), 38.

[6] One observer pointedly remarked that it was not the work of the *academy* that was discussed, as much as the work of the *academicians*. S. Krum, "Fizika na sessii Akademii nauk," *SORENA*, 1936, no. 5:105–16, on 106. Three years later, LFTI was transferred to the academy's jurisdiction.

[7] *Izvestiia Akademii nauk SSSR*, seriia fizicheskaia, 1936, nos. 1–2:5–409, on 5.

decade at the university.[8] That Tamm himself was in the process of building a school had been one of the reasons cited for his election to the academy in 1933.[9] Tamm's immediate superior, FIAN director and optics specialist S. I. Vavilov, was reporting on GOI (his second post), so Tamm's report on the state of nuclear theory would have to stand in for a more comprehensive assessment from Mandelstam. V. A. Fock (GOI) would also address the many-body problem in quantum mechanics, while Ia. I. Frenkel (1894–1952) (LFTI) would share his latest work on the solid-liquid phase transition.[10] Yet despite Joffe's (and Vavilov's) careful plans, Soviet physicists did not present a united front to party and state dignitaries nor to their fellow scientists. When *Pravda* applauded "the fresh gust of self-criticism that has burst its way into the walls of the academy," it was reporting more than the enactment of an ideological rite whose outcome was predetermined.[11] Joffe's younger colleagues, it turned out, took the exercise entirely in earnest—especially twenty-eight-year-old theorist Lev Landau (1908–68), who had begun his career at LFTI.[12]

The theorists made trouble because they felt their institutional status remained insecure. Despite the calculated deference shown to them by Vavilov, they had good reason to be concerned. Tamm's group would be formally dispersed to the various experimental laboratories in April 1938, albeit with continued permission to work on "general topics."[13] This was partly because Tamm's engineer brother had been arrested during the purges and partly because Tamm could not come up with a tenable theory of nuclear forces.[14] But a complaint about insufficient attention to school building also played a part. Only as Tamm began attending more closely to the demands of cosmic ray physicists, and as international praise began to accrue for his interpretation of Cherenkov radiation, did the decision begin to look premature.[15] For his part, Landau was already building a school of his own in Kharkov, eventually succeeding brilliantly—to this day, the designation "Landau school" requires no explanation among theoretical physicists.[16] But founding schools in *theoretical* physics relied upon a rather different understanding of the past of Russian physics and of the

---

[8] In a December 14, 1922, letter to his wife, Tamm noted that Mandelstam got a migraine if so much as forced to share a dinner table with even a polite Communist. L. I. Vernskii, "I. E. Tamm v dnevnikakh i pis'makh k Natalii Vasil'evne," in *Kapitsa, Tamm, Semenov v ocherkakh i pis'makh* (Moscow, 1998), 273–4.

[9] "Vybory chlenov-korrespondentov i pochetnykh chlenov," f. 2, op. 2 (1933), d. 9, l. 178, Russian Academy of Sciences (Arkhiv Rossiiskoi Akademii Nauk, hereafter cited as ARAN), Moscow.

[10] For Tamm's talk, see *Izvestiia AN SSSR*, seriia fizicheskaia, 1936, nos. 1–2:300–23; for Fock's, 351–62; for Frenkel's, 371–93.

[11] Quoted in Vizgin, "Martovskaia sessiia" (cit. n. 4), 71. For a close analysis of the rituals of criticism and self-criticism, see Alexei B. Kojevnikov, *Stalin's Great Science: The Times and Adventures of Soviet Physicists* (London, 2004), 186–216; Oleg Kharkhordin, *The Collective and the Individual in Russia* (Berkeley, Calif., 1999), 123–63.

[12] On the attitudes of the younger participants, see S. E. Frish, *Skvoz' prizmu vremeni* (Moscow, 1992), 217–8.

[13] I. E. Tamm, "Otchet laboratorii teoreticheskoi fiziki," 7 May 1938, f. 532 op. 1 d. 31 ll. 89–93 (including discussion), ARAN.

[14] The doomed Leonid Tamm's testimony is found in *Report of Court Proceedings in the Case of the Anti-Soviet Trotskyite Centre*, 23–30 Jan. 1937 (Moscow, 1937), 439–42.

[15] I. M. Frank and Ig. Tamm, "Coherent Visible Radiation of Fast Electrons Passing through Matter," *Doklady AN SSSR* 14, no. 3 (1937): 109–14; Ig. Tamm and S. Belenkii, "On the Soft Component of Cosmic Rays at Sea Level," *Journal of Physics USSR* 1, no. 3 (1939): 177–98. Tamm, Frank, and Cherenkov were awarded the Nobel Prize in 1958.

[16] E.g., Frank Wilczek, "Nobel Lecture: Asymptotic Freedom: From Paradox to Paradigm," *Reviews of Modern Physics* 77 (2005): 859. Cf. John Ziman, "Landau and His School," in *Of One Mind: The Collectivization of Science* (Woodbury, N.Y., 1995), 17–22.

relation of the Soviet physicist to his European counterparts. The March session was merely a marker of discontents and ambitions that stemmed from the early days of the Russian Revolution, whose generational effect had been especially strong among theorists born after 1890. It is also a point of entry for understanding how the unstable professional roles of Soviet theorists contributed to the renewal of intelligentsia sensibilities that, in part for reasons of disciplinary self-conception and probably in part from sheer historical ignorance, the theorists had largely suppressed during the formative years of their training.

In an era when abstract "theory" could be an ambiguous and occasionally dangerous social marker of one's area of expertise, theoretical physicists uniformly made theory credibly *Soviet* by endorsing a genetic fallacy among their colleagues: it had had no real place in tsarist Russia, it was "practically the same age as the revolution," and it had only gotten started "in the new conditions of scientific work created by the Soviet government."[17] Prepared by 1938 to treat Frenkel, Tamm, and Landau as nascent heads of Soviet schools, Vavilov deliberately cast theory as the labor of Soviet youth, for "it received almost no heritage, no tradition from the prerevolutionary past," when there had been "no theoretical school."[18] Since Soviet theory had blossomed in the 1920s and 1930s, the so-called golden age of modern physics, its vigorous international engagement with other practitioners of the new quantum mechanics made domestic institutionalization possible. The alacrity with which the first Soviet theorists had gained international regard during their travels abroad became central to their collective identity. The European moment was indeed constitutive because "this participation enabled the development of theoretical physics in the Soviet Union, [and] the founding of the prominent schools of L. D. Landau, I. E. Tamm, V. A. Fock, [and] Ia. I. Frenkel."[19]

Schools, in other words, were the form of life that let Soviet theorists to build and sustain informal and exclusive modes of social interaction amid recurrent cycles of egalitarian proletarianization and party intervention in the Soviet sciences during the Stalin era—processes in which the collective homogeneity of the physics community vis-à-vis interlopers could not be taken for granted.[20] In the smaller confines of Landau's elite seminar, wrote a postwar participant, "everyone felt as if they belonged to one faith," and that sense of social and intellectual unity was immensely valuable to its members.[21] At the same time, schools also came to provide a respectable outlet for the official reward mechanisms for individual distinction that won out

---

[17] E. V. Shpol'skii, "Fizika v SSSR 1917–1937," *Uspekhi fizicheskikh nauk* (hereafter cited as *UFN*) 18 (1937): 295–322, on 322; Shpol'skii, "Piat'desiat let sovetskoi fiziki," *UFN* 93 (1967): 197–276, on 213. (Shpol'skii was a classmate of Vavilov's in Moscow, and longtime editor of *UFN*.) Cf. S. Khaikin, "Dostizheniia v oblasti fiziki za 15 let," *Pod znamenem marksizma*, 1932, nos. 9–10:221–9; D. I. Blochinzew, "The Advance of Theoretical Physics in the Soviet Union in Twenty Years," *Physikalische Zeitschrift der Sowjetunion* 12 (1937): 542–3; Blokhintsev [Blochinzew], "Puti razvitiia teoreticheskoi fiziki v SSSR," *UFN* 33 (1947): 285–93, on 286; P. S. Kudriavtsev, "Iz istorii stanovleniia sovetskoi teoreticheskoi fiziki," in *Razvitie fiziki v Rossii*, vol. 2 (Moscow, 1970), 6–21, on 7.

[18] S. I. Vavilov, "Sovetskaia fizika—nauka molodezhi," *Priroda*, 1938, no. 9:37–8; Vavilov, "Fizika v Rossii i v SSSR," *Priroda*, 1932, nos. 11–12:989–1012, on 995.

[19] M. A. El'iashevich, "Ot vozniknoveniia kvantovykh predstavlenii do stanovleniia kvantovoi mekhaniki," *UFN* 122 (1977): 673–717, on 688–9.

[20] The agonistic milieu of early Soviet theory is detailed in Karl Hall, "Purely Practical Revolutionaries: A History of Stalinist Theoretical Physics, 1917–1941" (PhD diss., Harvard Univ., 1999).

[21] M. I. Kagan, "Shkola Landau: Chto ia o nei dumaiu. . ." *Priroda*, 1995, no. 3:76–90, on 85. Cf. B. L. Ioffe, "Landau's Theoretical Minimum, Landau's Seminar, ITEP in the Beginning of the 1950's" (preprint arXiv:hep-ph/0204295v1, April 2002), http://www.arXiv.org.

over the long term. It was an arena of professional individualization (*otlichie*) safely surrounded by strong norms of collective endeavor and thus well suited to mature Soviet society.[22]

Schools offered a legitimate institutional identity to this most cosmopolitan of sciences even when Soviet political autarky meant that the achievements of physicists were constantly being evaluated in patriotic terms. In the case of Landau, scorn for such provincialisms by no means meant that his research school lacked a local distinctiveness. But to understand why the disciplinary practices fostered by the Landau school were in a certain sense "Soviet," we must appeal to more than the reactive dynamic of state-sponsored rituals such as the March session. Recovering a more immediate, anthropological account of Soviet cultures of theory will thus be central to this essay.[23] To do this, we must also follow Landau to Zurich, Cambridge, and Copenhagen, the locales where he consolidated a truly Soviet identity and developed proactive agendas for *teoretika* (his neologism for theoretical physics played down the resonances with "theory" in the broader sense) back home. If Landau held out Niels Bohr as his model leader of a school, he did so not to emulate Bohr's theoretical practices in toto but to marginalize Soviet competitors to his own school. And in the mouths of Landau and his peers, Bohr's language of crisis and renunciation at the heart of quantum theory took on distinctive resonances in the context of massive Soviet industrialization circa 1930. In the failed revolution of quantum electrodynamics that followed, Soviet theorists encountered conceptual dilemmas that only served to exacerbate their dissatisfactions with local institutions. This made schools all the more important in developing "purely practical" procedures for Soviet theory, in which "practical" did not mean "applied" but rather "not overly concerned with the physics of principles," with all its attendant epistemological problems.[24] It would be matter of fact and pragmatic in technique without reducing to mere phenomenology, even if its abstruse subject matter looked at times to Soviet laypeople like the "abstract cult" of science once thought to have been characteristic of the prerevolutionary Russian scientist-*intelligent*.[25] The historical analogy was actually quite weak, yet it was their early Soviet conflation of revolutions—political and quantum—that had led the Soviets to treat the institutions of theory as a novel Soviet phenomenon. The collapse of the revolutionary analogy brought them back into history, and in finding ways to survive and thrive in the Stalinist order, they also became sometimes reluctant heirs of intelligentsia traditions that demanded an independent spirit "in matters large and small, in life and in science."[26] If rigorous disciplinary practices were always at the heart of the school, they were reinforced and rendered socially stable

---

[22] Kharkhordin, *Collective* (cit. n. 11), 337–40.

[23] Peter Galison and Andrew Warwick, eds., "Cultures of Theory," *Studies in History and Philosophy of Modern Physics* 29B (1998): 287–434.

[24] Landau once characterized his seminar method brusquely as "chisto delovoi" in response to a tendentious query. See "Protokol No. 9 zasedaniia Uchenogo Soveta Instituta fizicheskikh problem," 9 June 1951, f. 1, op. 1, d. 106, l. 20, Arkhiv Instituta Fizicheskikh Problem, Moscow.

[25] Boris Sokolov, *Nauka v sovetskoi rossii* (Berlin, 1921), 6. On mathematical abstraction as a purported political danger, see E. Kol'man, "Vreditel'stvo v nauke," *Bol'shevik*, 31 Jan. 1931, no. 2:75–6.

[26] See E. L. Feinberg's remarks on Tamm's relation to the prerevolutionary intelligentsia in "Epokha i lichnost'," in *Vospominaniia o I. E. Tamme*, 2nd ed., ed. E. L. Feinberg (Moscow, 1986), 225; echoed in Andrei Sakharov, *Memoirs* (New York, 1990), 128.

by a larger system of group values that Landau liked to call the "theoretical physics approach to life."[27]

BACK TO SCHOOL: POLEMICS AND PRECURSORS

March 1936 was less than two decades after the founding of LFTI, but in the foreshortened horizon of socialist construction, this was already more than a generation since Joffe himself had become the first post-October academician among the physicists. His report ranged from atomic physics to industrial laboratories and proudly noted the founding of daughter institutes in Kharkov, Tomsk, and Dnepropetrovsk. While the topical scope of the report does not concern us, the scope of Joffe's claims about LFTI's historical trajectory does, for although he did hearken back rather dismissively to the fin de siècle in Russian physics, he had good reason to paint LFTI as the engine that had propelled Soviet physics onto the world stage. However, Joffe's readiness to cast Soviet physics as nearly on par with the top two or three international counterparts invited Landau's derision, and the theorist took the floor at his most combative. (Only a very modest fraction of the hundreds of people in attendance could claim experience abroad comparable to Landau's.) Landau sweetly credited to Joffe much of the very existence of Soviet physics, for "in tsarist Russia up to the moment of the revolution physics practically didn't exist. Those little fragments of the Lebedev school that existed in Moscow could not pretend at the time to the title of real physics." So Joffe's subsequent contribution had been great indeed, but his rosy assessments were unsustainable, said Landau, declaring the actual situation "almost catastrophic" in comparison with the tasks at hand. He counted no more than a hundred independent researchers in Soviet physics, not all of whom evidenced much concern for training the next generation. Landau then launched into a pointed critique of the less-than-careful habits of research he saw prevailing in the profession, habits that led to a tremendous waste of resources. Such waste could have been prevented if senior colleagues such as Joffe, rather than designing entire experimental research agendas based on "extremely primitive" calculations, had consulted with tough-minded theorists in a timely way.[28]

The forty-year-old Tamm—also well traveled—was equally fervent, if generally more diplomatic, when his turn came to speak. He mounted a strong defense of basic research, maintaining that a proper "scientific base" would be the only guarantee of physicists' participation in socialist construction. Like Landau, however, he was not at all sanguine about the Russian past. He referred in passing to Lebedev and another contemporary as "isolated researchers who had not left behind schools, who had not founded a common scientific culture, who had not founded any scientific cadres, no opportunity to work as members of a strong collective, the fundamental requirement for any successful work."[29] These were harsh words, but the mixed audience took them as a healthy part of the exercise. Neither theorist hesitated to dispense with professional comity on this larger stage if he thought that displays of unity came at the cost of internal disciplinary reform. Landau even received some applause when he

---

[27] A. A. Abrikosov, "O L. D. Landau," in *Vospominaniia o L. D. Landau*, ed. I. M. Khalatnikov (Moscow, 1988), 33.
[28] L. D. Landau, "Vystuplenie," *Izv. AN*, nos. 1–2 (1936): 83–6.
[29] I. E. Tamm, ibid., 87–91.

condemned Joffe's high estimates of Soviet physics. It was commonly understood, even by those who did not entirely care for the "ugly" tone of Landau's criticism, that the debate was supposed to be "passionate" and not heed academic conventions.[30] Yet I single out these remarks because they did not accord well with the nascent patriotic trend—accelerated in the postwar era—to reclaim the past of Russian physics as a worthy predecessor to a vibrant Soviet present.[31] To do so meant, first and foremost, that Lebedev had to be retrieved, and in subsequent years he was—by one of the not-so-little "fragments," who dubbed him the "first organizer of collective scientific work" in Russian physics.[32] Especially among nontheorists, Landau and Tamm's tendentious dismissal of prerevolutionary history would rankle for a long time indeed.[33]

The mere fact that many Soviet contemporaries might have been inclined to treat the October Revolution as year zero scarcely comes as a surprise. But what were the disciplinary ramifications? As early as 1921, one can find Russian talk of a "new era" in atomic physics with reference to Rozhdestvenskii's spectroscopic work, and as we will see in the next section, quantum theory became a rich resource for talking about revolutions.[34] Geochemist and liberal V. I. Vernadskii perceptively surmised that the accelerating revival of Russian science already evident to him in 1922 would unavoidably come to be associated with the revolution, notwithstanding the diverse political views of the twenty scholars he named as instrumental in this process.[35] The eagerness of the Bolsheviks to transform the state's role as patron for science largely ensured this confluence of political and disciplinary temporalities. Yet we need to examine more closely the theorists' historical reasons for differentiating themselves in this fashion in the years after 1917, for the discrepancies can tell us much about the function of research schools in the Russian context. My aim is not to test the well-developed historiography on European research schools in the Russian case.[36] In any event, Russian historians of science have had an interest in research schools since the growth of science studies (*naukovedenie*) in the 1960s and 1970s, when it was hoped that better understanding of school formation could aid in the planning of science. In their work, these historians are often admirably sensitive to analytic distinctions, and they frequently invoke examples of Russian schools, yet seldom pause to ask whether

---

[30] Krum, "Fizika na sessii" (cit. n. 6), 110.

[31] Epitomized by *Razvitie fiziki v Rossii*, vol. 1 (Moscow, 1970).

[32] S. I. Vavilov, "Petr Nikolaevich Lebedev," *Liudi russkoi nauki* (Moscow, 1961), 277–84. Cf. P. P. Lazarev, *A. G. Stoletov, N. A. Umov, P. N. Lebedev, B. B. Golitsyn* (Leningrad, 1927); Shpol'skii, "Fizika v SSSR," 296, and Shpol'skii, "Piat'desiat let," 200 (both cit. n. 17); Iu. A. Khramov, *Nauchnye shkoly v fizike* (Kiev, 1987), 42–61; A. E. Ivanov, "Nauka," in A. N. Sakharov et al., ed., *Rossiia v nachale XX veka* (Moscow, 2002), 676. In a table on p. 59 of his book, Khramov identifies no fewer than twenty-three schools that purportedly evolved from the Lebedev school, which suggests that dilution of the concept may be as much a danger as fragmentation of the initial shared practices.

[33] L. V. Levshin, *Sergei Ivanovich Vavilov*, 2nd rev. ed. (Moscow, 2003), 60–1, berates the two theorists and restores Lebedev to his rightful place in the school pantheon.

[34] Sokolov, *Nauka v sovetskoi Rossii* (cit. n. 25), 59. On Rozhdestvenskii, see A. N. Osinovskii and A. F. Kononkov, *D. S. Rozhdestvenskii* (Moscow, 1974); S. E. Frish and A. I. Stozharov, eds., *Vospominaniia ob akademike D. S. Rozhdestvenskom: k 100-letiiu so dnia rozhdeniia* (Leningrad, 1976).

[35] See the entry of July 4, 1922, in his *Dnevniki: Mart 1921–avgust 1925*, ed. V. P. Volkov (Moscow, 1998), 67.

[36] Gerald L. Geison and Frederic L. Holmes, eds., *Research Schools: Historical Reappraisals*, Osiris 8 (1993); Lutz Danneberg, Wolfgang Höppner, and Ralf Klausnitzer, eds., *Stil, Schule, Disziplin: Analyse und Erprobung von Konzepten wissenschaftsgeschichtlicher Rekonstruktion* (Frankfurt am Main, 2005).

there was anything historically unusual about the Russian context.[37] In the present case, although Landau's students have reflected constructively on the admission criteria and sociology of this most durable of the Soviet schools, they take for granted the narrative of its origins and valuation.[38] Valuable published reflections on the less cohesive schools of Mandelstam and Tamm also betray no interest in historicizing the category.[39] But how did research schools become naturalized so thoroughly in twentieth-century Russia?

To understand the historical appeal of the "school" as institution, we must go back to revolutionary Russia and reflect further on the challenges facing an ambitious young physicist seeking to advance a discipline that had enjoyed only limited state support in the nineteenth century. How would a Russian physicist go about acquiring professional credibility internationally? What social groupings would be most effective professionally in a volatile political setting? How should textbooks be structured so that the mind of the student would be trained (*dressiruet'sia*) more reliably (and by implication, uniformly)? How might national identity enter into this most cosmopolitan of sciences? How would the physicist regard the legacy of imperial Russia? In important respects, the answers to these questions contribute to the evolving collective conception of the "school" in Russia.

In answer to the first question, the biographies of Frenkel, Tamm, Fock, Landau, and George Gamow (1904–68) make it clear that a period of study in central Europe was crucial for a Russian physicist in training. To use contemporary phrasing, these scholars were not possessed by that false sense of national pride that would treat as unnecessary the acquisition of experience in the West. To visit a German-speaking institution and to demonstrate the ease with which one could address problems set by a master physicist, thereby assuming a leadership role among his students—this was a crucial rite of passage for a truly talented Russian. Thus Landau's ability to feed *der Theorienfresser*, Wolfgang Pauli, set him apart in Zurich during extended visits in 1930 and 1931.[40] "Now there is some life in the institute," wrote Pauli's assistant, Rudolf Peierls, and "ardent discussions about everything and if you would come near my room you would hear a dreadful noise, because at any given time of the day there will be some people of different opinion trying to convince each other. All this, naturally, is due to Landau."[41]

What should the talented young physicist do upon returning home? Again drawing on contemporary formulations, it was important that small and fluid groups of students linked by common intellectual interests should have the opportunity to gather outside the usual constraints of official institutions—in a social unit that Russians call the circle, or *kruzhok*. The informal kruzhok could in turn serve as the kernel of

---

[37] S. R. Mikulinskii et al., eds., *Shkoly v nauke* (Moscow, 1977) (cf. partial exception in the essay of B. A. Starostin, which treats the eighteenth and early nineteenth centuries); Khramov, *Nauchnye shkoly* (cit. n. 32); D. Gouzévitch, "Nauchnaia shkola kak forma deiatel'nosti," *Voprosy istorii estestvoznaniia i tekhniki*, 2003, no. 1:64–93.

[38] Kagan, "Shkola Landau" (cit. n. 21); I. M. Khalatnikov, "Kak sozdavalos' shkola Landau" (1980), in Khalatnikov, *Vospomaniia o L. D. Landau* (cit. n. 27), 267–74. Much the same holds true for Khramov (*Nauchnye shkoly*, cit. n. 32), who treats Mandelstam, Landau, and Tamm among the theorists.

[39] B. M. Bolotovskii, "Shkola Tamma," in Feinberg, *Vospomaniia o I. E. Tamme* (cit. n. 26), 42–64; G. E. Gorelik, "Leonid Mandel'shtam i ego shkola," *Vestnik RAN* 74 (2004): 932–9.

[40] Rudolf Peierls to Evgeniia Kanegisser, 28 Jan. 1931, Rudolf Peierls Papers (hereafter cited as RPP), Department of Western Manuscripts, Bodleian Library, Oxford University, Oxford.

[41] Peierls to Genia, 6 Dec. 1930, ibid.

future research schools, insofar as it provided a fluid setting within which to form intellectual alliances and establish hierarchies of taste and talent. Convene a kruzhok in one's own apartment, if necessary, and when the opportunity arises, formalize it institutionally. At these meetings of the kruzhok, "participants exchange thoughts about topical problems of science, read reviews on the latest achievements of physics, [and] acquaint colleagues with the results of their researches." The subjects discussed ranged widely: "The kruzhok participants do not just discuss science at their meetings," writes one biographer, for "burning social questions also agitate them."[42] Although Leningrad may have dominated the academic scene in Landau's student days, Moscow later offered many opportunities for the scientific prodigy to transform his initial circle of students into "the center of an entire school of Russian physicists who have occupied chairs in universities and other higher educational institutions," as a colleague put it.[43]

The youthful reformer would demonstrate a flair for balancing the stubborn empiricism of many of his experimentalist colleagues and the historical dominance of Russian mathematics, achieving a "harmonious blending" of the two in his theoretical work. Undaunted by his own lack of teaching experience, the junior physicist could pursue an interest in textbook writing, with the spotty Russian publishing landscape permitting someone who was "very nearly a debutante" to intervene productively in university pedagogy—working from first principles, as it were. Deductive rather than inductive methods were to be preferred (as any reader of Landau's *Course of Theoretical Physics* would later confirm).

> We think that concision and precision of expression, unity of terminology, the strictest possible system in the assignment of material—these are the qualities needed for a textbook, especially in the exact sciences. These qualities are hardly reconcilable with a "historical" exposition with the aid of excerpts from books of different eras and of a different nature. We propose that the mind of the student is more reliably trained by strictly consistent and precise exposition of an elaborated series of ideas, than by a spotty and superficial historical survey.[44]

Authorship would be fluid, for the teacher's students could help assemble the materials for these textbooks under his strong editorial direction. When introducing the texts into the university curriculum, he would also enforce a strong hierarchy of examination standards by establishing a formal "minimum," a term now firmly associated with Landau.

Finally, how was the reform-minded physicist to sustain the coherence of the kruzhok amid the volatility of contemporary educational politics, given the state's recurrent suspicion of intellectuals? Learn to avoid becoming the creature of any particular patron, and do not treat rectors or institute directors as all-purpose fixers in daily affairs. Let the kruzhok be partisan only in its collective pride at being able to avoid such external entanglements. Treat the members of the kruzhok as an extended family, for whom there is no greater devotion than the life of science. In this combination

---

[42] See Anna Livanova, *Landau*, 2nd. ed. (Moscow, 1983), 23.
[43] See V. B. Berestetskii, "Lev Davidovich Landau," *UFN* 64 (1958): 615–23, on 616.
[44] See the introduction to the English edition in L. D. Landau and E. Lifshitz, *Mechanics*, trans. J. B. Sykes and J. S. Bell (Oxford, 1960), vii.

of imperatives, perhaps, we can see the social and intellectual roots of the Landau school.

If I have employed somewhat stilted characterizations of the historical elements of a nascent Landau school, it is because all of the factors described above rely on close paraphrasing and quotation of sentiments that predate Landau by decades. Although I could make the case that each of these challenges facing the young Russian physicist has direct analogies in the formation and identity of the Landau school, everything I have just recited was originally used to describe the career of the physicist Aleksandr Grigorievich Stoletov (1839–96).[45] And I am not referring simply to Soviet-era characterizations of a heroic Russian precursor molded to contemporary needs, for most of the crucial identifying markers—including the distinctive "minimum"—may be found in the obituary of Stoletov written by his biologist colleague Kliment Timiriazev.[46] Indeed, it is striking how many of Stoletov's colleagues agreed on the adoption of the kruzhok and then the school as crucial markers of his ability to bear European and Russian identities simultaneously.[47] Just as important, in the Soviet period all of these elements became standard tropes in Stoletov biographies.[48]

The point here is not to dilute the distinctiveness of the Landau school but rather to emphasize that the kruzhok provides a great deal of historical continuity in the dynamics of identity formation for the Russian scientist.[49] Most concretely in the case of Stoletov, but in many other manifestations as well, the kruzhok-to-school dynamic of the tsarist era provides many of the historical conditions for legitimation of subsequent Soviet institutions of theoretical physics. During the early Soviet period, the school was widely celebrated and increasingly emulated, and the figure of Pavlov

---

[45] Thus the previous three footnotes point to a later parallel, rather than the original source of the quotation (given below).

[46] K. A. Timiriazev, "Aleksandr Grigor'evich Stoletov," *Russkaia mysl'*, 1896, no. 11:262–80, on 265 (leader among students of master physicist, in this case Gustav Kirchhoff); 266 ("center of an entire school"); 267 (no false sense of national pride); 272 ("harmonious blending"); 273 (textbook authorship); 277 ("minimum").

[47] P. M. Pokrovskii, "Aleksandr Grigor'evich Stoletov," *Universitetskiia izvestiia* (Kiev) 36, 1896, no. 11:1–9; N. N. Shiller, "Kharakteristika lichnosti i nauchnykh trudov pokoinago professora," *Universitetskiia izvestiia* (Kiev) 36, no. 12 (1896): 1–10; A. Sokolov, "Aleksandr Grigr'evich Stoletov," *Zhurnal russkago fiziko-khimicheskago obshchestva* (hereafter cited as *ZhRFKhO*) 29, no. 2 (1897): 25–74, especially, 35–9.

[48] Lazarev, *A. G. Stoletov* (cit. n. 32), 12–3; M. Funder, "Rol' A. G. Stoletova v istorii russkoi fiziki," *Pod znamenem marksizma*, 1940, no. 2:165–80, on 167; V. Bolkhovitinov, *Aleksandr Grigor'evich Stoletov, 1839–1896* (Moscow, 1951), 210–45; A. I. Kompaneets, *Mirovozzrenie A. G. Stoletova* (Moscow, 1956), 34–5; A. A. Glagoleva-Arkad'eva, "Aleksandr Grigor'evich Stoletov," in *Liudi russkoi nauki* (cit. n. 32), 152–9; M. S. Sominskii, *Aleksandr Grigor'evich Stoletov* (Leningrad, 1970), 69–74, 163–6 (all on kruzhok and its institutionalization); Bolkhovitinov, *Aleksandr Grigor'evich Stoletov*, 148–9 ("burning social questions"), 151–2 ("exchange thoughts"), 153 (scientists as family); cf. V. M. Dukov, "Razvitie teorii elektromagnitnogo polia v trudakh russkikh fizikov do opytov Gertsa," *UFN* 49 (1953): 579. Stoletov's statement against historical exposition is quoted in Kompaneets, *Mirovozzrenie A. G. Stoletova*, 47. On the "scientists as family" trope, see also P. P. Lazarev, "O mezhdunarodnykh nauchnykh snosheniiakh," *Nauchnyi rabotnik*, 1926, no. 3:3–10. The kruzhok autonomy trope is found in I. M. Sechenov, *Avtobiograficheskie zapiski* (1904; repr. Moscow, 1952), 226, and is repeated by Lazarev in *A. G. Stoletov*, 26 (cit. n. 32).

[49] Daniel A. Alexandrov, "The Politics of Scientific '*Kruzhok*': Study Circles in Russian Science and Their Transformation in the 1920s," in E. I. Kolchinskii, ed., *Na perelome: Sovetskaia biologiia v 20-30-kh godakh* (St. Petersburg, 1997), 255–67; and Michael Gordin's essay in this volume, "The Heidelberg Circle: German Inflections on the Professionalization of Russian Chemistry in the 1860s." Cf. Barbara Walker, *Maximilian Voloshin and the Russian Literary Circle: Culture and Survival in Revolutionary Times* (Bloomington, Ind., 2005), 1–23.

likely provides the strongest element of continuity in this process.⁵⁰ One of his students emphasized that a common topic and a leader were not sufficient to constitute a school. "The term refers also to a definite body of views and ideas, to a whole complex of special methods of investigation. The latter are transmitted as a sort of scientific tradition, being sometimes handed down through a number of generations of scientific workers."⁵¹ The Soviet milieu in general and the physics community in particular thus offered a fertile environment linking the virtues of the kruzhok to school formation in historical terms.⁵²

Not so long ago, captive schools were still arrayed against free science, where alone "no dictum of authority oppresses."⁵³ As John Servos describes it, for much of the nineteenth and early twentieth century, European scientists most often used the term "school" in a derogatory fashion. It could signal dogma, unthinking emulation, even the stifling of genius. Dominance of a given research school could even impede the progress of a discipline. "Schools might train, but could they educate or liberate?"⁵⁴ Before 1900, the answer in Europe was usually negative. Had one asked Aleksandr Herzen at midcentury—recalling that he had earlier been trained in astronomy at Moscow University—his usage of "school" would likewise have been with this constricting definition.⁵⁵ Yet available evidence strongly suggests that Russian scientists nonetheless began to embrace the positive valence of the term comparatively early, and that this was subsequently taken up wholesale in Soviet usage. If Dmitrii Mendeleev reported from the famous 1860 Karlsruhe Congress about disagreements "among followers of different chemical schools" in the conventional agonistic sense, within a decade he was nominating A. M. Butlerov to a university chair by stressing the link between school formation and original research.⁵⁶ In contrast to German and English students of Gustav Kirchhoff, Stoletov praised his teacher's creation of a school in the 1870s, while by the early 1880s, physiologist I. M. Sechenov was already celebrating school formation in chemistry and zoology.⁵⁷ According to an early Soviet biographer, Sechenov's great service in the last quarter of the nineteenth century had been not so much introducing students to the latest developments in the West

---

⁵⁰ L. A. Andreyev, "The Great Teacher and Master of Science," *Scientific Monthly* 45, no. 2 (1937): 158–71; Y. P. Frolov, *Pavlov and His School: The Theory of Conditioned Reflexes* (London, 1938). On the workforce trained by Pavlov in the first phase of his career, see, especially, Daniel P. Todes, "Pavlov's Physiology Factory," *Isis* 88 (1997): 205–46.

⁵¹ Frolov, *Pavlov and His School* (cit. n. 50), 255.

⁵² A. K. Timiriazev, "Aleksandr Grigor'evich Stoletov—Osnovatel' russkoi fiziki," *UFN* 22 (1939): 369–83, on 370–1. This was the physicist son of K. A. Timiriazev and a thorn in Tamm's side at Moscow University.

⁵³ Emil du Bois-Reymond, *Kulturgeschichte und Naturwissenschaft* (Berlin, 1878).

⁵⁴ John Servos, "Research Schools and Their Histories," *Osiris* 8 (1993): 2–15, on 6.

⁵⁵ A. I. Gertsen, "Diletanty-romantiki," *Sobranie sochinenii*, vol. 3 (Moscow, 1954), 25, 32.

⁵⁶ D. I. Mendeleev, "Khimicheskii kongress v Karlsrue (Pis'mo k A. A. Voskresenskomu)," in *Sankt-Peterburgskie Vedomosti*, no. 238, 2 Nov. 1860; Mendeleev, "Predstavlenie v sovet universiteta" (1868), both in *Sochineniia*, vol. 15 (Moscow, 1949), 165, 295. For Soviet usage, cf. I. A. Kablukov, "Aleksandr Mikhailovich Butlerov—Osnovatel' russkoi khimicheskoi shkoly," *Zhurnal khimicheskoi promyshlennosti* 5 (1928): 903–9, drawing directly on reminiscences of Butlerov's students written in 1887; A. D. Petrov, "Peterburgskaia shkola A. M. Butlerova," in *Istoriia khimicheskikh nauk*, ed. Iu. I. Solov'ev (Moscow, 1961), 197–211; Iu. I. Solov'ev, *Istoriia khimii v Rossii: Nauchnye tsentry i osnovnye napravleniia issledovanii* (Moscow, 1985), 188–201. See also Gordin, "Heidelberg Circle" (cit. n. 49).

⁵⁷ A. G. Stoletov, "G. R. Kirkhgoff," *Priroda*, 1873, no. 2:174–99, on 175; I. M. Sechenov, "The Scientific Activity of the Russian Universities during the Last Twenty-Five Years," *Science* 3 (1883): 756–9, a translation of an article in *Vestnik evropy*.

but his ability to coax them toward independent approaches to scientific problems. This constituted the foundation of the "Russian physiological school." In Russian usage, school formation thus became a sign of scientific progress vis-à-vis western Europe, rather than a marker of restricted outlook.[58] It was a highly functional form of local cosmopolitanism.

In Sechenov's day, one increasingly saw distinctive research methods (*priemy issledovaniia*) characteristic of "scientists of Russian nationality," and there were many domains that had recently acquired their own "Russian schools" (meaning both ethnic Russian scientists and other scientists working on Russian soil). Already before the fin de siècle, schools were becoming constitutive of Russian science.[59] After the 1905 revolution, the kruzhok also enjoyed renewed popularity in a slightly more tolerant university setting.[60] V. A. Steklov (1864–1926), who after 1906 became the dominant mathematician in St. Petersburg with an interest in mathematical physics, is known to have been an active supporter of science kruzhki.[61] While this did nothing for Russian theoretical physics per se, it did ensure comparative neutrality when young physicists bent on theorization sought to avoid the gaze of the senior physics professors by convening their own kruzhok.

By far the most important school precursor for Soviet physicists was Lebedev, best known for his experimental measurement of light pressure in the 1890s. The son of a Moscow merchant, he effectively became Stoletov's successor in Moscow and was among the key faculty presiding at the opening of Moscow University's Physics Institute in 1904. Initially staffed by three people, his laboratory grew to nearly thirty members by the end of the decade. Although Lebedev was known as a rather indifferent lecturer, he showed great skill at reconciling laboratory research and training without compromising the former. He frequently wrote articles in German and encouraged his students to publish abroad. The expansion of his research agendas, however, came along comparatively narrow experimental lines: beyond visible light, he wanted to develop techniques for measuring any ponderomotive forces exerted on solids or gases by other wave phenomena. In that respect, his enlistment of protégés in an increasingly complex division of labor is mainly a story of professionalization, and evidence of a Lebedev school should simply be taken as a symptom of that diverse Pan-European process.[62]

Russian society drew Lebedev into a larger role, however. When the government overreacted to student demonstrations in 1911, the rectors at Moscow University resigned their administrative positions in protest at this heavy-handed disruption of university life. The Ministry of Education then unexpectedly fired them as professors;

---

[58] M. N. Shaternikov, "Biograficheskii ocherk I. M. Sechenova," in I. M. Sechenov, *Izbrannye trudy* (Moscow, 1935), xv, quoted in M. G. Iaroshevskii, "Logika razvitiia nauki i nauchnaia shkola," in Mikulinskii et al., *Shkoly v nauke* (cit. n. 37), 29.

[59] V. I. Modestov, *Russkaia nauka v poslednie dvadtsat' piat' let* (Odessa, 1890), 10. My thanks to Tatiana Khripachenko for help with this source.

[60] Samuel D. Kassow, *Students, Professors, and the State in Tsarist Russia* (Berkeley, Calif., 1989), 302–3.

[61] I. I. Markush, "K voprosu o sozdanii peterburgskoi–leningradskoi shkoly matematicheskoi fiziki V. A. Steklova," *Istoriia i metodologiia estestvennykh nauk* 16 (1974): 141–53, on 144.

[62] P. Forman, J. L. Heilbron, and S. Weart, "Physics around 1900: Personnel, Funding, and Productivity of the Academic Establishments," *Historical Studies in the Physical Sciences* 5 (1975): 1–185; A. M. Korzukhina, *Ot prosveshcheniia k nauke: Fizika v Moskovskom i S.-Peterburgskom universitetakh vo vtoroi polovine XIX v.–nachale XX v.* (Dubna, 2006), 63–86. St. Petersburg University's physics institute also opened around this time.

Lebedev and many other colleagues demanded that they be reinstated. After the government refused to relent, he and about a third of the faculty resigned, crippling the university. This was not an act of political engagement as such, for Lebedev and most of his colleagues had little sympathy for student agendas and were scarcely eager to acknowledge the failure of the liberal tactic of compromise since 1905. They simply wanted the university to be "above politics."[63] That is why Lebedev found it rhetorically convenient to invoke the temporally distant Mikhail Lomonosov as the proxy for his contemporary dilemmas some months after his resignation. The physicist, by now ailing with the heart disease that would soon end his life, published a memorial essay in *Russkie vedomosti* marking the 200th anniversary of Lomonosov's birth.[64] According to Lebedev, Lomonosov had never been dissuaded from his main aim, "educating Russian society and forcing it to reflect upon new ideas that were alien to her." Lebedev believed that those few Russian scientists who had achieved something worthy of note had done so more despite the conditions offered by Russia than because of them. What he called the "Russian scientific school" had continually been forced to struggle for its very existence.[65] Via the school, a political problem could be recast as a cultural one, and divisive state objectives could be displaced on to unifying national ones. Cohort solidarity could benefit science, thus benefiting the common good. But that was a strategy often better suited to the canny "nonparty" Vavilov than to the splenetic "Trotskyite" Landau under high Stalinism. Lebedev's students—several of whom were among the politically (and disciplinarily) cautious colleagues the former Menshevik Tamm found himself despising when he returned to Moscow to teach in 1922—not only celebrated Lebedev's genuine achievements but also benefited indirectly from the canonization of his school during the early Soviet period.[66] Their living witness to "enduring scientific schools" was further proof that Soviet Russia was no longer a "barbarous country with low culture."[67]

Not only did the absence of a school signal the absence of a prerevolutionary history for Soviet theorists. It also gave them license to reinvent the kruzhok in Soviet life.

THE JAZZ BAND PLAYS AT THE REVOLUTION

There actually was a prerevolutionary theoretical physicist in Russia who offered a small measure of continuity between pre- and postwar experience, fostering the

---

[63] Kassow, *Students, Professors* (cit. n. 60), 353–66. An inordinate number of those resigning came from the natural science faculty.

[64] P. N. Lebedev, "Pamiati pervogo russkogo uchenogo 1711–1911 gg.," *Sobranie sochinenii*, ed. T. P. Kravets (Moscow, 1963), 350–8.

[65] Ibid., 357. This was in contrast to the school formed by his revered teacher, August Kundt. Lebedev, "August Kundt" (1894), in ibid., 49–67, on 61–6.

[66] T. P. Kravets, "P. N. Lebedev i sozdannaia im fizicheskaia shkola" (1913), in *Ot N'iutona do Vavilova: Ocherki i vospominania* (Leningrad, 1967), 321–7; Lazarev, *A. G. Stoletov* (cit. n. 32), 38; Vavilov, "Fizika v Rossii" (cit. n. 18), 995; Shpol'skii, "Fizika v SSSR" (cit. n. 17), 296; O. A. Lezhneva, "Issledovaniia po istorii fiziki," in *Razvitie fiziki v SSSR*, vol. 2 (Moscow, 1967), 345; Khramov, *Nauchnye shkoly* (cit. n. 32), 46–61.

[67] Lazarev, "O mezhdunarodnykh nauchnykh snosheniiakh" (cit. n. 48), 4, likewise writing about Lomonosov and the dilemmas of Russia's lone geniuses under the tsars—and the latter-day benefits of the Soviets.

school ideal in his own quirky way: Paul Ehrenfest (1880–1933)—Pavel Sigismundovich to his Russian colleagues.[68] Although he eventually became the successor to H. A. Lorentz's chair at Leiden in 1912, as a Viennese Jew his initial professional prospects had been rather poor. He and his Russian wife, Tatiana Afanasieva, had decided to move to St. Petersburg in 1907 for family reasons, and Ehrenfest scraped by on editorial assignments, winning only one short-term teaching contract at the Polytechnic Institute. It was in Russia that he and Joffe, also Jewish and also facing obstacles to obtaining a permanent position in the capital, became fast friends. Although both men were trained at German-speaking universities, each took his Russian experience as constitutive in forming a scientific identity vis-à-vis an initially ambivalent professional world.[69]

The two men had first met briefly in Munich, where Joffe was studying with Wilhelm Röntgen (1902–6). A major factor in the later tension between Joffe and Landau regarding the professionalization of theory was Joffe's own youthful confidence as an experimentalist who could keep up with the latest results of atomic theory and productively query colleagues such as Ehrenfest. After assuming the theory chair in Munich, Arnold Sommerfeld had been eager to get up to speed with experimental developments in Röntgen's institute, and he determined to spend an hour or so in "training" each day. He latched on to Joffe as his interlocutor, and Joffe proposed not long afterward that they gather at a local café and hold the discussions there in a less formal atmosphere: a Bavarian kruzhok. Joffe and Ernst Wagner then took the initiative, inviting a mixture of theoretical and experimental physicists, crystallographers, and physical chemists to the kruzhok, with Sommerfeld student Peter Debye eventually taking the lead in Sommerfeld's (and later Joffe's) absence.[70]

In St. Petersburg, Joffe and Ehrenfest joined with Rozhdestvenskii to start their own kruzhok on much the same model, having been urged to do so by Tatiana's aunt, who simply assumed that the kruzhok was the natural social milieu for restless young scientific intelligentsia.[71] Mathematicians were always welcome, assuming they were ready to engage Ehrenfest's interests in statistical mechanics, for instance. The kruzhok was formed largely to bypass disapproving university superiors and to discuss the latest developments in atomic physics.[72] When word of the mass resignations in Moscow came to the capital, it was many of these same figures who used their control of the council of the physics division of the Russian Physico-Chemical Society to call for their St. Petersburg colleagues not to become involved. Members of the kruzhok were the most vocal in protesting this minority position, forcing the society to pass a resolution that included the following: "Anyone who holds dear

---

[68] Martin J. Klein, *Paul Ehrenfest: The Making of a Theoretical Physicist* (Amsterdam, 1970); V. Ia. Frenkel', *Paul' Erenfest* (Moscow, 1971).

[69] Ehrenfest to Joffe, 20 Aug. 1910; Joffe to Ehrenfest, 25 Aug. 1910, in *Erenfest-Ioffe nauchnaia perepiska 1907-1933gg.*, 2nd ed., ed. V. Ia. Frenkel' (Leningrad, 1990), 66–7, 270–1.

[70] A. F. Joffe, *Vstrechi s fizikami* (Leningrad, 1983), 34; Wagner to Joffe, 4 Jan. 1908, ibid., 138.

[71] Klein, *Paul Ehrenfest* (cit. n. 68), 85.

[72] By this time, even senior physicist I. I. Borgman (1849–1914), former rector and object of their disdain, had his own formal physics kruzhok for the advanced undergraduates to report on current topics, so there was no special social frisson to the Ehrenfest kruzhok. They were in direct competition to advance more ambitious research agendas for their young peers. Cf. "Fizicheskii kruzhok pri un-te za 1911," *Otchet o sostoianii i deiatel'nosti imperatorskogo S.-Peterburgskogo universiteta za 1911 god* (St. Petersburg, 1912), 217–9.

the development of physics in Russia has followed with a feeling of deep satisfaction as a model school of physicists has grown and developed in Moscow ... The Moscow school represents a viable developing organism" engaged in all the activities necessary to train independent researchers. "We cannot reconcile ourselves to the thought that the oldest Russian university is depriving itself of a school of physicists so exceptional in its significance."[73]

Although this availed the members of the kruzhok nothing, it did seal Joffe's determination to build a research school of his own. Having defended his Russian doctoral dissertation in 1915 (with Steklov as one of the opponents), Joffe lost a bid for a university professorship but finally won a chair at the Polytechnic later that year. The following year, the kruzhok was institutionalized as an interdisciplinary seminar led by Joffe, and an inordinately high percentage of physicists who gained positions of influence in the Soviet era counted this as a rite of passage—Frenkel included. The kruzhok also gave Ehrenfest a chance to further the training of a university student named Yuri Krutkov (1890–1952), who then followed him to Leiden and later published the earliest survey of quantum theory in Russian.[74] Krutkov is generally reckoned the first Russian theoretical physicist, albeit without benefit of the "school" moniker, as the epigraph by Tamm implies.[75] Daniel Alexandrov has rightly observed that this kruzhok served in turn as a model for the proliferation of similar informal institutions during the early Soviet era.[76] Indeed, when Joffe subsequently felt the bite of young Soviet theorists eager to put him in his place, Ehrenfest calmed him with reminders of their own youthful behavior toward senior professors while running the St. Petersburg kruzhok.[77]

War and revolution cut Ehrenfest off from Russia for the better part of six years, but he was drawn back by the chance to interact with young Soviet physicists. After renewing correspondence in 1920, he became the sole foreigner to attend the 1924 meeting of the Russian Association of Physicists.[78] As he told Joffe afterward, "Both in Germany and in Holland I am coming closer and closer to the conclusion that it's infinitely easier for me to make intellectual contact with Russians than with non-Russians."[79] Soviet students took note, swarming the diminutive theorist after his presentation on the correspondence principle and the quantum theory of light, at which he melodramatically declared in his stilted Russian that the battle between wave and corpuscular theories "takes place in the heart of every physicist."[80] That the battle had famously driven Bohr, Hendrik Kramers, and John Slater to speculate that both energy conservation and causality might only be preserved at the statistical level did not discourage Ehrenfest's audience. "In these young people we

[73] M. S. Sominskii, *Abram Fedorovich Ioffe* (Moscow, 1964), 176–8.
[74] Iu. Krutkov, "O teorii kvantov," *ZhRFKhO* 48 (1916): 43–76.
[75] V. Ia. Frenkel, "Iurii Aleksandrovich Krutkov," *UFN* 102 (1970): 639–54.
[76] Alexandrov, "Politics of Scientific '*Kruzhok*'" (cit. n. 49).
[77] Joffe to Ehrenfest, 27 Dec. 1932, *Erenfest-Ioffe nauchnaia perepiska* (cit. n. 69), 298–301; Ehrenfest to Joffe, 21 Dec. 1932, Ehrenfest Scientific Correspondence (hereafter cited as ESC), Museum Boerhaave, Leiden.
[78] Joffe to Ehrenfest, 18 June 1920, and Ehrenfest to Joffe, 6 Sept. 1920, *Erenfest-Ioffe nauchnaia perepiska* (cit. n. 69), 274–5, 138–42; Joffe to NTO VSNKh, before 17 July 1924, in *Organizatsiia nauki v pervye gody Sovetskoi vlasti (1917–1925): Sbornik dokumentov* (Leningrad, 1968), 392; Sominskii, *Abram Fedorovich Ioffe* (cit. n. 73), 398–9.
[79] Ehrenfest to Joffe, 19 Oct. 1924, *Erenfest-Ioffe nauchnaia perepiska* (cit. n. 69), 174.
[80] See the brief resumé in P. S. Ehrenfest, "Teoriia kvantov," *ZhRFKhO* 56 (1924): 449–50.

have the future scientific workers," proclaimed a junior physicist at the scene, "and their attitude toward the purely theoretical problems touched upon at the congress clearly demonstrates that they are interested not only in practice but in genuine *living physical science*." The dialectical appeal of Bohr's correspondence principle offered even more, however, for it suggested to the young Russian that "there is nothing to be concerned about by the contradiction of modern physics, which is undergoing at present in quantum theory and relativity theory a revolution equal in its significance to the world revolution [sic] that took place seven years ago."[81] Some months later, writing about magnetism just before the appearance of Pauli's seminal papers formulating the exclusion principle, Tamm likewise painted modern physics as characterized by "the accumulation of deep internal contradictions in new fruitful theories."[82]

Soviet enthusiasm for the contradictions of modern physics in the early 1920s could not be taken for granted—resources were too scarce. While in the German-speaking universities in 1921 "the interest of the greatest portion of the younger physicists [was] concentrated on either atomic theory or quantum theory," the same could not be said for their Soviet counterparts.[83] The dominance of the mathematicians not only limited the ranks of the physicists in Petrograd and Moscow but also made it more likely that mathematically adept students with a bent for applied problems would choose mechanical engineering over physics.[84] In Moscow, specialization in the exact sciences was regarded as "without prospects" early in the decade, even though the physicists were generally credited by militant students with being "closer to industry, to workers, to the organs of Soviet power, [and] joined us a lot earlier than the mathematicians living in their calculations or the flower-loving botanists."[85] Those few who did opt to study physics lagged considerably behind their counterparts in Petrograd when it came to relativity and quantum theory.[86]

Moscow University appointed its first professor of theoretical physics, S. A. Boguslavskii, in the spring of 1919, although the civil war kept him from assuming his post until 1921. Trained under the Göttingen crystallographer Woldemar Voigt, Boguslavskii was best known for his attempt to give pyroelectricity a basis in quantum

---

[81] D. Galanin, "Osnovnye voprosy sovremennoi fiziki," *Krasnaia molodezh'*, Oct. 1924, no. 3:146–54, on 147, 154 (emphasis in original). D. D. Galanin worked with Tamm on a modest research project in the spring of 1923 at Sverdlov Communist University, f. 5221, op. 4, d. 68, l. 93, State Archive of the Russian Federation (Gosudarstvennyi Arkhiv Rossiiskoi Federatsii, hereafter cited as GARF), Moscow. This is likely the father of an accomplished postwar student of quantum electrodynamics, A. D. Galanin.

[82] I. E. Tamm, "Magnetizm i stroenie atomov," *UFN* 5, nos. 1–2 (1925): 105–37, translated as "Magnetism and the Structure of Atoms," *Soviet Physics Uspekhi* 36 (April 1993): 246.

[83] Hans Thirring, "Ziele und Methoden der theoretischen Physik," *Die Naturwissenschaften* 9 (1921): 1023–8, on 1026.

[84] Although faculty numbers at Moscow University were nearly the same, four times as many students pursued advanced degrees in pure and applied mathematics as in physics. See *Otchet o sostoianii i deistviiakh imperatorskago Moskovskago Universiteta za 1914 god* (Moscow, 1915); *Otchet o sostoianii i deistviiakh Moskovskago Universiteta za 1916 god* (Moscow, 1917). If anything, Petrograd favored mathematics even more.

[85] N. M. Beskin, "Vospominaniia o Moskovskom fizmate nachala 20-kh godov," *Istoriko-matematicheskie issledovaniia* 34 (1993): 164; Semen Rin, "Revoliutsiia na fizmate," *Krasnoe studenchestvo*, 7 Nov. 1927, nos. 4–5:148.

[86] P. S. Kudriavtsev, "Prepodavanie fiziki v moskovskom universitete v 1922–1926 gg.," *IMEN* 17 (1975): 126.

theory.[87] Although Boguslavskii took an active interest in atomic theory,[88] both Tamm and Mandelstam held him in low regard.[89] As it turned out, they were the primary beneficiaries when the unfortunate Boguslavskii took ill and died of tuberculosis in the fall of 1923, leaving the modest "office of theoretical physics with laboratory" in disarray.[90] Ehrenfest was invited to take the chair but declined in deference to Lorentz.[91] Mandelstam was finally lured back to Moscow with an appeal to form a kruzhok, bringing together "people who desire and who are able to work, to put an end to the countless intrigues that have permeated the grounds of the [university's physics] institute."[92] Tamm, who had been surviving since 1922 on modest teaching assignments at Sverdlov Communist University and the second Moscow University (run directly by the Commissariat of Education), as well as consulting for the state electrical power supplier, promptly accepted Mandelstam's invitation to join the staff.[93] With this foothold, he became the leading advocate of modern physics as he remained more politically engaged than the reticent Mandelstam. For Tamm, political conservatism and disciplinary conservatism went hand in hand, which is why he told his wife in 1922 that he was becoming more radical and would keep doing so ("leveiu i budu levet'"), because the "struggle for the sovietization of higher education" demanded no less.[94] Even further to the left, the new party monopolists had their reasons for defending the "theoretical science" of the physicists, indirectly benefiting the modernizers. For it was here that "one can specialize in those problems which have cardinal significance for our worldview and which find application in party propagandistic work (structure of matter, theory of relativity, astronomy)."[95] Theoretical physicists happily exploited this conflation of tasks. Despite dropping any party affiliation soon after October 1917, Tamm retained a reputation as a Bolshevik among the faculty until the late 1920s, when party watchdogs began detecting signs of "wavering."[96]

This level of apparent politicization was not necessarily at odds with professionalization for someone such as Tamm in the early years of Soviet power. Like Frenkel just old enough to have wrestled with the "accursed questions" of the Russian intelligentsia during war and revolution, Tamm cast them aside and turned to what he likewise called the "accursed questions" of fundamental physical theory, generally

---

[87] S. Boguslawski [Boguslavskii], "Pyroelektrizität auf Grund der Quantentheorie," *Physikalische Zeitschrift* 15 (1914): 569–72.

[88] See S. Boguslavskii, "K voprosu o stroenii atomnogo iadra," *ZhRFKhO* 52 (1922): 73–84; Boguslavskii, "O vnutrennem iavlenii Zeemana," ibid., 89–97; Boguslavskii, "K istolkovaniiu uravneniia, opredeliaiushchego chastotu sveta izluchaemogo model'iu atoma Bora," ibid., 85–7; Boguslavskii, "On the Problem of Two Moving Charges in Its Connexions with the Atomic Theory," *Philosophical Magazine* 45 (1923): 145–60.

[89] I. E. Tamm to N. V. Tamm, 11 Nov. 1922, in *Priroda*, July 1995, no. 7:150.

[90] *Otchet I-go Moskovskogo Gosudarstvennogo Universiteta za 1923 g.* (Moscow, 1924), 134–6; D. D. Gulo, A. F. Kononkov, and B. I. Spasskii, "Nauchnye trudy S. A. Boguslavskogo," in *Voprosy istorii fiziko-matematicheskikh nauk*, ed. K. A. Rybnikov et al. (Moscow, 1963), 332–45.

[91] Ehrenfest to Joffe, 6 July 1924, *Erenfest-Ioffe nauchnaia perepiska* (cit. n. 69), 164–5.

[92] G. S. Landsberg to L. I. Mandelstam, 18 June 1924, f. 1622, op. 1, d. 75, l. 1, ARAN.

[93] On Tamm at Sverdlov, see f. 5221, op. 4, d. 83, l. 46, GARF. Tamm and Mandelstam had worked together in Odessa from 1920 to 1922.

[94] Tamm to his wife, 6 Nov. 1922, in Vernskii, "I. E. Tamm v dnevnikakh i pis'makh" (cit. n. 8), 269.

[95] A. Vyshinskii, A. Timiriazev, and V. Egorshin, "Chto takoe fizmat?" *Izvestiia*, 5 June 1927, 7.

[96] I. M. Frank, "Otryvki vospominanii raznykh let," in Feinberg, *Vospominaniia o I. E. Tamme* (cit. n. 26), 259; A. A. Maksimov, "Kharakteristika sostava osnovnykh professorskikh grupp fizicheskogo razdela fizmata," f. 225, op. 1, d. 23, Moscow State University Archive, Moscow.

working hard to ensure that the latter would *not* have to speak to extradisciplinary issues at every turn.[97] Theirs was not the classic October resolution of "internal conflict" for the tsarist-era scientist-intelligent finally given "the opportunity to infuse his scientific work into the construction of a new socialist society."[98] Or rather, there was little perceived conflict to begin with, because the second term of the equation was largely taken for granted among those who reached maturity after 1917. Tamm and his younger peers, while supremely confident in the efficacy of the new quantum theories they came to master in the course of the decade, did not subscribe to the boundary-dissolving cultural logic of the Bolsheviks. To be sure, that they were increasingly drawn into work "on the cultural front" was testimony to the gradual sovietization of intellectual life, but it was more than that.[99] Although suspicious in the early 1920s of the few Communist physicists as "new monopolists" on the left, and contemptuous of the more opportunistic Old Regime physicists on the right, they came to see the politically insecure and often professionally undistinguished "gray" middle as equally worrisome for theory.[100] Especially in the hands of its quantum theoretic flag bearers, the disciplinary question of the place of atomic physics became entangled at times in the recurrent political question of who was to dominate Soviet culture—more ideologically militant "proletarian" intellectuals, or "fellow travelers," intellectuals who were unquestionably "Soviet" but invested in professional values. No other options were available. As fellow travelers, these theorists' challenge was at once intellectual and social in nature: to make quantum physics central to the profession, and to secure an institutional role for theory in the process.

The rejection of generational prerogatives that F. Scott Fitzgerald would identify with the Jazz Age found analogous expression among these young Soviets as they entered Leningrad University in the early 1920s. Several of them formed the so-called Jazz Band as the (decidedly nonmusical) vehicle for this iconoclastic social dynamic. The atmosphere of "militant informality" that they cultivated in this kruzhok has been colorfully described by Matvei Bronshtein's biographers.[101] Centered around "Jonny" (Gamow), "Dimus" (Dmitrii Ivanenko, 1904–94), and "Dau" (Landau), membership in this cohort seems to have been based in equal measure on talent for modern physics and capacity for scathing irony, with no subject exempt from heated debate, summary scorn, or rollicking amusement.[102] As a biographer of Landau depicts it, "[T]hey spared no one, including one another. Salvos of criticism were aimed in all directions, and the more gladly, the more important the 'cause.'"[103]

It was in this spirit that the Jazz Band regularly issued an informal manuscript titled *Physikalische Dummheiten*, devoted to uncovering mistaken calculations and

---

[97] See the youthful reflections reproduced by Tamm's grandson in L. I. Vernskii, "V kabinete i vne ego," Feinberg, *Vospominaniia o I. E. Tamme* (cit. n. 26), 82, 86. Cf. V. Ya. Frenkel, *Yakov Ilich Frenkel: His Work, Life and Letters*, trans. A. S. Silbergleit (Basel, 1996), 20–1.

[98] The reference is to biochemist, tsarist-era radical, and Bolshevik favorite A. N. Bakh (b. 1857), "Kak i kogda ia stal marksistom," *Front nauki i tekhniki*, 1933, no. 2:123–4.

[99] Katerina Clark, "The 'Quiet Revolution' in Soviet Intellectual Life," in *Russia in the Era of NEP: Explorations in Soviet Society and Culture*, ed. S. Fitzpatrick, A. Rabinowitch, and R. Stites (Bloomington, Ind., 1991), 210–30.

[100] I. E. Tamm to N. V. Tamm, [Fall] 1922 and 6 Nov. 1922, reproduced in *Priroda*, July 1995, no. 7:148–9. Cf. Sokolov, *Nauka v sovetskoi Rossii* (cit. n. 25), 19–23.

[101] G. E. Gorelik and V. Ia. Frenkel, *Matvei Petrovich Bronshtein* (Moscow, 1990), 25–9.

[102] Compare Landau's remarks to a Danish student group in March 1931, as related by Hendrik Casimir, *Haphazard Reality: Half a Century of Science* (New York, 1983), 113.

[103] Livanova, *Landau* (cit. n. 42), 23.

misguided assertions in the leading physics journals of the day. Its contents were often announced without warning at the university physics seminar, to the occasional discomfiture of senior physicists in attendance. When Joffe sniffed about "half-mad Schrödingerites who don't understand why the rest of physics and letters other than Ψ still exist," it was members of the Jazz Band he had in mind.[104] There is perhaps no better testimony to the unease this prevalent kruzhok dynamic engendered among cultural authorities than the bookish desire to condemn "kruzhok cliquishness" by definition, so to speak, as "a tendency toward the formation of tight and closed groups within the greater collective or organization, leading to splitting, disorganization, and to isolation from the masses."[105] The cliquishness of these new adepts of quantum mechanics would only be increased by the opportunity to obtain further training abroad.

The international successes of quantum mechanics in the mid-1920s fired revolutionary expectations for the future of theory, which Soviet theorists saw as threatened by the indifference of mediocre fellow physicists who neither properly appreciated its potential nor even acknowledged a viable role for theory.[106] I say "revolutionary" advisedly, because the term carried so much cultural baggage into these disciplinary disputes, yet "revolutionary" was still embraced knowingly by these same theorists. Without intellectual ratification (and occasional resistance) from renowned outsiders such as Einstein and Bohr, however, they would not have made much headway in staking their sometimes convoluted claims to cultural authority at home. The language of crisis and revolution that they invoked was partly an echo of Bohr's institute and the anomalies of the old quantum theory circa 1923, taken over to the problems of relativistic quantum mechanics in the late 1920s. Following Suman Seth, we can regard this as a rhetorical moment in the continual reconfiguration of the international theoretical physics community, with portions seeing the "physics of principles" (e.g., Copenhagen), while others saw the "physics of problems" (e.g., Arnold Sommerfeld's school), with little sense of accompanying crisis.[107] If histories of modern theory in Europe have frequently adopted the crisis language of the former to characterize the processes of conceptual change involved, the background of recurrent political crises in Soviet Russia also hints at how the physics of problems could gain traction in the wake of failed "revolutions" in the physics of principles—not just between groups but within them.[108]

Those few Soviet physicists with precious experience abroad were often in a stronger position to advocate science as culture within the Bolshevik enlightenment

---

[104] Joffe to Ehrenfest, 26 May 1928, *Erenfest-Ioffe nauchnaia perepiska* (cit. n. 69), 281.

[105] S. v. "kruzhkovshchina," in *Malaia sovetskaia entsiklopediia* (1929), as noted in B. F. Egorov, "Russkie kruzhki," *Iz istorii russkoi kul'tury*, vol. 5 (Moscow, 1996), 504–17, on 513.

[106] This was not a uniquely Soviet problem as there was no one "natural" trajectory for the institutionalization of (quantum) theory. In the closest parallel, French physics was also slow to create chairs for theoretical physics, as described by Dominique Pestre, *Physique et physiciens en France, 1918–1940* (Montreux, 1984), chap. 4.

[107] Suman Seth, "Crisis and the Construction of Modern Theoretical Physics," *British Journal for the History of Science* 40 (2007): 25–51.

[108] Much this sort of piecewise resolution may be seen in quantum electrodynamics in the late 1940s, as described in S. S. Schweber, *QED and the Men Who Made It* (Princeton, N.J. 1994). For an explicit focus on problem solving as collective process, see also David Kaiser, *Drawing Theories Apart: The Dispersion of Feynman Diagrams in Postwar Physics* (Chicago, 2005); David Kaiser, Kenji Ito, and Karl Hall, "Spreading the Tools of Theory: Feynman Diagrams in the United States, Japan, and the Soviet Union," *Social Studies of Science* 34 (2004): 879–922.

project, countering or reappropriating the stronger utilitarian strain in Bolshevik political economy. They could do so in part because so many of them shared with party leaders a picture of Russia's "backwardness" that made the approach of socialism hinge upon wresting the fruits of (Western) science from the hands of "civilized barbarians" and placing science "on the mantel of cultural rebirth."[109] Once the Soviet state had begun investing considerable sums in these domains, there was no more treasured endorsement among the Bolsheviks than, for instance, a German scholar's comment that "all competent critics testify that the new Russia stands behind no other civilized land [*Kulturland*]" in the natural sciences and medicine.[110] In such an environment, the rare young physicist such as Gamow who "bore the banner of Soviet science abroad in honorable fashion" had every reason to expect that he would attract "the attention of the cultural world" at home.[111] As the Soviet Union undertook a massive industrialization effort in the late 1920s and early 1930s, even so abstract an endeavor as nuclear physics could become the focus of Soviet physicists' bids for public esteem and state support—in the militaristic language of the day, they proclaimed the "storming of the atom."[112]

Although Soviet theorists made no crucial contributions to early quantum mechanics (1925–27), once Gamow made the first successful application of the theory to the atomic nucleus in 1928, it did not take long for his achievement to become a Soviet virtue.[113] For in his work, theory had "decisively broken with stubborn conservatism and classicism no matter what."[114] Gamow was even heralded in *Pravda* for raising the profile of Soviet science.[115] He wasted no time capitalizing on his newfound celebrity. Just as he was preparing to submit his paper, he appealed to Bohr (whom he had not met) for an invitation to Copenhagen, an invitation soon forthcoming.[116] Glad to come in closer contact "with young Russian science," Bohr enthused to Ehrenfest that Gamow "has seized on the problem of nuclear structure with the highest

---

[109] L. B. Kamenev, "Rabochaia revoliutsiia i nauka," *Narodnoe Prosveshchenie*, 1925, no. 9:23–5 ("civilized barbarians" was adopted from an unidentified Russian scientist); A. I. Rykov is quoted in I. Brusilovskii, "Nauchnyi front," *Krasnaia gazeta*, 22 March 1928, 2. For Lenin's obsession with this topic, see, e.g., the compilation *V. I. Lenin o kul'ture* (Moscow, 1980). On American debates about Russian backwardness, see David C. Engerman, *Modernization from the Other Shore: American Intellectuals and the Romance of Russian Development* (Cambridge, Mass., 2003).

[110] Eduard Meyer, "Eindrücke von der Jubiläumsfeier der russischen Akademie der Wissenschaften," *Deutsche Rundschau* 52, no. 205 (1925): 101–18, on 112.

[111] V. E. L'vov, "Mirovoe otkrytie sovetskogo uchenogo," *Krasnaia gazeta*, 17 May 1929, 4.

[112] Fizik [Physicist], "Shturm atomov," *Krasnaia gazeta*, 28 Feb. 1928, 5. Certain details suggest the story was planted by Joffe. Frenkel later repeated the trope: "We are now on the eve of the last decisive storming of the problem of matter, and in all likelihood we will soon be able to celebrate final victory." Ia. I. Frenkel, "Proiskhozhdenie i razvitie volnovoi mekhaniki," *Priroda*, 1930, no. 1:3–32, on 32.

[113] George Gamow, "Zur Quantentheorie des Atomkernes," *Zeitschrift für Physik* 51 (1928): 204–12; Ronald W. Gurney and Edward U. Condon, "Wave Mechanics and Radioactive Disintegration," *Nature* 122 (1928): 439; Roger H. Steuwer, "Gamow's Theory of Alpha-Decay," in *The Kaleidoscope of Science*, ed. Edna Ullmann-Margalit (Dordrecht, 1986), 147–86.

[114] Vavilov, "Fizika v Rossii" (cit. n.18), 1004. The official history of Soviet theory prepared for the fiftieth anniversary of the revolution—whose editorial board included Tamm, M. A. Leontovich, and Landau student A. A. Abrikosov—simply bypassed the modest first decade of "revolutionary" theory by starting with quantum field theory, at which point V. A. Fock's contributions became indispensable. See *Razvitie fiziki v SSSR*, vol. 1 (Moscow, 1967).

[115] Demian Bednyi, "Do atomov dobralis'," *Pravda*, 28 Nov. 1928; cf. George Gamow, *My World Line: An Informal Autobiography* (New York, 1970), 74–5.

[116] Gamow to Bohr, 28 July 1928, Bohr Scientific Correspondence (hereafter cited as BSC), Niels Bohr Archive, Copenhagen.

imaginative capacity."[117] Within a short time, Gamow found his way into Bohr's inner circle.[118]

Gamow's time in Copenhagen and Cambridge opened the way for other visitors, including Landau. On Frenkel's recommendation, the Commissariat of Education dispatched Landau to Germany in the fall of 1929. By the following spring, Landau had obtained funding from the Rockefeller Foundation to extend his stay in Europe for another year, thus providing him with the opportunity to search out all that was new and noteworthy in Western physics.[119] In short order, he visited Berlin and Leipzig but finally settled in Zurich under the watchful eye of Wolfgang Pauli. In certain respects, Pauli, at an earlier age, had been no less brash toward his elders than Landau had toward his and thus was unlikely to be cowed or offended by his visitor's antics. The cozy confines of the Eidgenössische Technische Hochschule indeed did little to dampen the revolutionary sensibilities of a twenty-one-year-old scholar given to "very extreme views about everything, not only physics."[120] Mildly flattered by the Swiss government's unwillingness to extend his visa for more than a brief stay, Landau noted with some amusement that a generation earlier his compatriot Lenin had been judged less potentially troublesome.[121]

Landau sought notoriety of another sort, however. The frisson of student nonconformity, combined with professional approbation for his first physics papers on quantum mechanics while still at the university, instilled in him a fierce determination to fashion a distinct persona for himself among the "bourgeois" European physicists whom he simultaneously revered and despised.[122] His new friend Peierls (Pauli's assistant) came to see in Landau a helpful "litmus test" for drawing out the true sentiments of everyone he encountered.[123] The twenty-three-year-old Berliner and son of an Allgemeine Elektrizitäts-Gesellschaft managing director marveled at Landau's capacity to shock bourgeois people. It was not simply that he spouted revolutionary views—that was not uncommon among European intellectuals.

> But he discusses not at all with reasons; he simply puts against them a system which is as closed and "obvious" as theirs. He does not try to be objective but is as subjective as they are, only with different prejudices and this amazes them awfully much. It is really an interesting experiment to notice the effect he makes on different people . . . If somebody is not shocked, he is surely in a very safe [psychological] position.[124]

---

[117] Bohr to Ehrenfest, 28 Oct. 1928, BSC.
[118] See the contemporary letter of Nevill Mott to his mother, reproduced in Nevill Mott, *A Life in Science* (London, 1986), 28.
[119] Landau to Peierls, 17 May 1930, RPP; V. I. Frenkel' and P. Josephson, "Sovetskie fiziki–stipendiaty Rokfellerovsko go fonda," UFN 160, no. 11 (1990): 103–43, on 130.
[120] Rudolf Peierls interview by J. L. Heilbron, 18 June 1963, AHQP, Niels Bohr Library, American Institute of Physics, College Park, Maryland.
[121] Rudolf Peierls, *Bird of Passage* (Princeton, N.J. 1984), 49.
[122] In Copenhagen, Landau's dogmatic pronouncements on all topics were soon lampooned; see Heisenberg to Pauli, 12 March 1931, in Wolfgang Pauli, *Wissenschaftlicher Briefwechsel*, vol. 2, ed. K. von Meyenn (Berlin, 1985), 66. The unusually informal social atmosphere of Bohr's institute made it perhaps the only place in central European academia where scientists such as Landau and Gamow could give free rein to these unconventional behaviors. See Finn Aaserud, *Redirecting Science: Niels Bohr, Philanthropy, and the Rise of Nuclear Physics* (Cambridge, UK, 1990), 6–15, for a colorful evocation of the Copenhagen setting.
[123] Peierls to Genia, 6 Jan. 1931, RPP.
[124] Peierls to Genia, 21 Dec. 1930, RPP.

These antics extended to the physics colloquium, where Landau summarily condemned Edward Milne's cosmological "kinematic relativity" as "horrific nonsense," injecting his favorite epithet "pathological" in every phrase, again discomfiting the "bigwigs" in attendance.[125]

Convinced that proponents of the existing theory had not yet provided the answer, Landau kept asking, "[W]*ho* is going to make quantum electrodynamics?"[126] During a fruitful year of intellectual and geographic peregrination, Landau finally seized upon a conceptual conflict that seemed to suit his purposes. The end result was a contentious paper with Peierls that proclaimed limits upon the applicable domain of quantum mechanics and, while not actually ushering in the new quantum electrodynamics, still celebrated the failure of the "old" theory.[127] Landau and Peierls's hasty assertions that there would be "no physical quantities and no measurements in the sense of wave mechanics" in the future relativistic theory provoked Bohr and Léon Rosenfeld to produce in response a seminal paper clarifying the problem of measurement in quantum electrodynamics.[128] Although it did sow the seeds for many future anecdotes about him in the physics community, this dispute did not initiate the conceptual rupture Landau had sought. On the contrary, it helped put measurement on an even firmer footing within the prevailing formalism. It is not my purpose here to give a proper account of the "triumph" of Bohr's views on field measurement, and the Bohr-Rosenfeld paper will consequently get short shrift.[129] Yet there were profound temperamental differences between Bohr and Landau in their approaches to physics, differences whose consequences for the practice of theoretical physics are largely obscured in standard accounts of how Bohr and Rosenfeld disposed of the criticisms of Landau and Peierls.[130]

The restless Landau had already left for Leipzig by the time the final draft was officially submitted for publication, but he found the city's denizens too "boring" to justify lingering there for long.[131] An element of social insecurity may have been

---

[125] Peierls to Genia, 11 Feb. 1931, RPP.

[126] H. Casimir interview by Thomas S. Kuhn, 5 July 1963, AHQP. Léon Rosenfeld was in attendance and injected this recollection.

[127] L. D. Landau and R. Peierls, "Erweiterung des Unbestimmtheitsprinzips für die relativistische Quantentheorie," *Zeitschrift für Physik* 69 (1931): 56–69.

[128] Niels Bohr and Léon Rosenfeld, "Zur Frage der Messbarkeit der elektromagnetischen Feldgrössen," *Konigelige danske videnskabernes selskab, mathematisk-fysiske meddelelser* 12, no. 8 (1933); English translation by Aage Petersen in R. S. Cohen and J. J. Stachel, *Selected Papers of Léon Rosenfeld* (Boston, 1979), 357–400.

[129] For authoritative treatments of the Bohr-Rosenfeld paper, see Olivier Darrigol, "Cohérence et complétude de la mécanique quantique: L'exemple de 'Bohr-Rosenfeld,'" *Revue d'histoire des sciences* 44, no. 2 (1991): 137–79; Jørgen Kalckar, introduction to Niels Bohr, *Collected Works*, vol. 7 (Amsterdam, 1996), 3–40; Theodore R. Talbot, "Bohr and Rosenfeld's Foundations for Quantum Electrodynamics" (PhD diss., Columbia Univ., 1978).

[130] Léon Rosenfeld, "On Quantum Electrodynamics," in *Niels Bohr and the Development of Physics*, ed. Wolfgang Pauli (New York, 1955), 70–95; Rosenfeld, "Niels Bohr in the Thirties," in *Niels Bohr: His Life and Work as Seen by His Friends and Colleagues*, ed. S. Rozental (Amsterdam, 1967), 125–7; Jørgen Kalckar, "Measurability Problems in the Quantum Theory of Fields," in *Foundations of Quantum Mechanics* (Proceedings of the International School of Physics Enrico Fermi), ed. B. d'Espagnat (New York, 1971), 127–69; Max Jammer, *The Philosophy of Quantum Mechanics* (New York, 1974), 142–5; Arthur I. Miller, "Measurement Problems in Quantum Field Theory in the 1930's," in *Sixty-Two Years of Uncertainty*, ed. Arthur I. Miller (New York, 1990), 139–52; Abraham Pais, *Niels Bohr's Times, in Physics, Philosophy, and Polity* (Oxford, 1991), 358–64.

[131] Landau to Peierls, 6 Feb. 1930, RPP.

involved. When Peierls visited Leipzig not long afterward, he soon picked up on Heisenberg's drive to cultivate a school in the Sommerfeld mold, noting his ability to farm out topics in productive fashion, impressed at his competitiveness in sports—perhaps the off-putting aspect for the unathletic Landau—and quite conscious that Heisenberg "does all these things not like a professor but like a boy [*mal'chik*]."[132] What eluded the peripatetic Soviet, however, was a sense of where next to focus his energies, despite repeated epistolary prompting from Peierls. With the commencement of his Rockefeller fellowship, Landau at last made his way to Cambridge, hoping for inspiration from Paul Dirac. At Cambridge, however, he was disturbed to find the Englishman deeply immersed in an attempt to formulate a unified theory of particles based on proton-electron annihilation, a theory then hampered by puzzling negative-energy electron states.[133] "Not that he is the only one who seems to me to hope to reach quantum electrodynamics along old pathways," Landau informed Peierls, "but something is amiss with him. I have the very strong feeling that it has to do with serious symptoms of contamination (perhaps better named Einstein Sickness)."[134] The young Soviet clearly entertained no desire to pursue the "dream of philosophers" that Dirac invoked in his search for aesthetic unity in physical theory.

It was Peierls, well versed in observability problems from his student days in Munich, who at last persuaded Landau to join the search for conceptual breakdown in earnest.[135] "With your light quantum clarifications I believe I am now fully in agreement," Landau informed Peierls at summer's end. "I would only formulate them somewhat differently." Actually reaching accord on these formulations would occupy the better part of six months once their colleagues got wind of the project, but Landau was finally convinced that publication was necessary to correct previous mistaken notions concerning the interaction of photons with the electromagnetic field. As he told Peierls, "These matters also seem to me independent in a certain sense from the tragedies of contemporary quantum electrodynamics, and must for example come out of any future electrodynamics where $m \to \infty$."[136]

It was upon his return to Copenhagen for a lengthier stay beginning in September 1930 that Landau determined to make his way to center stage at the Institut for Teoretisk Fysik, immersing himself in study of the Danish language and quickly picking up the equivalents of "inferior" and "philistine," the better to express his contempt for the bourgeois establishment.[137] Bohr's young assistant, Hendrik Casimir, concisely conveyed the impression made by the visiting Soviet: "Landau, a young Russian, is terribly clever. Enormously disagreeable but not unamusing. Bohr is also very fond of him."[138] In this last, Landau was fortunate as he had shown very little regard for Bohr's own current research concerns, while Bohr in turn was not much taken with

---

[132] Peierls to Genia, 26 July 1931, RPP. In the German-speaking world, *Knabenphysik* had become a common designation for the early days of quantum mechanics in Göttingen.

[133] Helge Kragh, *Dirac: A Scientific Biography* (Cambridge, UK, 1990), 94–105; Schweber, *QED* (cit. n. 108), 56–69.

[134] Landau to Peierls, 9 Aug. 1930, RPP.

[135] E.g., Peierls gave a report on "Neuere Arbeiten zur physikalischen Deutung der Quantenmechanik (prinzipielle Unschärfe der Beobachtung)" on February 3, 1928, as cited in Pauli, *Wissenschaftlicher Briefwechsel* (cit. n. 122), 11n9.

[136] Landau to Peierls, 9 Aug. 1930, RPP.

[137] Casimir, *Haphazard Reality* (cit. n. 102), 105. On Landau's aggressive behavior, see also the interview of Christian Møller by Thomas S. Kuhn, 29 July 1963, AHQP.

[138] Casimir to S. Goudsmit, 27 Sept. 1930, Goudsmit Correspondence, AHQP.

"narrow" applications of quantum mechanics such as diamagnetism (where Landau made his first breakthrough). Upon arriving in Copenhagen Landau immediately asked, "What's Bohr doing? What's Bohr doing?" When told that Bohr was discussing examples of complementarity, Landau promptly exclaimed, "Oh, yes, but that's not physics."[139] What did count as physics for both of them could be found in the formal content of the uncertainty relations. The interpretation of their proper application to the electromagnetic field struck both men as central to determining how much of the quantum mechanical formalism would survive in a future theory incorporating relativistic processes.

Landau aimed to call the entire formalism into question, to the consternation of Bohr and the edifying distraction of his peers. Casimir, a delighted witness to many of these protracted exchanges, exclaimed that "a Bohr-Landau discussion on any subject in or outside physics is always a splendid spectacle."[140]

According to Bohr and Rosenfeld, it turned out that Landau and Peierls had neglected to explain the connection between their uncertainty relations and the commutation relations for the electromagnetic fields. And quite apart from the woes of quantum theory, they had not taken into account the statistical fluctuations in a classical field measured over finite distances. Bohr and Rosenfeld went back and carefully defined all the field quantities in operational terms, demonstrating that the errors in momentum measurement could be just balanced out by a combination of extended test charges and field fluctuations. They concluded that the classical field of the macroscopic test body does not impair the microscopic field measurements. It is, instead, "an essential feature of the ultimate adaptation of quantum field theory to the measurability problem."

The irony is that the close attention to *Gedanken* procedures that enabled Bohr and Rosenfeld to make the field-theoretic concepts operationally meaningful also took them further from experimentally realizable configurations. Their extended test charges occupied the entire measurement cell, and consistency notwithstanding, there was little in their formulation to guide the working experimentalist. Years later, Peierls, while deferring to the fundamental correctness of the Bohr-Rosenfeld results, still claimed that when one looked closely at the demands they were making upon the actual experimental configuration, "the resulting operations look quite unlike any kind of measurement that an experimentalist would design."[141] It is safe to say that Landau shared this view and had every reason to foster relations with experimentalists premised on the utility of theoretical insights for realizable experiments in the laboratory.

It is certainly not the case that Landau was ever inclined to repudiate his and Peierls's work after the appearance of the Bohr-Rosenfeld paper. On the contrary, at an international conference in Kiev nearly three decades later, he still alluded to it proudly as a groundbreaking work.[142] When Bohr visited the Soviet Union for the

---

[139] Casimir interview (cit. n. 126). The characterization of Bohr's concerns at the time is Casimir's, while Landau's response is recalled by Rosenfeld, who sat in on the interview.

[140] Casimir to Ehrenfest, 15 Dec. 1930, ESC.

[141] R. E. Peierls, "Field Theory Since Maxwell," in C. Domb, ed., *Clerk Maxwell and Modern Science* (London, 1963), 36.

[142] L. D. Landau, "On Analytical Properties of Vertex Parts in Quantum Field Theory," *Ninth International Annual Conference on High Energy Physics*, vol. 2 (Moscow, 1960), 97. Cf. Landau, "Fundamental Problems," *Collected Papers of L. D. Landau* (New York, 1965), 801.

last time in 1961, Landau modestly confessed before flocks of admiring students that it was Bohr himself who had taught him to understand the uncertainty principle in quantum mechanics.[143] Whatever the truth of that assessment, there is no evidence Landau ever acknowledged the Bohr-Rosenfeld paper publicly, and even his students were occasionally puzzled to note that he never broached the subject of measurement.[144] Throughout his career he consistently maintained a position first articulated in the official report he gave on his travels in June 1931:

> My scientific work developed in several different directions during this period. A crucial problem of modern theoretical physics occupied an essential locus in this . . . the problem of the unification into a single whole of the two most general modern theories: the principle of relativity and quantum theory. This problem has led to immense complications, rendering its precise resolution an issue only for the extremely distant future.[145]

The pragmatic solutions to the dilemmas of quantum electrodynamics achieved in the 1940s never entirely satisfied Landau, for whom the "extremely distant future" turned out to extend well beyond his own career. There would be no revolution.

MATTERS OF STATE, AND SOME POPULAR INTEREST

Among Soviet theoretical physicists in the interwar period, it was the Dane Niels Bohr who served as the most important reference point for these cross-cultural encounters, and his influence reverberated more broadly in Soviet intellectual culture. As with Ehrenfest, it was of no small importance that Bohr did not hail from France, Germany, or Great Britain as that made it a good deal more difficult for philosopher-courtiers to project Soviet political rivalries on to Denmark's most famous scientist. Although he was at the center of numerous Soviet disputes about the Copenhagen interpretation of quantum mechanics, Bohr was exceedingly difficult to cast as a puppet of larger foreign powers, and that lent a unique value to his presence in Russia.[146] I would suggest that the ways in which he was seen to be seeking understanding of the Soviet Union in the context of interwar European politics were linked with Soviet theoretical physicists' attempts to stabilize their own roles within the Soviet system of science.

When a world-renowned physicist such as Bohr joined "the best minds of humanity" by bearing witness to the Soviet experiment, it was more than a mere propaganda coup for party-state officials.[147] It was also an important contribution to the efforts of Soviet theoretical physicists to advance concrete research agendas, both within a heterogeneous physics community and within this broader Soviet discourse of civilization and backwardness.

When Tamm first met Bohr at a conference (in Leiden), the Dane proved to be an enigmatic exception to the younger Russian's growing peer confidence; this per-

---

[143] Livanova, *Landau* (cit. n. 42), 31.

[144] A student who knew Landau from 1951 onward testifies that he never heard his teacher discuss the foundations of quantum mechanics. I. E. Dzialoshinskii, "Landau glazami uchenika," in Khalatnikov, *Vospominaniia o L. D. Landau* (cit. n. 27), 119–20.

[145] V. I. Frenkel' and P. Josephson, "Sovetskie fiziki–stipendiaty Rokfellerovskogo fonda," *UFN* 160, no. 11 (1911): 103-43, on 131.

[146] On Soviet philosophical disputes about quantum mechanics, see Loren R. Graham, *Science, Philosophy, and Human Behavior in the Soviet Union* (New York, 1987).

[147] A. F. Joffe, "K priezdu Nil'sa Bora v Sovetskii Soiuz," *Izvestiia*, 4 May 1934, 6.

ception of Bohr was almost as visceral as it was intellectual. Tamm confided to his wife that Bohr "makes a completely overwhelming impression—the heavy face, as if hewn from a rock, the deep-set, I would say, not entirely normal eyes." For the Soviet theorist still looking to make his mark, and not much given to philosophical reflection about modern physics at a moment when Soviet philosophers would happily have celebrated Bohr's own predilections in a Soviet physicist, Bohr provided a sometimes problematic but also invaluable mediating role for Tamm and his colleagues at home. In Tamm's eyes, the earnest Bohr fostered the impression that "it's not easy bearing the burden of genius."[148] This was not the institution builder described by Finn Aaserud,[149] but rather the sage presiding over the Copenhagen "Mecca" of theoretical physics (Tamm's term).[150] Quite apart from the actual intellectual stimulus that Bohr's obsession with the foundational concepts of physics provided to his colleagues, I would suggest that the public image of the barely comprehensible genius provided a useful foil for Soviet theorists bent on portraying themselves as tough-minded pragmatists by comparison. This is despite the fact that their "pragmatism" made sense only if construed in fairly narrow disciplinary terms and that in the near term their pragmatism was making little direct contribution to the economic aspects of socialist construction.

Bohr had been a foreign member of the Soviet Academy of Sciences since 1924 and the object of repeated invitations to visit.[151] In 1934, Bohr finally made his first trip to the Soviet Union in the company of his assistant, Léon Rosenfeld. Joffe provided fulsome greetings in the pages of *Izvestiia*, proclaiming the arrival of the top theorist a joyful occasion for Bohr to witness the enthusiasm and heroic efforts of the proletariat to build a happy life on earth.[152] At his initial destination, in Leningrad, Bohr witnessed the May Day celebrations and pronounced himself impressed by the martial splendor. As tanks rumbled past, a local reporter wrote in rapture that "the orchestra can no longer be heard. The square is filled with the music of metal in motion."[153] Such panegyrics to technological accomplishment were nonetheless rather incidental to the Danish physicist, who proceeded to Moscow and then on to Kharkov, where he was slated to attend a conference hosted by Landau.[154] While the substantive lure of the visit for Bohr may have been the chance to argue with the combative Landau, the urban milieu of the Kharkov physics institute made no less an impression on foreign physicists.[155]

The British science journalist J. G. Crowther reported that the public aspects of the conference were a success, including a lecture by Bohr on causality, which was

---

[148] Tamm to his wife, 11 March 1928, *Priroda,* July 1995, no. 7:158.

[149] Aaserud, *Redirecting Science* (cit. n. 122).

[150] I. E. Tamm to N. V. Tamm, 12 June 1931, papers held by the Tamm Family, Moscow.

[151] P. P. Lazarev et al., "Zapiska ob uchenykh trudakh Nil'sa Bora (Niels Bohr)," *Izvestiia rossiiskoi akademii nauk* (1924): 458–9.

[152] Joffe, "K priezdu Nil'sa Bora" (cit. n. 147).

[153] *Leningradskaia Pravda,* 4 May 1934, quoted in V. Ia. Frenkel, "Nil's Bor i sovetskie fiziki," in *Nil's Bor i nauka XX veka: Sbornik nauchnykh trudov* (Kiev, 1988), 20–5, on 21.

[154] On Bohr's visit and the impression it made on his hosts, see also J. G. Crowther, *Soviet Science* (London, 1936), 112–25. Further sources may be found in the exhaustive study by Peder J. E. Kragh, "Niels Bohr and the Soviet Union between the Two World Wars: Resources at the Niels Bohr Archive" (master's thesis, Copenhagen Univ., 2003).

[155] In an impromptu *Baedeker* guide for the trip, the orthographically challenged George Gamow noted, "If you are not expetially interested in facturies nothing to beee seen in this place except of Landau." Gamow to the Bohrs, 14 April 1934, Niels Bohr Archive, Copenhagen.

Figure 1. Conference on theoretical physics in Kharkov, May 1934. From left to right: D. D. Ivanenko, L. Tisza (partially obscured), L. Rosenfeld, (figure obscured), Iu. B. Rumer, N. Bohr, J. G. Crowther, L. D. Landau, M. Plesset, Ia. I. Frenkel, I. Waller, E. J. Williams, W. Gordon, V. A. Fock, I. E. Tamm. Landau and Ivanenko purposely wore more casual attire to establish their disdain for conventional social mores and to emphasize generational distinctions. (Photo courtesy of the Niels Bohr Archive, Copenhagen.)

attended by top functionaries such as the Ukrainian commissar of enlightenment (a chemist by training). This "was not regarded merely as a meeting of persons engaged in researches of no interest to others," claimed Crowther. "The arrangements showed that theoretical physics, however recondite, was considered to be a matter of State, and some popular interest."[156] One of the participants obligingly reported on the proceedings in the pages of *Nature*. Although he dwelt primarily on the technical substance of the conference, he also remarked on the "youthful enthusiasm" of the staff and expressed admiration for the Dzerzhinskii School for Orphans (run by the notorious State Political Directorate, or GPU), which (among other things) provided its charges with vocational training for the manufacture of Leica lenses.[157]

Bohr's experience also points to the need to incorporate the Soviet side of these interactions. Tamm, a passionate socialist, spent much time cloistered with Bohr during the 1934 journey, and they by no means limited their conversations to physics. For Tamm, this encounter marked a retreat from the factionalist instincts honed in the politics of his youth, since he had been inclined to follow the Soviet government's position that all the states of Europe were effectively arrayed in opposition to Soviet

---

[156] Crowther, *Soviet Science*, 119 (cit. n. 154).
[157] "International Congress on Theoretical Physics at Kharkov," *Nature* 134 (21 July 1934): 109–10. The author may well have been Crowther, but judging from the technical content, Welsh physicist E. J. Williams seems the more likely candidate.

Communism, even if some of them proclaimed socialist platforms. But as Tamm reported to Paul Dirac in his slightly stilted English, "I mentioned to Bohr occasionally the discussion we have had with you last autumn about Bohr's theory of the absence of any intellectual and psychological difference between nations; he argued on the point for about an hour and, as you did rightly foretell, totally convinced me."[158] Bohr was arguing for alliances in opposition to Fascism, but it is interesting that the Danish scientist had to work so hard to persuade his Soviet counterpart to assess the political subjects of the Soviet experiment in universal terms. If it was the case that "under the unifying bonds of science [diverse European scholars] found themselves in complete harmony," there were those among Tamm's contemporaries who realized that this commonplace could not be taken for granted.[159]

Bohr visited the Soviet Union again in 1937 and in 1961. This is not the occasion to delve into the personal relations between Landau and Bohr, but I would argue that their considerable mutual affection and respect had very little to do with compatible temperaments or intellectual styles and a great deal to do with Bohr's peculiar status as an approving foreigner. For the young Landau, these encounters were crucial in the formation of his personal and professional identity as a Soviet. For public purposes, Landau adopted Bohr (and to a lesser extent, Wolfgang Pauli) as his teachers but not from any great urge to emulate Bohr's concern for foundational questions. Landau wanted instead to disassociate himself conspicuously from his onetime Soviet mentors, especially Frenkel. Conveniently, neither the Dane nor the Swiss stemmed from one of the great powers of European politics in the 1930s, which meant that Landau could identify himself as a disciple without readily falling prey to increasingly patriotic talk at home about "national" schools in the sciences. And at the same time, he could work assiduously to build his own distinctive school, very much in keeping with Soviet expectations of cultural autarky.[160] Landau's close friend Iurii Rumer painted the Soviets as "inheritors of the culture of the dying classes [i.e., of Western capitalism]," obliged to proclaim the "tremendous revolutionary teaching" of twentieth-century physics.[161] As the prospect of conceptual revolution receded in the 1930s, however, "purely practical" theory came to thrive in the nascent Landau school. Schools provided the social mechanism to sustain "local" disciplinary affinities with theorists like Bohr in a heterogeneous global physics community. At the same time they fostered specific collective relations among Soviet compatriots that kept theory credible as Landau retreated from Bohr-style crisis talk to develop a training regimen more in keeping with the Sommerfeld-style physics of problems.

Bohr's own desire to explore the philosophical implications of modern physics well beyond the realm of quantum mechanics also had domestic utility for Landau, who had social as well as epistemological reasons for maintaining firmer disciplinary boundaries. That is why he found it expedient at a moment of institutional crisis in 1935 to include a brief attack on Bohr in a polemic for *Izvestiia*, chiding him

---

[158] Tamm to Dirac, 13 May 1934 (with minor spelling corrections), Dirac Scientific Correspondence, Florida State University, Tallahassee, or Churchill College, Cambridge, UK. Tamm repeated the anecdote in his obituary of Bohr, "Nil's Bor—Velikii fizik XX veka," 80 *UFN* (1963): 191–5.

[159] F. A. Paneth, "The Mendeléeff Centenary and Scientific Progress in the U.S.S.R.," *Nature* 134 (24 Nov. 1934): 799–801.

[160] On Landau's pedagogical efforts, see Karl Hall, "'Think Less about Foundations': A Short Course on the *Course of Theoretical Physics* of Landau and Lifshitz," in *Training Scientists, Crafting Science*, ed. David Kaiser (Cambridge, Mass., 2005), 253–86.

[161] Iu. Rumer, "Teoriia otnositel'nosti," *Izvestiia*, 22 Oct. 1935, 3.

for the August 1932 lecture on "Light and Life," in which the Danish physicist had attempted to extend some of the notions of complementarity to the life sciences.[162] Landau expressed amazement that a man who understood physics so profoundly could say such foolish things about biology. But again, this wasn't so much a matter of Landau's engaging Bohr in a scientific debate as it was an attempt to hold up a particular image of a Western scientist whose stature was already acknowledged by Landau's domestic critics and thus could not easily be dismissed by them. Yet it was also a useful means of establishing differences that pointed to peculiarly Soviet disciplinary virtues without acceding to know-nothing demands for a sui generis "proletarian science." When philosopher-courtiers later started faulting Landau, Frenkel, Fock, and Tamm as disciples of the "idealist" Bohr, they failed to realize that these missionaries had no need of the Copenhagen theology as it was popularly understood.[163] Fervency and technique would suffice—only Fock would later develop a taste for exegesis, and Bohr found him heterodox.[164]

CONCLUSION

After Landau moved to Kharkov in 1932 he began to concentrate increasingly on the physics of the solid state and on the creation of a pedagogical framework for a distinctive school. What I call the "principled phenomenology" that became the basis for Landau's research program with students achieved its coherence within the school framework as a technique-oriented dialectic of global principles versus local models. The famous Theoretical Minimum quickly became the standard for membership in the Landau school, with early students such as A. S. Kompaneets, E. M. Lifschitz, A. I. Akhiezer, I. Ia. Pomeranchuk, and I. M. Khalatnikov among the successful aspirants.

With his 1941 theory of superfluid helium—neither a fundamental microscopic theory nor an ad hoc model—Landau sealed his reputation among his peers. As his first students began to come into their own as professional theorists, he was finally selected to the Academy of Sciences in 1946, with his growing "scientific school" as an important factor in his favor. As successive volumes of the *Course of Theoretical Physics* made their way into print at home and abroad, textbooks also became approved markers of school building. By the time Landau was awarded the Nobel Prize in 1962, kruzhok sensibilities had little to do with the officially sanctioned role of the school in Soviet science. The Landau school was most concretely institutionalized in 1965 in a theory institute founded outside Moscow by his students. Though small by Soviet standards, it remained one of the most productive institutes in the academy on a per capita basis until the demise of the Soviet Union.

Both Landau and his associates recognized some of the latent tensions between the school as a marker of unofficial science (i.e., as kruzhok legacy) and as an epitome of official science. Through the next two generations the moral qualities of this social cohort remained very much alive to its members, who saw in scientific schools "one

---

[162] L. D. Landau, "Burzhuaziia i sovremennaia fizika," *Izvestiia*, 23 Nov. 1935; Niels Bohr, "Light and Life," *Nature* 131 (1933): 421–3, 457–9.

[163] J. L. Heilbron, "The Earliest Missionaries of the Copenhagen Spirit," *Rev. Hist. Sci.* 38, nos. 3–4 (1985): 195–230.

[164] Graham, *Science* (cit. n. 146), 337–43.

of the highest humanistic achievements of civilization."[165] In the 1990s, after many key members of the Landau school had dispersed to take positions outside Russia, Khalatnikov depicted the school as "something typical of this country. Indeed, this is a unique phenomenon which cannot be found anywhere else in the world."[166] In his eyes, its peculiar esprit de corps constituted a tradition foreign to the individualism of physicists in the West.

What for Khalatnikov was a historically peculiar set of professional virtues, however, was also less obviously a product of deep ambivalence toward Soviet power among Landau and his students in the years after his arrest in 1938. The experience of the young A. I. Larkin (1932–2005) is revealing in this regard. His youthful determination in the early post-Stalin era to "be obligated to no one, belong to nothing" was driven by disgust at the pervasive role of party institutions in daily life, and the tremendous prestige and comparative autonomy of Soviet physics in the 1950s made it an appealing career outlet. Larkin was somewhat shaken upon witnessing his teacher A. B. Migdal's readiness to shrug off a French colleague's membership in the Communist Party. (Migdal had taken his doctorate with Landau and was perhaps the most successful disciple in fostering a school of his own.) Neither entirely facetious nor fully in earnest, Migdal explained to Larkin that in this world one had to have friends, and the French Communists were very friendly people. As Larkin relates it, he thenceforth resolved that his party would be the Migdal school and the Landau school.[167] The resilience of this kind of self-identification was both intellectual and political. The social dynamic that made the Landau school a productive institution for criticizing, testing, and extending contemporary theoretical practices was in no small part the outcome of similar longstanding intelligentsia dilemmas of collective identity.

[165] V. F. Klepikov, "Shkola fiziki I zhizni," *A. i. Akhiezer: Ocherki i vospominaniia*, ed. V. G. Bar'iakhtar (Kharkiv, 2003), 177.
[166] I. M. Khalatnikov, "Our History," Landau Institute for Theoretical Physics (1966), http://www.itp.ac.ru.
[167] A. I. Larkin, "A. B. Migdal v moei zhizni," *Vospominaniia ob akademike A. B. Migdale,* ed. N. O. Agasian (Moscow, 2003), 40–8.

Imagining the Cosmos:
Utopians, Mystics, and the Popular Culture of Spaceflight in Revolutionary Russia

By Asif A. Siddiqi*

ABSTRACT

This essay investigates the explosive Soviet interest in space travel during the New Economic Policy (NEP) era of the 1920s, as expressed through amateur societies, the press, literature, painting, film, and other popular culture. In recovering an obscured history of the roots of Russian cosmonautics, it shows how the cause of space exploration in early twentieth-century Russia originally stemmed from two ideological strands: technological utopianism and the mystical occult tradition of Cosmism. The former (seemingly modern, urban, international, materialist) alternately clashed and meshed with the latter (superficially archaic, pastoral, Russian, spiritual), creating an often contradictory but urgent language of space enthusiasm. Cosmic activists, who saw themselves as part of a new Soviet intelligentsia, actively used both ideals to communicate their views directly to the public. The essay argues that despite superficial differences, technological utopianism and Cosmism shared much of the same iconography, language, and goals, particularly the imperative to transform and control the natural world. In other words, the modern rocket with its new Communist cosmonaut was conceived as much in a leap of faith as in a reach for reason.

> By taking a pair of steps, I crossed over the threshold from one epoch to another, into the space [era].[1]
> —Mikhail Popov, organizer of the world's first interplanetary exhibition, on what it felt like to step into the display hall, 1927

Space achievements represented an important marker of Soviet claims to global preeminence during the cold war. In books, movies, posters, and songs, Soviet authorities sang the glories of their space program; cosmonauts and artifacts toured the world using rhetoric that conflated mastery of space with mastery of nature. During and after the cold war, both Russian and Western historians underlined the connection between the Soviet space program and Marxist fascinations with technol-

---

*Dept. of History, Fordham University, 441 E. Fordham Road, Bronx, NY 10458; siddiqi@fordham.edu.

The author wishes to thank Michael Hagemeister as well as the participants of the weekly seminar at the Shelby Cullom Davis Center at Princeton University for their useful comments.

[1] Sergei Samoilovich, *Grazhdanin vselennoi (Cherty zhizni i deiatel' nosti Konstantina Eduardovicha Tsiolkovskogo)* (Kaluga, 1969), 181.

ogy.² These accounts located the social and cultural origins of the Soviet space program as part of the project of modernization, secularism, and "progress." When the first young hero cosmonauts flew into space in the early 1960s, Soviet commentators repeatedly depicted them as emblematic of a modern and technologically sophisticated Russia, overtaking the West. Furthermore, unlike American astronauts who thanked God for their successes, Soviet cosmonauts were explicitly atheistic; one of the first cosmonauts, the young Gherman Titov, famously declared on a visit to the United States that during his seventeen orbits of the Earth, he had seen "no God or angels," adding that "no God helped build our rocket."[3] And in the 1970s, when the Soviets launched their first cargo ship to a space station, they named it simply *Progress*.

Through lenses of modernity, secularism, and progress, historians typically traced back the history of the Soviet space program to the "patriarch" of Soviet cosmonautics, Konstantin Tsiolkovskii, who in 1903 produced the first mathematical substantiations that spaceflight was possible. According to this deeply engrained story, the Bolsheviks recognized the value of his work after the Russian Revolution, honored him with many awards, and made him a national treasure. To the Bolsheviks, Tsiolkovskii's ideas were a perfect vehicle for catapulting Russia into the modern technological age of Ford and Taylor. Soon, inspired by Tsiolkovskii, young men and women joined together to build rockets. The Soviet government supported them and, in 1933, sponsored the creation of a national institute to build rockets. The intellectual and engineering groundwork that they created eventually bore fruit a quarter century later with the launch of *Sputnik*, the world's first artificial satellite.[4] The received story, built on a series of willful distortions, masked a set of complex social and cultural processes, particularly the ways in which social and cultural factors outside state sponsorship—besides popular Marxist rhetoric about the role of technology—enabled the project of space exploration in the Soviet Union.

In the late 1920s and early 1930s, Tsiolkovskii's ideas on space exploration fed enormous popular interest in the cause of cosmic travel in the Soviet Union. With little or no support from the state, amateur and technically minded enthusiasts formed short-lived societies to discuss their interests and exchange information.[5] Some put up impressive exhibitions displaying the visions of the major prognosticators of the day such as Tsiolkovskii, the American Robert Goddard, and the Romanian-German Herman Oberth. In the popular media, advocates wrote about the power of technology to improve and remake Russian society. On the cultural front, the science fiction of Aleksei Tolstoi, the paintings of the Suprematists and the Amaravella collective, and Iakov Protazanov's famous interplanetary movie *Aelita* all engaged mystical and spiritual ideas of the place of humanity in the cosmos. These embryonic artistic,

---

[2] William Shelton, *Soviet Space Exploration: The First Decade* (New York, 1968); James E. Oberg, *Red Star in Orbit* (New York, 1981); William P. Barry, "The Missile Design Bureaux and Soviet Piloted Space Policy, 1953–1974" (DPhil diss., Univ. of Oxford, 1995); David Easton Potts, "Soviet Man in Space: Politics and Technology from Stalin to Gorbachev (Volumes 1 and 2)" (PhD diss., Georgetown Univ., 1992).

[3] "Titov, Denying God, Puts His Faith in People," *New York Times*, 7 May 1962.

[4] Nicholas Daniloff, *The Kremlin and the Cosmos* (New York, 1972); James Harford, *Korolev: How One Man Masterminded the Soviet Drive to Beat America to the Moon* (New York, 1997).

[5] For an international perspective, see Asif A. Siddiqi, "Nauka za stenami akademii: K. E. Tsiolkovskii i ego al'ternativnaia set' neformal'noi nauchnoi kommunikatsii," *Voprosy istorii estestvoznaniia i tekhniki*, 2005, no. 4:137–54.

philosophical, and cultural explorations were important not only because they underlined an interest in the power of modern science but also because they disseminated ideas about space travel that were not simply about technology or modernization.

In a number of important ways, the space enthusiasts represented a counterexample to the more prominent elements of Soviet scientific and technical intelligentsia of the period. The two groups shared a few common traits. Both possessed a reverence for knowledge about the natural and material world. They exhibited an ambivalence between reason and faith, the former represented by an aspiration for modernization and the latter by a weakness for mysticism. Finally, although few of the space enthusiasts were revolutionary in the way that many Russian intelligentsia self-identified, the space obsessed saw themselves as the vanguard of a new era; the resistance they faced from public quarters for their utopian leanings emboldened their self-image as revolutionary and iconoclastic actors.

Yet two major characteristics distinguished the *kosmopolitov* from the burgeoning Soviet scientific and technical intelligentsia. First, the space obsessed could claim no formal education in the natural sciences; their "higher" knowledge was often the result of informal schooling or, at best, mediocre institutions. Second, they embraced an antielitist stance that led them to actively engage with the popular culture of the day. In fact, their very embrace of more popular and populist forms of communication contributed as much to their estrangement from the orthodox scientific community as their lack of formal educational identifiers, the autodidact Tsiolkovskii being the quintessential embodiment of this alienation. Revisiting the noise that these space enthusiasts generated—which spanned the revolutionary divide of 1917—opens a critical window into the discursive strategies used by marginal scientific actors in revolutionary Russia to advance seemingly outlandish scientific ideas. Theirs was the curious case of a demographic who strongly identified with the mainstream scientific and technical intelligentsia while being almost completely alienated from them.

Their cause, space exploration, was a small but important part of the wild cultural explorations of the New Economic Policy (NEP) era of the 1920s; it stemmed from both ideological oppositions and unions. Two intellectual strands contributed to the birth and sustenance of the 1920s space fad: technological utopianism and the mystical tradition of Cosmism. The former (seemingly modern, urban, international, materialist) clashed and meshed with the latter (superficially archaic, pastoral, Russian, spiritual), creating a complex ideological context for popular interest in spaceflight. If the language of technological utopianism has retained its place in the received history of Russian space travel, the role of Cosmism has been all but obscured. Recovering the "hidden" history of the Cosmist roots of Soviet space travel underscores how advocates of interplanetary flight from the early Bolshevik era navigated the entire spectrum between extreme technology fetishism (such as the amateur student societies) and extreme occult fascinations (the Biocosmists). The most important bridge between these two seemingly contradictory worldviews was Konstantin Tsiolkovskii, the patriarch of Russian space travel.

TECHNOLOGICAL UTOPIANISM

Russian utopian thought, which has a history long predating Bolshevism, Marxism, and indeed the nineteenth century, encompassed everything from overtly secular ideas to explicitly theological conceptions, and from monarchist ideals to anarchist visions. Already before the revolution of 1917, Russian utopian philosophy incorporated both Marxist notions and twentieth-century modernist ideals of science and technology. The revolution, however, allowed technological utopian visions to move from the wisp of dreams to the arena of *possibility*. After 1917, an ostensibly secular brand of millenarianism entered the picture.

The richest expressions of this meeting of sensibilities between utopia, technology, and possibility occurred during the NEP years, when the country moved through a rapid economic recovery that fostered what Sheila Fitzpatrick called "an upsurge of optimism among the Bolshevik leaders."[6] Notwithstanding harsh conditions in the cities, the urban population continued to grow through the 1920s due to peasant migration into the cities and massive demobilization following the end of the civil war. Despite one million unemployed in 1924, wages finally began to rise the same year, and the standard of living for the average factory worker—someone like the tireless space crusader Fridrikh Tsander—began to improve noticeably. With urban renewal accelerating and the first fruits of the revolution appearing, people conjured up old dreams of utopia in new and experimental ways. In his indispensable study *Revolutionary Dreams: Utopian Vision and Experimental Life in the Russian Revolution*, Richard Stites has described the many ways in which a wide spectrum of actors, from the poorest peasant to the most influential member of the intelligentsia, invoked, debated over, wrote about, and often rejected utopia.[7] From ritual to religion, mannerisms to machines, and art to architecture, utopian thought pervaded Soviet society at all levels. The utopian discussions of the period were not monolithic; in fact, their very contradictions and illogic often gave the social experimentation a rich and expansive tenor.

In the 1920s, technology played a major role in the social conjuring, debating, and enabling of utopias. Prominent voices of the scientific and technical intelligentsia, as well as Bolshevik leaders, engaged in this discourse, and indeed, their pronouncements reflected the same types of tensions between naiveté and pragmatism emblematic of broader NEP culture. Lenin's fascination with the rapid electrification of Russia, industrial Taylorism, and the construction of modernized railroads in Russia were certainly all practical, but they also carried with them an underlying idea that technology itself was a possible panacea.[8] Beyond his oft-quoted phrase "communism equals Soviet power plus the electrification of the entire country," Lenin had an almost evangelical view of the role of electricity, and technology in general, as if it had the power to transform nation and culture. H. G. Wells, after interviewing Lenin in 1920, wrote, "Lenin, [who] like a good orthodox Marxist, denounces all 'Utopians,' has succumbed at last to Utopia, the Utopia of electricians."[9] Even as Lenin underlined "the need to dream," he was also unforgiving of those who shied away from the harsh realities of practical action. In the prerevolutionary days, Lenin

---

[6] Sheila Fitzpatrick, *The Russian Revolution*, 2nd ed. (Oxford, 1994), 113.

[7] Richard Stites, *Revolutionary Dreams: Utopian Vision and Experimental Life in the Russian Revolution* (New York, 1989). See also Paul Josephson, "'Projects of the Century' in Soviet History: Large-Scale Technologies from Lenin to Gorbachev," *Technology and Culture* 36 (1995): 519–59.

[8] For Lenin's personal role in plans for Soviet electrification, Taylorism, and railroads, see Jonathan Coopersmith, *The Electrification of Russia, 1880–1926* (Ithaca, N.Y., 1992), 153–5; Anthony Heywood, *Modernising Lenin's Russia: Economic Reconstruction, Foreign Trade, and the Railway* (Cambridge, UK, 1999); and Kendall E. Bailes, "Alexei Gastev and the Soviet Controversy over Taylorism, 1918–24," *Soviet Studies* 29 (1977): 373–94.

[9] H. G. Wells, *Russia in the Shadows* (New York, 1921), 158–9.

had been consistently critical of utopian socialists as well as the Populists for their unrealistic goals.[10]

Lev Trotskii (Leon Trotsky), another hardheaded revolutionary few would characterize as being impractical, also spoke rather uncritically of the powers of science and technology. In his 1923 tract *Literature and Revolution*, Trotskii wrote that because of the revolution, "[t]he shell of life will hardly have time to form before it will burst open again under the pressure of new technical and cultural achievements." Under the twin spells of science and utopia, Trotskii conjectured that advances in medicine would create a new "superman," able to "rise to the heights of an Aristotle, a Goethe, or a Marx."[11] Maksim Gor'kii (Gorky), one of the most important cultural commentators of the day, who held Konstantin Tsiolkovskii in very high esteem, frequently spoke of technology as miraculous and a panacea to the world's ills; he coined the phrase "an area of miracles" to speak of the power of science.[12]

Stites and others have pointed to the Russian interest in aviation, which held a much broader fascination for the Soviet populace in the 1920s than did spaceflight, as reflective of "a kinetic metaphor for liberation." Aviation represented a mixture of modernity and liberation that proved irresistible to many leading Bolsheviks. They appropriated its symbolic meanings to encourage and inculcate ideas about a new world and used it to bridge the literal and metaphorical gaps between urban and rural masses.[13] Yet although flight served as a metaphor for liberation, and perhaps even emancipation, it had some basis in the reality of the 1920s; both in Soviet Russia and the rest of the developed world, most urban citizens had seen pictures or drawings of airplanes if not an actual machine flying over their heads.

The dream of spaceflight in the 1920s differed in two significant ways from the concurrent interest in aviation. First, spaceflight, which was also about liberation from the Earth, pushed the physical limits of emancipation beyond conception, past the boundaries of the visible skies. Second, spaceflight was entirely a discourse of fantasy: voyages beyond the atmosphere had no precedent or template. Liberation and fantasy in one shape or other are common to most utopian dreams, but by extending liberation (into space) and pushing utopian speculations beyond reality (into fantasy), the spaceflight discourse was infused with a "universal" (in both senses of the word) appeal that aviation lacked. For a brief period in the 1920s, spaceflight was the most potent manifestation of the "fantasy of liberation" and indeed may be seen as a "liberation of fantasy." The speculations about spaceflight would not have been possible without the promise of new twentieth-century technology that made the utopias of liberation and fantasy attainable. As one single force—a combination of technology, fantasy, and liberation—spaceflight promised what aviation could only offer in

---

[10] The "need to dream" quotation is from V. I. Lenin, *Polnoe sobranie sochinenii: Izdanie piatoe*, vol. 6 (Moscow, 1959), 171–2. For Lenin and utopianism, see Stites, *Revolutionary Dreams* (cit. n. 7), 41–6; Robert C. Tucker, "Lenin's Bolshevism as a Culture in the Making," in *Bolshevik Culture: Experiment and Order in the Russian Revolution*, ed. Abbott Gleason, Peter Kenez, and Richard Stites (Bloomington, Ind., 1985), 25–38; Rodney Barfield, "Lenin's Utopianism: State and Revolution," *Slavic Review* 21 (March 1971): 45–56.

[11] Leon Trotsky, *Literature and Revolution* (Ann Arbor, Mich., 1975).

[12] Bernice Glatzer Rosenthal, "Political Implications of the Occult Revival," in *The Occult in Russian and Soviet Culture*, ed. Bernice Glatzer Rosenthal (Ithaca, N.Y., 1997), 390.

[13] Richard Stites, "Utopias in the Air and on the Ground: Futuristic Dreaming in the Russian Revolution," *Russian History/Histoire Russe* 11, nos. 2–3 (1984): 236–57.

part: total liberation from the signifiers of the past—social injustice, imperfection, gravity, and ultimately, the Earth.

COSMISM

Technology, fantasy, and liberation also figured prominently in a parallel set of ideas known as Russian Cosmism that has fed into a nationalist discourse in present-day Russia.[14] In the early twentieth century, Cosmism resonated strongly in some Russian intellectual circles as a corpus of philosophical thought about the evolution of both humanity and the universe and the relationship between the two.[15] The philosophy influenced many famous Russian intellectuals in the 1920s. They included Bolshevik ideologues, scientists, writers, philosophers, poets, artists, and architects, who gathered in Moscow and Kaluga, Tsiolkovskii's hometown, to discuss its attributes. Cosmism's intellectual foundations comprised a hodgepodge of Eastern and Western philosophical traditions, theosophy, Pan-Slavism, and Russian Orthodox thinking.[16] The outcome was a nationalist and often reactionary philosophy that continues to attract the attention of many Russian intellectuals.

Tsiolkovskii served as a key contributor to the canon of Cosmism, but the most important worldview that fed into twentieth-century Cosmism stemmed from the writings of Nikolai Fedorovich Fedorov (1828–1903), the eccentric philosopher whose works influenced many, including Dostoevskii (Fyodor Dostoevsky), Gor'kii, and Tolstoi (Leo Tolstoy).[17] While working as a librarian at the Rumiantsev Library in Moscow, Fedorov developed his infamous *Filosofiia obshchego dela* (Philosophy of the common task), the most enduring and notorious of his many works.[18] Described by one Western biographer as "one of the most profound, comprehensive, and original ideas in the history of Russian speculation," Fedorov's doctrine, published after his death in 1906, was about "the common task" of all humanity, to resurrect the dead.[19] Fedorov's mission stemmed from a distinctly theocratic view of the universe in which he saw Christianity as primarily a religion of resurrection, an idea that attracted

---

[14] For links between modern Russian Cosmism and post-Soviet Russian nationalism, see James P. Scanlan, ed., *Russian Thought after Communism: The Recovery of a Philosophical Heritage* (Armonk, N.Y., 1994), 26–8.

[15] For only a sampling of works on Russian Cosmism since the late 1980s, see L. V. Fesenkova, ed., *Russkii kosmizm i sovremennost'* (Moscow, 1990); Svetlana Semenova, "Russkii kosmizm," *Svobodnaia mysl'*, 1992, no. 17:81–97; Semenova and A. G. Gacheva, eds., *Russkii kosmizm: Antologiia filosofskoi mysli* (Moscow, 1993); O. D. Kurakina, *Russkii kosmizm kak sotsiokul'turnyi fenomenon* (Moscow, 1993).

[16] For the best English-language meditation on Russian Cosmism as a historical process, see Michael Hagemeister, "Russian Cosmism in the 1920s and Today," in Rosenthal, *Occult in Russian and Soviet Culture* (cit. n. 12), 185–202.

[17] Those said to be influenced by Fedorov included writers (Dostoevskii, Gor'kii, Odoevskii, Pasternak, Platonov, Tolstoi), political thinkers (Bogdanov, Lunacharskii), poets (Khlebnikov, Maiakovskii, Zabolotskii), painters (Filonov), architects (Mel'nikov), heliobiologists (Chizhevskii), and scientists (Tsiolkovskii, Vernadskii). For a description of the Moscow-based Fedorovtsy (supporters of N. F. Fedorov) in the 1920s, see Michael Hagemeister, *Nikolaj Fedorov: Studien zu Leben, Werk und Wirkung* (Munich, 1989), 343–62.

[18] Fedorov devotees independently printed and distributed 480 copies of the original in 1906. A second volume was issued in 1913. His writings have been collected in A. G. Gacheva and Svetlana Semenova, eds., *N. F. Fedorov: Sobranie sochinenii v chetyrekh tomakh*, 5 vols. (Vols. 1–4 and supplement) (Moscow, 1995–2000).

[19] George M. Young Jr., *Nikolai F. Fedorov: An Introduction* (Belmont, Mass., 1979), 7.

both Dostoevskii and Tolstoi. He believed that humanity's moral task was to emulate Christ and make bodily resurrection possible. Mass resurrection would finally eliminate the artificial boundaries among the "brotherhood" of humanity, that is, between previous and current generations. In other words, none of the ills of society could be solved without devising a solution to the inevitability of death. He argued that using all of the resources at its disposal, including science and technology, humanity should engage in a quest to reassemble the corporeal particles lost in the "disintegration" of human death. In an ideal utopian setting ("as it ought to be"), Fedorov believed that there would be no birth and no death, only the progressive reanimation of the deceased millions from history.[20]

Two aspects of Fedorov's "philosophy of the common task" related to Cosmism in general and to voyages into space in particular. First, to achieve his ultimate goal of "liberation from death," Fedorov called for restructuring human society and its natural environment, which for him included not only the Earth but the entire universe. In the early postrevolutionary era, the idea of "regulating nature" by taking absolute control over it resonated deeply with the scientific and technical intelligentsia, who, infected by Bolshevik claims of remaking the social universe, were also interested in remaking the natural one.[21] Second, Fedorov believed that humans from Earth would have to travel into the cosmos—to the Moon, the planets, and stars—to recover disintegrated particles of deceased human beings that are spread throughout the universe. Once the bodies of the deceased were reconstituted (in forms that might not resemble humans), the resurrected would then settle throughout the universe. In his *Philosophy of the Common Task*, Fedorov wrote, "[The] conquest of the Path to Space is an absolute imperative, imposed on us as a duty in preparation for the Resurrection. We must take possession of new regions of Space because there is not enough space on Earth to allow the co-existence of all the resurrected generations."[22]

Fedorov's ideas of restructuring humanity and the cosmos, especially the supreme role of science and technology in this transformation, anticipated Tsiolkovskii's writings, which are sprinkled with the Promethean urge to remake everything that surrounds us. Many historians have claimed that Fedorov inculcated Tsiolkovskii with his ideas about space travel. During his brief stay in Moscow as a teen in the 1870s, Tsiolkovskii had indeed met daily with Fedorov, who worked at a Moscow library. Fedorov played a critical role in supporting the young student in his struggle to learn more about the natural sciences. As Tsiolkovskii later remembered, "It is no exaggeration to say that for me he took the place of university professors."[23] Yet, those who suggest that Fedorov may have influenced Tsiolkovskii to take up the cause of spaceflight are certainly mistaken. Throughout his life, Tsiolkovskii himself maintained

[20] Summarized from Fedorov works collected in vols. 1 and 2 of Gacheva and Semenova, *N. F. Fedorov* (cit. n. 18).

[21] The famous Russian geochemist Vladimir Vernadskii, who shared these views (although he probably never heard of Fedorov), headed the Commission for the Study of the Natural Productive Forces (KEPS), a body whose goals encompassed such transformative projects as harnessing solar and electromagnetic forces for the good of Russian society. Kendall E. Bailes, *Science and Russian Culture in an Age of Revolutions: V. I. Vernadsky and His Scientific School, 1863–1945* (Bloomington, Ind., 1990). Remarkably, Bailes never once mentions Vernadskii's interest in Cosmism. For Vernadskii and Cosmism, see G. P. Aksenov, "O nauchnom odinochestve Vernadskogo," *Voprosy filosofii*, 1993, no. 6:74–87.

[22] Jean Clair, "From Humboldt to Hubble," in *Cosmos: From Romanticism to the Avant Garde*, ed. Jean Clair (Munich, 1999), 25; Young, *Nikolai F. Fedorov* (cit. n. 19), 182–3.

[23] Konstantin Altaiskii, "Moskovskaia iunost' Tsiolkovskogo," *Moskva*, 1966, no. 9:176–92, on 181.

that during his tenure of study under Fedorov, the two never discussed space travel although both had independently begun thinking of the possibility by this time.[24]

In parallel with his more technical writings, Tsiolkovskii issued numerous short monographs, beginning in the late nineteenth century, that touched on the philosophy of cosmic travel. These two strands, the technical and the philosophical, intertwined and influenced each other throughout his life, and although his philosophical writings are less well known than his technical ones, they form a corpus of work that exceeds in size his combined works on aeronautics, rocketry, and space travel.[25] Tsiolkovskii brought a messianic and transformative vision to the cause of spaceflight that mimicked some of Fedorov's ideas about immortality and cosmic unity. He also drew upon occult thought rooting back to German philosopher Carl du Prel, who was famous for drawing a link between cosmic and biological evolution, that is, that Darwinian natural selection acted on planetary bodies just as they acted on living organisms.[26] In Tsiolkovskii's worldview, the occult, theories of evolution, and Christianity existed without contradiction. At a fundamental level, Tsiolkovskii was a religious thinker whose life was an attempt to reconcile the scientific views of nature that seemed to contradict his strong faith in Christ. As such, he expended a great deal of energy explaining biblical events with the aid of contemporary science.

Like Fedorov, Tsiolkovskii believed that humanity's place in the universe depended on two related ideas, monism and panpsychism. He described both of these concepts in *Monizm vselennoi* (Monism of the universe), a brochure he self-published in 1925 that would be his most complete statement of cosmic philosophy. According to his version of monism, all matter in the universe, including organic matter, is made out of a single substance, has the same structure, and obeys the same set of laws. He explained panpsychism as the belief that all matter is made up of "atoms of ether," even smaller than "regular" atoms, which are in and of themselves *living organisms* or "happy atoms."[27] When these atoms combine in different ways, they produce different living beings with differing abilities. Because these ether atoms are indestructible,

---

[24] The legend that Fedorov pointed Tsiolkovskii in the direction of space travel probably originated from scientist Viktor Shlovskii in his "'K' in 'Kosmonavtika ot A do Ia,'" *Literaturnaia gazeta*, 7 April 1971. See also V. E. L'vov, *Zagadochnyi starik: Povesti* (Leningrad, 1977). Many Western and Russian authors, without any evidence, make a direct causal connection between Fedorov and Tsiolkovskii. Michael Holquist, "Konstantin Tsiolkovsky: Science Fiction and Philosophy in the History of Soviet Space Exploration," in *Intersections: Fantasy and Science Fiction*, ed. George E. Slusser and Eric S. Rabkin (Carbondale, Ill., 1987), 74–86; Holquist, "The Philosophical Bases of Soviet Space Exploration," *Key Reporter* 50 (Winter 1985–86): 2–4; and Vladimir V. Lytkin, "Tsiolkovsky's Inspiration," *Ad Astra*, Nov.–Dec. 1998, 34–9.

[25] Especially through the 1920s, during the height of the "space fad," Tsiolkovskii's output on philosophical topics increased dramatically. He self-published such works as *The Wealth of the Universe* (1920), *The Origins of Life on Earth* (1922), *Monism of the Universe* (1925), *Reason for Space* (1925), *The Future of Earth and Humanity* (1928), *The Will of the Universe: Unknown Intelligent Forces* (1928), *Love for Oneself or the Source of Egoism* (1928), *Intellect and Passion* (1928), *The Social Organization of Humanity* (1928), and *The Goal of Stellar Voyages* (1929). All of these works, and others unpublished during his lifetime, have been compiled into one volume: L. V. Golovanov and E. A. Timoshenkova, eds., *K. E. Tsiolkovskii: Genii sredi liudi* (Moscow, 2002). For the best analysis of Tsiolkovskii's philosophical works, see V. S. Avduevskii, ed., *K. E. Tsiolkovskii: Kosmicheskaia filosofiia* (Moscow, 2001), 370–472.

[26] For a sympathetic analysis of the differences between Fedorov and Tsiolkovskii's philosophies, see V. V. Kaziutinskii, "Kosmizm i kosmicheskaia filosofiia," in *Osvoenie aerokosmicheskogo prostranstva: Proshloe, nastoiashchee, budushchee*, ed. B. V. Raushenbakh (Moscow, 1997), 139–44.

[27] K. E. Tsiolkovskii, *Monizm vselennoi* (Kaluga, 1925). Tsiolkovskii's ideas were not original; they were heavily influenced by the ideas of such German thinkers as Gottfried Leibniz and Ernst Haeckel.

there is no such thing as true death as the atoms can be reconstituted in different combinations from the one that gave life to a specific human being.[28]

For all their "progressive" ideas about the role of science and technology and human expansion of space, Fedorov and Tsiolkovskii had a darker side to their vision. Fedorov's "common task" had a distinctly totalitarian tinge as it did not allow choice in the equation, that is, he argued humans would *have* to participate in his project without exception. Tsiolkovskii's view of the search for human perfection also reflected his firm belief in eugenics; he advocated the extermination of imperfect plants and animal life and called for a "battle against the procreation of defective people and animals."[29] In a piece finished in 1918, he wrote:

> I do not desire to live the life of the lowest races [such as] the life of a negro or an Indian. Therefore, the benefit of any atom, even the atom of a Papuan, requires the extinction also of the lowest races of humanity, and in an extreme measure the most imperfect individuals in the races.[30]

This view of space travel, which combined the search for human perfection, racial purity, and occult thinking, provided the fundamental impetus to Tsiolkovskii's more mathematically inclined meditations on rocket flight into outer space. Tsiolkovskii seamlessly combined his fascinations with technology and the occult into a fully formed weltanschauung. Yet to much of his audience in the 1920s—especially those young and technology-minded students who were inspired to dream of space travel—his goal of space travel fit nicely with prevailing Bolshevik rhetoric connecting technology with modernity. The technophiles, in fact, believed that by avoiding Tsiolkovskii's mystical invocations, they could construct a vision of space travel that directly countered antiquated notions of the cosmos as part of an epistemology of superstition and folktales. In forming societies to argue their cause, they saw in space travel a vehicle for creating a new world of machines and men.

TECHNOLOGICAL UTOPIANISM: THE COSMIC SOCIETIES

Most of the men and women who organized cosmic societies in the 1920s did so without any material support or encouragement from the state. They did, however, absorb official discourses on the role of technology as a panacea for all social ills in new, postrevolutionary Russia. Space advocates saw in space exploration (and its corollary, rocketry) a manifestation of the cold hard power of rationality, science, and mathematics to move society ahead on the path of "progress" and "modernization."

Several technology-enraptured (and short-lived) societies coalesced during the period of the space fad. Of these, the most important and influential was the Moscow-based Society for the Study of Interplanetary Communications (Obshchestva Izuche-

---

[28] In *Volia vselennoi* [Will of the universe], a brochure published in 1928 in Kaluga, Tsiolkovskii wrote, "Death is one of the illusions of a weak human mind. There is no death, for the existence of an atom in inorganic matter is not marked by memory and time—it is as if the latter does not exist at all." K. E. Tsiolkovskii, "Volia vselennoi," in Golovanov and Timoshenkova, *K. E. Tsiolkovskii* (cit. n. 25), 228–9.

[29] K. E. Tsiokovskii, "Liubov' k samomu sebe, ili istinnoe sebialiuboe," in Golovanov and Timoshenkova, *K. E. Tsiolkovskii* (cit. n. 25), 378–402, on 401.

[30] K. Tsiolkovskii, "Etika ili estestvennye osnovy nravstvennosti," in Avduevskii, *K. E. Tsiolkovskii* (cit. n. 25), 82.

niia Mezhplanetnykh Soobshchenii, OIMS), formed in 1924. It was not only the first group in the world to effectively organize for the cause of space exploration but also the first to build a domestic and international network around the idea. The history of the organization, a combination of serendipity, willful devotion, and eventual loss of momentum due to indifference from the state, illustrates the ways in which technological utopianism inspired a few to bring an esoteric idea to many.[31]

The society emerged during the first intense wave of public fascination with spaceflight in the spring of 1924, set off by a story in the newspaper *Izvestiia* under the headline "Is Utopia Really Possible?" about the recently published meditations on spaceflight written by the foreigners Oberth and Goddard.[32] Spurred to promote a Russian source for such ideas, the sixty-six-year-old Tsiolkovskii immediately republished his own prerevolutionary works on spaceflight. Almost overnight the Soviet media began to devote considerable attention to the cosmos. News and rumors of Oberth and Goddard's exploits, the publication of Aleksei Tolstoi's new space fiction novel *Aelita*, and the "Great Mars Opposition" of August 1924—when Mars and Earth were closer to each other than in hundreds of years—fed an explosion of public interest in space. In one lengthy *Pravda* article ("Voyage into Cosmic Space"), the author narrated the new history of space exploration, harking back to Leonardo da Vinci, Cyrano de Bergerac, Jules Verne, and H. G. Wells. The history naturally culminated with the works of Tsiolkovskii, Oberth, and Goddard. Palpably excited by the optimism of the times, the writer concluded, "[W]ithin a few years, hundreds of heavenly ships will push into the starry cosmos."[33]

The media frenzy over space exploration in early 1924 might have faded away had it not been for some resourceful young men and women. In April 1924, about a dozen students at the prestigious Zhukovskii Military Air Engineering Academy's Military-Science Society (VNO) set up a Section on Reactive Motion to exchange ideas about rockets.[34] In compiling a list of goals, the section touched on all the key strategies that would characterize the ensuing space fad, from its technical side (building rockets), to outreach (lectures, publications, and bookstores), to building a community (by interesting others in the same topics), to opening a channel to the West (by collecting media from overseas), to acknowledging the artistic medium as a possible way to educate and popularize (by branching into film).

The section first organized a public lecture. One of the section leaders, Morris Leiteizen, whose father was a famous prerevolutionary Bolshevik, asked a family friend, Mikhail Lapirov-Skoblo, to do the honors. Lapirov-Skoblo, thirty-five years old and a rising member of the reconstituted postrevolutionary technical intelligentsia, had been briefly acquainted with Lenin. After the latter's death, he served as deputy chairman of the Scientific-Technical Department of the Supreme Council of the People's Economy (VSNKh, or Vesenkha), a body tasked with supervisory duties

---

[31] For a detailed account of the society, see Asif A. Siddiqi, "Making Spaceflight Modern: A Cultural History of the World's First Space Advocacy Group," in *The Societal Impact of Spaceflight*, ed. Steven J. Dick and Roger D. Launius (Washington, D.C., 2007), 513–37.

[32] "Novosti nauki i tekhniki: Neuzheli ne utopiia?" *Izvestiia VTsIK*, 2 Oct. 1923.

[33] M. Ia. Lapirov-Skoblo, "Puteshestviia v mezhplanetnye prostranstva," *Pravda*, 15 April 1924. For Goddard's prominent role in the space fad, see Asif A. Siddiqi, "Deep Impact: Robert Goddard and the Soviet 'Space Fad' of the 1920s," *History and Technology* 20, no. 2 (2004): 97–113.

[34] The leading VNO student members included V. P. Kaperskii, M. G. Leiteizen, A. I. Makarevskii, M. A. Rezunov, and N. A. Sokolov-Sokolenok., r. 4, op. 14, d. 197, ll. 32–3, Archive of the Russian Academy of Sciences (hereafter cited as ARAN), Moscow.

over applied research and development in Soviet industry. He also headed *Pravda*'s department of science and technology.³⁵ Tsiolkovskii's recently published works so inspired Lapirov-Skoblo that he wrote the first well-researched expositions on space travel in the postrevolutionary era in *Pravda* and other publications.³⁶

Lapirov-Skoblo's lecture, held on the evening of Friday, May 30, 1924, was a resounding success. Tickets sold out two days earlier; on the day of the talk, the organizers were forced to call for the police to control the mass of people who wanted to attend. Attendees eagerly bought up all the utopian literature on space travel on display—H. G. Wells's *War of the Worlds*, Russian science fiction from Aleksei Tolstoi and Aleksandr Beliaev, and books by the popular science writer Iakov Perel'man. Lapirov-Skoblo's lecture, titled "Interplanetary Communications (How Modern Science and Technology Solves This Question)," may have been the first exposition on space exploration in Russia open to the general public. His lecture was a typical example of the rhetoric of the technological utopian space advocates; he linked the idea of spaceflight with both modern technology and the future of a new Bolshevik Russia, a nation he believed had left behind its roots in tradition, backwardness, and peasant life. He concluded by calling on the Soviet populace to build rocket engines to "transform into reality the centuries-old dream of flight into space."³⁷

Following Lapirov-Skoblo's talk, section members invited the audience to sign up to form the core of a public society, thus opening up membership to laypeople outside the Zhukovskii Academy. While the complete list of 179 names has been lost, the surviving pages give a sense of these people. Of the 121 names preserved, 104 were men. The majority of the members (68) were young, between the ages of twenty and thirty. In terms of professions, a total of 96 members, that is, roughly 80 percent, were evenly split between students and workers. A smaller number identified themselves as "scientific workers," "writers," or "scientists and inventors."³⁸

Grigorii Kramarov, elected to chair the new society's "presidium," recalled forty years later that no one had any illusions that the Soviet Union would soon be sending men into space. He remembered that "in the work of the society [we] all saw one more possibility to aid the Motherland, to aid in the building of socialism." Instead of building rockets, the society would bring science and technology to the masses. Its members were "convinced that the society's work would contribute to the preparation of cadres, who in the future would create the economic and scientific and technical base for solving the greatest problems."³⁹ They paid lip service to the notion that technology would improve social conditions in revolutionary Russia. In a speech to factory workers, Fridrikh Tsander, one of the principal activists in the society, spoke of the many benefits to be gained from space travel: of "senior citizens [who] will

---

³⁵ When Lenin supervised the formation of the State Commission for Electrification of Russia (GEOLRO) in 1920, he tapped Lapirov-Skoblo to represent the Vesenkha on GOELRO. For a biography, see r. 14, op. 14, d. 197, ll. 30–30b, ARAN.

³⁶ For his other articles, both titled "Puteshestviia v mezhplanetnye prostranstva," see *Molodaia gvardiia*, 1924, no. 5, and *Khochu vse znat'*, 1924, no. 3:140.

³⁷ For the transcript, see r. 4, op. 14, d. 194, ll. 49–62, ARAN. For recollections of attendees, see r. 4, op. 14, d. 197, ll. 35–8, ARAN; G. Kramarov, *Na zare kosmonavtiki: K 40-letiiu osnovaniia pervogo v mire obshchestva mezhplanetnykh soobshchenii* (Moscow, 1965), 25–8.

³⁸ R. 4, op. 14, d. 196, ll. 6–21, ARAN; V. M. Komarov and I. N. Tarasenko, "20 iunia—50 let so vremeni sozdaniia v moskve obshchestva izucheniia mezhplanetnykh soobshchenii (1924g.)," *Iz istorii aviatsii i kosmonavtiki* 22 (1974): 75–82; Kramarov, *Na zare kosmonavtiki* (cit. n. 37), 28.

³⁹ Kramarov, *Na zare kosmonavtiki* (cit. n. 37), 50.

find it much easier to maintain health in [space]," of the "inhabitants of Mars . . . [whose] inventions could help us to a great extent to become happy and well off," and of "[a]stronomy, [which] more than the other sciences, calls upon man to unite for a longer and happier life."[40] When critics attacked their views for being utopian, the members responded by calling their opponents "conservative," thus locating supporters and detractors of space exploration within a binary world; one was either modern ("with science and technology") or traditional (against "progress").[41]

Throughout 1924, the society held numerous lectures and debates in Moscow, Leningrad, Kharkov, Ryazan, Tula, Saratov, and elsewhere, introducing the idea of space exploration to a huge audience beyond technology fetishists. But despite their many successes—including one near-riotous event in October 1924, when the Moscow horse militia had to be called out to control unruly crowds interested in rumors of a rocket launch to the Moon that year—lack of state support proved to be the society's undoing. In late 1924, when the society petitioned the administrative department of the Moscow city council to register the organization officially, the city council rejected the application on the grounds that the society had "insufficient scientific strength among its members."[42] The society's members also had to deal with less committed members, who were unable to sustain interest in the face of both the widespread poverty of the times and the possibility that space exploration was decades away. Society head Kramarov remembered that the most common question from the audience after each lecture was "How quickly would flight to the planets be accomplished?"[43] When it became clear that travel into space was years, if not decades, away, the "accidental members" dispersed quickly, leaving only a handful of the truly dedicated to pursue the cause. Eventually, even the faithful had to come down to earth; most, such as Tsander, had little time to devote to activities that did not provide money for living. Valentin Chernov, for example, remembered later that his job as a violinist forced him to abandon the society.[44] Like many utopians, the society was unable to sustain a vision beyond the short term.

TECHNOLOGICAL UTOPIANISM: THE MEDIA

Dissemination of celebratory ideas about space travel during the NEP era depended greatly on the existence of vibrant popular scientific media, which directly equated technology with modernization and societal benefit.[45] The journal *Khochu vse znat'* (I want to know all), published by the Leningrad-based newspaper *Rabochei gazety*

---

[40] See F. A. Tsander, "Doklad inzhenera F. A. Tsandera a svoem izobretenii," in *Iz nauchnogo naslediia* (Moscow, 1967), 10–4.

[41] V. Chernov, "Raketa na lunu," r. 4, op. 14, d. 194, ll. 1–3, ARAN.

[42] R. 4, op. 14, d. 197, l. 19, ARAN. Tsander later confirmed that the lack of "scientific workers" among members of the "board of directors,"—i.e., Tsander, Leiteizen, Kaperskii, Rezunov, Chernov, Serebrennikov, and Kramarov—was a source of dissension that contributed to the society's dissolution.

[43] Kramarov, *Na zare kosmonavtiki* (cit. n. 37), 56.

[44] Ibid., 51–2. Tsander, in his autobiography, notes that "the lack of published material and of spare time did not permit us to work intensively." "Autobiography of Friedrich Arturovich Tsander, Mechanical Engineer," in N. A. Rynin, *Rockets*, vol. 2, no. 4, of *Interplanetary Flight and Communication*, trans. T. Pelz (Jerusalem, 1971), 187.

[45] For a view on the role of popular science in the postrevolutionary period, see James T. Andrews, *Science for the Masses: The Bolshevik State, Public Science, and the Popular Imagination in Soviet Russia, 1917–1934* (College Station, Tex., 2003).

(Working gazette), for example, set out to "[help] readers in developing a material understanding of the world" and to "familiarize readers with the newest achievements in modern science and technology" that would benefit the revolution.[46] Publishers, both private and public, found that scientific titles were particularly popular among urban masses. Jeffrey Brooks notes that "[p]ublishers had difficulty keeping up with the demand for works in popular science," which "comprised a fifth of [all] titles published from 1921–27."[47] By the mid-1920s, biweekly and monthly journals devoted to popular science were ubiquitous on newsstands and included both new and older publications.[48] The partially independent Leningrad-based publisher P. P. Soikin, which published the journals *Mir prikliuchenii* (World of adventure), *Priroda i liudi* (Nature and people), and *Vestnik znaniia* (Journal of knowledge), played an influential role in the popularization of science through the second and third decades of the twentieth century. Having published Lenin's first legal work in prerevolutionary times, Soikin remained one of the few imperial-era publishing concerns allowed to operate during the NEP years.[49] Although private publishers were producing only about 5 percent of all copies of books in 1925, Soikin carved out a dominating niche in the popular science market that remained unchallenged until complete nationalization of the press in the post-NEP era. Circulation of *Vestnik znaniia*, one of Soikin's most popular monthlies, for example, increased from 25,000 in 1925 to 75,000 by 1931.[50] Such publications were widely available via bookstores such as Leningrad's Nauka i Znanie (Science and Knowledge), one of the largest in the city, which catered exclusively to scientific and applied scientific titles. Its catalog in 1928 boasted around 7,000 titles from "all branches of [scientific and technical] knowledge."[51]

Space and space-related topics constituted a significant, although by no means major, slice of the popular science literature. Based upon an in-depth search through the popular science literature in early twentieth-century Russia, my research suggests that the number of articles on spaceflight published between 1923 and 1932 (inclusive), the key years spanning the space fad, amounted to nearly 250 articles and more than thirty books. Compared with the other pressing topics of the day, this output did not represent a great number, but that so many works on space exploration were published on such an arcane subject is in and of itself a striking result.[52] By comparison, in the United States, only *two* nonfiction monographs on spaceflight appeared in the

---

[46] Advertisement for *Khochu vse znat'* in inside cover of various issues of *Vestnik znaniia*.

[47] Jeffrey Brooks, "The Breakdown in Production and Distribution of Printed Material, 1917–1927," in Gleason, Kenez, and Stites, *Bolshevik Culture* (cit. n. 10), 159, 168–9.

[48] Popular science journals included *Bor'ba mirov* (The world's struggle), *Khochu vse znat'* (I want to know all), *Krasnaia nov'* (Red virgin soil), *Mir prikliuchenii* (World of adventure), *Nauka i tekhnika* (Science and technology), *Pioner* (Pioneer), *Priroda i liudi* (Nature and people), *Tekhnika i zhizn'* (Technology and life), *Tekhnika-molodezhi* (Technology for youth), *Vestnik znaniia* (Journal of knowledge), *V masterskoi prirody* (In nature's workshop), *Vsemirnyi sledopyt'* (World pathfinder), and *Znanie-sila* (Knowledge is power).

[49] A. M. Admiral'skii and S. V. Belov, *Rytsar' knigi: Ocherki zhizni i deiatel'nosti P. P. Soikina* (Leningrad, 1970).

[50] Publication runs are from the back pages of *Vestnik znaniia* in 1925 and 1931.

[51] From commercial advertisements in the back covers of various popular science magazines in 1928.

[52] Asif A. Siddiqi, "The Rockets' Red Glare: Spaceflight and the Russian Imagination, 1857–1957" (PhD diss., Carnegie Mellon Univ., 2004).

same period. Only in Germany, the single Western nation with a vocal spaceflight community, were there comparable levels of media attention.[53]

The content of popular science media suggests that readers were not merely passive receptors of information on spaceflight. Brooks has noted that Soviet newspapers during the NEP era contained three spheres of discourse: explanatory, interactive, and informational.[54] The dialogue over spaceflight in popular science journals echoed these divisions. Both *Nauka i tekhnika* (Science and technology) and *Vestnik znaniia* had forums for interacting with readers. The former, under the banner "Correspondence with Readers," published more than two dozen responses to readers' letters per issue. Inquiries and comments came from all over the country: Moscow, Rostov-on-Don, Voronezh, Leningrad, Krasnodar, Voznesensk, Kharkov, Grozny, Kiev, Taganrog, Donbass, and elsewhere. *Vestnik znaniia* had a similar section titled "Living Communication," which published numerous editorial responses to readers' letters on various topics. The transformative, beneficial, and modernizing aspects of space travel were rarely, if ever, questioned in the exchange of ideas.

Many readers asked where to get materials on space, a service that the journals provided repeatedly, pointing out not only articles on space published in the journals' own pages but also those published elsewhere.[55] Some of the responses provided information while others clarified ambiguous topics. To comrade A. Semenov from Leningrad, for example, *Nauka i tekhnika* used a drawing to illustrate the changing distances between the planets. In some cases, the journal editors displayed a distinctly pedantic attitude to its readers, implying that lack of scientific and technical knowledge about space travel was indicative of ignorance about the modern world. For example, *Nauka i tekhnika* chastised comrade Pavliuchenko from Aleksandrovka for his "bewilderment" about movement through space in the absence of matter to push against.[56] On occasion, the journals acknowledged the "many numbers of questions to the Editors" on the topic; *Vestnik znaniia* claimed that numerous readers of the journal were dissatisfied with short articles on space travel and demanded complete books on the topic.[57] Some readers' communications required special attention. *Vestnik znaniia* returned comrade Iosifov's manuscript, "The Importance of the Planet and Its Satellites in the Solar System," with several points explaining why his conclusions were "absolutely incorrect." In the same readers' section, comrade Goldenveizer conjectured about the unpleasant sensations space travelers might experience in a vessel, some of which had been discussed by Tsiolkovskii, Noordung,

---

[53] For the two American monographs, see Robert H. Goddard, *A Method of Reaching Extreme Altitudes*, Smithsonian Miscellaneous Collections, vol. 71, no. 2 (Washington, D.C., 1919); David Lasser, *The Conquest of Space* (New York, 1931). For the German space fad, see Michael J. Neufeld, "Weimar Culture and Futuristic Technology: The Rocketry and Spaceflight Fad in Germany, 1923–1933," *Tech. Cult.* 31 (Oct. 1992): 725–52.

[54] Jeffrey Brooks, "The Press and Its Message: Images of America in the 1920s and 1930s," in *Russia in the Era of NEP: Explorations in Soviet Society and Culture*, ed. Sheila Fitzpatrick, Alexander Rabinowitch, and Richard Stites (Bloomington, Ind., 1991), 231–52.

[55] "Pred'iaviteliiu bileta avio-loterei ser. 008, no. 10220 (Baku)," *Nauka i tekhnika*, 19 Aug. 1927, no. 34:35; Ia. I. Perel'man, "Mezhplanetnye polety," *Vestnik znaniia*, 1928, no. 4:254; and "Tov. Miklashevskomu (Moskva)," *Nauka i tekhnika*, 9 June 1928, no. 23:31.

[56] "Tov. A. Semenovu (Leningrad)," *Nauka i tekhnika*, 18 Nov. 1927, no. 47:28; "Tov. Pavliuchenko (d. Aleksandrovka)," *Nauka i tekhnika*, 7 April 1928, no. 14:30. For an answer to a similar question, see "L'vovu," *Vestnik znaniia*, 25 Jan. 1931, no. 2:127.

[57] "Ot redkatskii," *Vestnik znaniia*, 1928, no. 11:551.

and others.⁵⁸ In one case, when a reader anticipated Fermi's paradox in relation to the possibility of space travel, *Vestnik znaniia* devoted a full article with responses from prominent writers, including Tsiolkovskii and Iakov Perel'man, to the question "Is Interplanetary Communications Possible?"⁵⁹ According to the writers, the answer was a resounding "yes," but only because modern science and technology would make it possible.

THE COSMOPOLITANS

Beyond societies and publishing, space advocates of the 1920s also used the medium of the *vystavka*, or "display," to publicize their cause. Through exhibits, enthusiasts were able to let their visions run free in more creative ways than was possible via lectures or publications. By exposing the possibility of space travel for the first time to thousands, they served a very important role during the space fad. Unlike the technological utopians who organized or wrote, exhibition organizers represented a constituency that embraced certain mystical ideas about spaceflight. In their lexicon, Tsiolkovskii assumed near-messianic status in a cause that was equal amounts of fetishizing of technology and speculation about human evolution. Recovering the history of the exhibitions underscores how in the 1920s, the line between lunar aspirations and lunacy was often invisible and that the lexicon of technological utopians was frequently indistinguishable from those who were mystically minded.

In 1925, a group of spaceflight enthusiasts organized a small exhibition of spaceflight-related artifacts in Kiev.⁶⁰ Although the exhibit remained open for less than three months, its success prompted one of its organizers, Aleksandr Fedorov, to join with the Moscow-based Association of Inventors (Assotsiatsiia Izobretatelei-Izobretateliam, AIIZ) to open the world's first international exhibition on space travel in 1927.⁶¹ The AIIZ, a forum for amateur enthusiasts to discuss their interests in science and technology, had recently created the Sector for Propaganda and Popularization of Astronautics to promote the cause of spaceflight.⁶² The sector's leading members included a motley crew of self-described inventors: a pilot, a former convict, a student, a technician, a librarian, and Fedorov.⁶³ Obsessed with Tsiolkovskii, the idiosyncratic Fedorov found a shared cause in his fellow exhibition organizers, who, like Fedorov, seemed to see the old man in overtly evangelical terms. In one letter to Tsiolkovskii, Fedorov wrote that he considered himself "fortunate to work under the leadership

---

⁵⁸ "I. T. Iosifovu" and "Podp. Goldenveizeru," in *Vestnik znaniia*, 10 Oct. 1931, no. 19:1004.

⁵⁹ "Vozmozhny li mezhplanetnye soobshcheniia?" *Vestnik znaniia*, 1930, no. 4:152–3. Fermi's paradox describes the seeming contradiction of our galaxy being more than a billion years old—and therefore possibly full of alien life—but humanity's having no contact with them.

⁶⁰ The only detailed documentary evidence on the exhibition are three letters from Fedorov to Tsiolkovskii describing the works of the Kiev Society, written in August–September 1925. Fedorov to Tsiolkovskii, 16 Aug. 1925, Kiev, r. 4, op. 14, d. 195, ll. 10–2, ARAN.

⁶¹ The literal translation of AIIZ is "Association of Inventor-to-Inventor," but the society was commonly known as the Association of Inventors.

⁶² R. 4, op. 14, d. 198, l. 41, ARAN. The precise word they used was *zvezdoplavaniia*, which literally translates as "stellar dynamics" in the same way that *vozdukhoplavaniia* means "aerodynamics." The closest English word is "astronautics," a term that Belgian writer J. J. Rosny invented. Other sections in the AIIZ included one for "culture-propaganda," one for language, and one for developing a universal language.

⁶³ The "organizational committee" of the AIIZ's astronautics sector included G. A. Polevoi (pilot), I. S. Beliaev (former convict), A. S. Suvorov (student), Z. G. Piatetskii (technician), and O. V. Kholoptseva (librarian). R. 4, op. 14, d. 198, ll. 1–2, ARAN.

of creative great ideas, a thinker of our times and a preacher of great inconceivable truths!"[64]

Having organized the previous exhibition in Kiev, Fedorov suggested to the sector that it host a major exhibition of space artifacts in Moscow. The idea was to construct models of rockets and spacecraft conceived by the leading Soviet and foreign theoreticians of the day and display them with information for curious visitors. The association planned to display many of Tsiolkovskii's publications on spaceflight in one place—a library of sorts that they called the "smithy of all inventors."[65] The exhibition, which would also commemorate the tenth anniversary of the great October Revolution, would be augmented by a publicity blitz on space travel, including lectures in dozens of locales in and around the capital city. The ragtag band of organizers united in their zealous belief in the power of "invention" and "inventors" and held up Tsiolkovskii as some sort of "prophet" of a new era, "superior even to Edison."[66]

Although Soviet-era accounts focused solely on the organizers' fascination with modern technology, Fedorov and his associates were inspired not only by the products of modern engineering but also by a mystic calling. They referred to themselves as "cosmopolitans" (*kosmopolitov*), a word derived from the term *cosmopolite* ("citizen of the world"), and their cause as "cosmopolitanism" (*kosmopolizma*).[67] Unlike many other technically minded popularizers of space exploration in the 1920s who carefully ignored Tsiolkovskii's spiritually oriented works about Cosmism and human destiny, the exhibition organizers embraced them, deifying Tsiolkovskii as a preacher, a visionary, the father of cosmopolitanism. They embraced the "master's" vision of animate matter and monism and believed in the importance of their efforts as part of a big evolutionary leap for all of humanity. In several effusive communications to Tsiolkovskii (the "first honorary captain rocket-mobilist") in late 1927, the organizers referenced Leibniz's worldview on monism and underscored the power of inventors to "find the resources for human immortality"—the foundation of the Cosmist view of the universe.[68] Their rationale for space exploration had as much to do with equating technology with modernization as with a self-important and mystical notion of human destiny that harked back to the nineteenth century.

The exhibition, unimaginatively named the "World's First Exhibition of Models of Interplanetary Apparatus, Mechanisms, Instruments, and Historical Materials," opened on April 24, 1927, not far from what is now Maiakovskii Square at number 68 (now 28) Tverskaia Street, one of Moscow's biggest thoroughfares. Open to the public for two months, the exhibition had an elaborately designed entrance with a huge display of an imagined planetary landscape, designed and built by Arkhipov,

---

[64] Fedorov to Tsiolkovskii, 7 Sept. 1926, Moscow, f. 555, op. 4, d. 641, ll. 1–5, ARAN.

[65] Efofbi [O. V. Khloptseva, pseud.] and Polevoi to Tsiolkovskii, 5 Feb. 1927, Moscow, f. 555, op. 3, d. 198, ll. 6–8, ARAN.

[66] Efofbi to Tsiolkovskii, 3 Dec. 1928, Moscow, f. 555, op. 3, d. 199, ll. 5–6, ARAN.

[67] AIIZ to Tsiolkovskii, 21 Jan. 1927, Moscow, f. 555, op. 3, d. 198, ll. 1–1ob, ARAN. Although nearly identical, the word *kosmopolizma* differed in meaning and etymology from the pejorative term *kosmopolitizm* that party ideologues used in the late 1940s to describe a "decadent" and "bourgeois" lifestyle during the late Stalin years. The latter word was first introduced into public discourse in January 1949. *Kosmopolitov* was probably derived from the early seventeenth-century French word *cosmopolite*, as in a "citizen of the world."

[68] Efofbi to Tsiolkovskii, 7 Dec. 1927, Moscow, f. 555, op. 3, d. 198, ll. 34–34ob, ARAN; AIIZ to Tsiolkovskii, 18 Dec. 1927, Moscow, f. 555, op. 3, d. 198, ll. 38–38ob, ARAN.

Figure 1. An image from the "World's First Exhibition of Models of Interplanetary Apparatus, Mechanisms, Instruments and Historical Materials" held in Moscow in 1927 shows the area devoted to the "patriarch" of Soviet space exploration, Konstantin Tsiolkovskii. A bust, specially commissioned for the event, was adorned with Tsiolkovskii's publications and models. The arrangement communicates an obvious conflation between the organizers' feelings about Tsiolkovskii's scientific contributions and their attitude toward the old man as a prophet with almost mystical qualities. (Reprinted with the permission of Ron Miller.)

placed behind a large pane of glass. Part of the display, somewhat incorrectly called "Lunar Panorama," showed a hypothetical planet with orange soil and blue vegetation crisscrossed by straight canals. A giant silver rocket descended from the starry sky while a voyager in a spacesuit (made of plywood) stood at the edge of a crater. Organizer Mikhail Popov described the feeling of entering the exhibition: "By taking a pair of steps, I crossed over the threshold of one epoch to another, into the space [era]."[69]

Although state organs ignored the show, it succeeded resoundingly with the public. According to the organizers, in two months, between 10,000 and 12,000 people visited the exhibition. Visitors included schoolchildren, workers, service employees, artists, scientists, policemen, and such luminaries as poet Vladimir Maiakovskii.[70] Visitors, who were invited to record their impressions in a book of comments, were both effusive and candid. One person, who signed as "Gorev," wrote, "Our mind is not accustomed to all the 'wonderful and unknown' which literally was [sic] seen and heard, as if in a dream, yet we understand that this is not a fantasy but a completely feasible idea supported by the achievements of science and engineering." Another person, an artist from the Third State Cinematographic Studio, recommended that "[i]t would be desirable that our inventors achieve the first landing on the moon." One of the most captivated visitors was S. G. Vortkin, a reporter from the most important workers' news daily, *Rabochaia moskva*, who wrote, "I am going to accompany you on the first flight. I am quite serious about this. As soon as I heard what you had done, I tried in every way to make certain that you would take me with you. Please do not refuse my request."[71]

SPACEFLIGHT IN ART AND CULTURE

The degree of popular Soviet fascination with space in the 1920s is also underlined by how deeply it resonated in the various art forms of the day. From literature to film to painting to poetry to architecture to language, clusters of artists produced works that reflected their belief that cosmic travel was an inevitable part of their future. A small sampling of this vast output—Tolstoi's novel *Aelita*, Protazanov's movie of the same title, Malevich's Suprematist paintings, and the Amaravella group's artwork—highlights some of the key dimensions of this cultural discourse. On the surface, artists with a spiritual-flavored view of the cosmos may have been disengaged from the modernist technologically minded utopians, but in fact they were linked by a network united in the cause of space exploration. And like their more "scientifically minded" space-enthusiast colleagues, the artists produced their populist work largely isolated from the elite Soviet scientific and technical intelligentsia of the NEP era.

Literature

The most widely disseminated media for communicating ideas about space exploration was *nauchno-fantastika* (literally, "scientific-fantasy"). Although many historians have explored the various dimensions of Soviet science fiction in the early

---

[69] Samoilovich, *Grazhdanin vselennoi* (cit. n. 1), 181.
[70] "Vospominaniia Z. G. Piatetskogo," r. 4, op. 14, d. 198, l. 38, ARAN; "Vospominaniia O. V. Kholoptsevoi," r. 4, op. 14, d. 198, l. 11, ARAN.
[71] Comments from Rynin, *Rockets* (cit. n. 44), 205–6.

decades of the twentieth century, its use of *space* as a plot or philosophical device has remained largely unscrutinized.⁷² Space fiction, which constituted about one-fifth of all Soviet science fiction in the postrevolutionary period to World War II, was remarkable for its disproportionate social resonance given the subgenre's low numbers. To some degree, most of the space-related works reflected the same characteristics of the broader science fiction literature, that is, almost all such works were technologically optimistic and can be divided into adventure stories (*krasnyi pinkerton,* or "red detective") and future utopias. Richard Stites's claim that "[Soviet s]cience fiction was a striking example of revolutionary discourse because of its total vision of communist life and its treatment of 'revolutionary dreams'" was also true for the smaller subset of space fiction.⁷³ Although the stories were less about social than technological revolution, the prevailing mood of revolution allowed the latter to be conflated with the former.

The most famous Soviet science fiction novel of the 1920s, Aleksei Tolstoi's *Aelita: Zakat Marsa* (Aelita: Sunset of Mars), first published in serialized form in 1922–23, remains the most famous *space* fiction work of the period.⁷⁴ It also perfectly encapsulated the contradictory themes of space advocacy in the 1920s. In the story, an engineer and a soldier voyage to Mars, where the latter incites a proletarian revolution among the bourgeois Martians. Aelita is the queen of Mars who falls in love with the Red Army soldier. On one level, the novel incorporates many elements of postrevolutionary utopian science fiction: a bourgeois enemy, a socialist revolution, modern science and technology, adventure and romance borrowed from Edgar Rice Burroughs, and utopian dreaming. Yet *Aelita*'s narrative also has hints of mysticism, especially ideas infused with theosophy and ancient anthroposophic ideas, not dissimilar to Fedorov and Tsiolkovskii's Cosmist views of the universe.⁷⁵ Defending his position from critics who blamed him for being too "emotional" in the novel, Tolstoi wrote, "Art—an artistic creation—appears momentarily like a dream. It has no place for logic, because its goal is not to find a cause for some sort of event, but to give in all its fullness a living piece of cosmos."⁷⁶ His use of the lexicon of panpsychism suggests a link to the mystical side of Tsiolkovskii and the Cosmists.⁷⁷

Aelita, despite its invocation of space travel, or maybe because of its Cosmist overtones, was a novel less about looking forward than looking to the past. Although regarded as the most important Soviet science fiction novel of the period, *Aelita*, Halina Stephan rightly claims, "concluded rather than inaugurated a literary tradition." Yet, the technologically minded spaceflight enthusiasts of Tolstoi's day avoided the mysticism and found it futuristic since the novel was the first of the period that used a rocket for interplanetary travel. Members of the Moscow Society for the Study of

---

⁷² For general reviews of early Soviet science fiction, see Darko Suvin, "The Utopian Tradition of Russian Science Fiction," *Modern Language Review* 66 (1971): 139–59; A. F. Britikov, *Russkii Sovetskii nauchno-fantasticheskii roman* (Leningrad, 1970); Patrick L. McGuire, *Red Stars: Political Aspects of Soviet Science Fiction* (Ann Arbor, Mich., 1985).

⁷³ Stites, *Revolutionary Dreams* (cit. n. 7), 167–8.

⁷⁴ The novel was originally published in three serialized parts in the journal *Krasnaia nov'*. In 1923, it was published as a stand-alone novel as *Aelita* (*Zakat Marsa*) (Moscow, 1923).

⁷⁵ Halina Stephan makes a similar point. Stephan, "Aleksei Tolstoi's *Aelita* and the Inauguration of Soviet Science Fiction," *Canadian-American Slavic Studies* 18 (1984): 63–75.

⁷⁶ Tolstoi quoted in ibid., 72–3.

⁷⁷ See also Ian Christie, "Down to Earth: *Aelita* Relocated," in *Inside the Film Factory: New Approaches to Russian and Soviet Cinema*, ed. Richard Taylor and Ian Christie (London, 1991), 97–8; Rosenthal, introduction to *Occult in Russian and Soviet Culture* (cit. n. 12), 25.

Interplanetary Communications were so taken by Tolstoi's use of the rocket that they considered using the story to develop a film script—a project that was brought to fruition by others.[78]

Film

The movie version of Tolstoi's *Aelita* appeared soon after publication of the print version and was directed by Iakov Protazanov, the Russian film director of pre-revolutionary fame.[79] Released officially in September 1924 at the peak of the space fad, *Aelita* has since been hailed as the most important Soviet science fiction movie of the interwar era. It also contributed enormously to the popularization of spaceflight in Soviet culture in the 1920s. For example, interest in the movie after its release drove up attendance numbers at interplanetary talks sponsored by space societies such as the OIMS. The film also established a new standard for Soviet cinema, if not in quality, then certainly in popularity and hype. Weeks of intense advertising campaigns in *Pravda* and *Kino-gazeta* (Movie gazette) preceded its release, while airplanes dropped thousands of leaflets announcing the opening over Voronezh.[80] Tickets for the opening shows sold out, and the size of the crowd on opening night prevented even Protazanov from attending.

Protazanov, who, like Tolstoi, had only recently returned to the Soviet Union from exile, engineered a significant transformation in Tolstoi's relatively conventional novel, producing a remarkable movie that not only mirrored and telescoped many prevailing social concerns of the NEP-era in movie form but also critiqued Tolstoi's novel itself. With the help of scriptwriters Aleksei Faiko and Fedor Otsep, Protazanov reimagined Tolstoi's original account of the voyage to Mars as a dream in the mind of the protagonist Los'.[81] The so-called revolution on Mars—which occupies only one-fourth of the film—is riddled with ambiguities that do not demarcate strictly along bipolar lines (capitalist-communist, benevolent-exploitative); nothing is really what it seems. Here, Los' is not simply a one-dimensional caricature of the new Soviet man but rather a man living in and mirroring the contradictory realities of NEP life.

---

[78] Leiteizen to Tsiolkovskii, 4 May 1924, Moscow, f. 555, op. 4, d. 356, ll. 2–3, ARAN. In addition to *Aelita*, Aleksandr Bogdanov's *Krasnaia zvezda* (Red star) enraptured space enthusiasts of the period. Less about spaceflight than about an idealized Communist utopia on the planet Mars, the novel has also been seen by some scholars as a warning on how socialism might take on distinctly totalitarian tones if sufficiently militarized. The Society for the Study of Interplanetary Communications evidently established communication with Bogdanov in 1924, interested in his idea of using atomic power to propel spaceships. Space enthusiasts were less likely to explore Bogdanov's philosophical arguments than his technological vision; both parties shared a view of technology as autonomous, positive, and liberating. Loren R. Graham, "Bogdanov's Inner Message," in *Red Star: The First Bolshevik Utopia*, ed. Loren R. Graham and Richard Stites (Bloomington, Ind., 1984), 241–53.

[79] M. Aleinikov, *Iakov Protazanov: O tvorcheskom puti rezhisera*, 2nd ed. (Moscow, 1957); Aleinikov, *Iakov Protazanov* (Moscow, 1961); Ian Christie and Julian Graffy, eds., *Protazanov and the Continuity of Russian Cinema* (London, 1993); Denise J. Youngblood, "The Return of the Native: Yakov Protazanov and Soviet Cinema," in Taylor and Christie, *Inside the Film Factory* (cit. n. 77), 103–23.

[80] The movie was produced by a new multinational company, Mezhrabpom-Rus', a joint Russian-German company that combined Mezhrabpom (International Workers' Aid), a Berlin-based relief organization and Rus', a Russian production company formed in 1918. Richard Taylor, *The Politics of the Soviet Cinema, 1917–1929* (Cambridge, UK, 1979), 74.

[81] Faiko and Otsep made changes to the original plot with Tolstoi's agreement. Aleinnikov, *Iakov Protazanov* (cit. n. 79), 32. Most Western sources incorrectly list his name as "Otsen" instead of the correct "Otsep."

In the movie *Aelita*, Protazanov sought to produce an "impartial" work, so the negative response surprised him. By and large, the state media criticized the film. In fact, the movie caused so much controversy that as late as 1928, newspapers and journals were still engaged in attacking the movie for being "alien to the working class," for its "petty bourgeois ending" because Los' returns to the domesticities of marriage, and for being "too Western."[82] Although many critics wrote off *Aelita* as a misstep in Protazanov's long career, it was an incredibly popular film; it did, after all, feature evocative acting, exotic scenes in interplanetary space, a glamorous princess, and women in provocative costumes. Grigorii Kramarov, the head of the OIMS, later underscored how "the book and film played a significant role in strengthening interest towards interplanetary communications and contributed to the development of activities of our Society."[83] Among those deeply affected by the hoopla over *Aelita* was ten-year-old Vladimir Chelomei; forty-five years later, as general designer of the Soviet space program, he named a new project of his, a huge space complex to send the first Soviet cosmonauts to Mars, *Aelita*.[84]

Art

Besides *Aelita*, both the novel and the film, other Russian works of art crossed the lines dividing technology and mysticism. Some scholars have claimed connections between the Russian avant-garde and Cosmism, arguing that the universal views of Nikolai Fedorov deeply influenced artistic personalities such as Vasilii Kandinskii, Kazimir Malevich, and Pavel Filonov.[85] But these connections were neither monolithic nor consistent. No single movement encapsulated the contradictions of the Soviet space fad better than did the Suprematists. Mentored by one of the legendary artists of the Russian avant-garde, Malevich, the Suprematists exemplified the duality and ambiguity of the space fad, cutting across not only mysticism (Cosmism) and science (space technology) but also the time and politics of the imperial and Bolshevik eras.

Suprematism as an organized movement of Russian and Soviet artists developed in the mid-1910s by extending and rejecting many of the foundations of Cubism. It reached its peak right after the October Revolution and then expanded into other media (principally architecture) in the early 1920s before losing direction late in the decade. Malevich had unveiled Suprematism at an exhibition of futurist art in 1915, with works that in their geometric shapes and colors completely dispensed with representations of conventional space and perspective. The paintings acquired a peculiarly compelling nature by the juxtaposition of colors and shapes that conveyed a continuum of space and time rather than self-contained and defined objects or ideas.

---

[82] Youngblood, "Return of the Native" (cit. n. 79), 111–2; Youngblood, *Soviet Cinema in the Silent Era, 1918–1935* (Ann Arbor, Mich., 1985), 30–2. For a list of reviews, see Aleinikov, *Iakov Protazanov* (cit. n. 79), 408. For Protazanov's comment about being "impartial," see p. 31 of Aleinikov's study.

[83] Kramarov, *Na zare kosmonavtiki* (cit. n. 37), 19–20.

[84] Asif A. Siddiqi, *The Soviet Space Race with Apollo* (Gainesville, Fla., 2003), 745–54.

[85] Iurii Linnik, *Russkii kosmizm i russkii avangard* (Petrozavodsk, 1995); Michael Holquist, "Tsiolkovsky as a Moment in the Prehistory of the Avant-Garde," in *Laboratory of Dreams: The Russian Avant-Garde and Cultural Experiment*, ed. John E. Bowlt and Olga Matich (Stanford, Calif., 1996), 100–17.

Malevich himself called his work the "nonobjective world," that is, a perception of the environment's distilled spaciousness.[86]

Such an approach naturally led many Suprematist artists to eulogize first aviation and then ultimately the cosmos as the ultimate environment of spaciousness. In their paintings, such as Boris Ender's *Cosmic Landscape* (1923), space—both cosmic and otherwise—became an integral part of the composition instead of "filler" in more traditional artistic creations. Malevich expressed interest in the most modern frontiers of art and science and technology, and he spent many years in pursuit of what he called the "science of art." He firmly believed in the power of technological "progress" and, like many other intellectuals of the day, supported the perfection of nature via artificial means. Malevich wrote, "I shall make my whole state comfortable and convenient, and, what is more, I shall convert other states and eventually the whole globe to my comfort and convenience."[87] His writings show an undeniably technologically utopian gloss, sprinkled with flirtations with anarchist ideas. Some scholars have suggested that Malevich, like many other Russian intellectuals, was captivated by mysticism and theosophy. For example, Igor Kazus claimed Malevich was "the first Russian artist to take note of [Fedorov's views of the universe, and] placed [them] at the base of Suprematism." [88] Malevich's many writings and works, however, suggest that his works were attempts to merge some of the disparate ideological underpinnings of modernity and spiritualism, that is, technological utopianism and mysticism.

Malevich's interest in spatial ideas beyond Earth first manifested themselves after 1916. As he wrote to a friend, "Earth has been abandoned like a worm-eaten house. And an aspiration towards space is in fact lodged in man and his consciousness, a longing to break away from the globe of the earth."[89] Paintings at the time show geometric forms (usually squares or rectangles) with hollowed-out spaces and stretched drops of color, drenched in white light that highlighted things unimaginable on Earth, that is, without reference to any form of nature. There was literally no up or down. Malevich's engagement with spatial ideas in the cosmic sense reached a zenith in 1917–18, during the height of the revolutionary years and just after the first major references to space travel appeared in the media. In 1919, he explicitly articulated the notion that Suprematism itself could be part of the project of space exploration:

> Between [Earth and the Moon], a new Suprematist satellite can be constructed, equipped with every component, which will move along an orbit shaping its new track. . . . I have ripped through the blue lampshade of the constraints of color. I have come out into the white. Follow me, comrade aviators! Swim into the abyss. I have set up the semaphores

---

[86] Larissa A. Zhadova, *Malevich: Suprematism and Revolution in Russian Art, 1910–1930* (London, 1982), 49–50.

[87] Serge Fauchereau, *Malevich* (New York, 1993), 27.

[88] Quotation from Igor A. Kazus, "The Idea of Cosmic Architecture and the Russian Avant-Garde of the Early Twentieth Century," in Clair, *Cosmos* (cit. n. 22), 194. John Golding also notes that "Malevich had . . . fallen under the spell of other occultists and pseudo-scientists fascinated with ideas about the fourth dimension, which had already been disseminated by the turn of the century." Golding, *Paths to the Absolute: Mondrian, Malevich, Kandinsky, Pollock, Newman, Rothko, and Still* (Princeton, N.J., 2000), 62. See also Igor A. Kazus, "Cosmic Architecture and the Russian Avant-Garde," *Project Russia* 15 (undated): 81–8. For a compelling and convincing counterargument, see Zhadova, *Malevich* (cit. n. 86), 59.

[89] Quoted in Zhadova, *Malevich* (cit. n. 86), 124n39.

of Suprematism. I have overcome the lining of the colored sky. . . . Swim! The white free abyss, infinity is before you.[90]

Some of Malevich's paintings from this period, such as *Suprematism* (1917) and *Drawing* (1918), depict objects not dissimilar to what we might today call space stations or futuristic cities in the cosmos. Malevich, of course, never alluded to them as such, and most certainly would not have known about such things given that few people in the world had yet articulated similar ideas in print. Yet the paintings show a remarkable understanding of the basic concepts of space travel, particularly the idea of space stations, and predate similar artistic visions that were common in Soviet popular science journals and pulp fiction of the 1920s. Malevich's fascination with the cosmos peaked around 1918 with his attempts to achieve an absolute spaciousness with pure whiteness, a white light of infinity that he represented in perhaps his most extreme avant-garde experiment, *White Square on White* (1918).

Like Malevich's works, many of his protégés' works hinted at a Fedorovian or Cosmist view of space. The case of the Society of Easel Painters (OST), which included a number of Malevich protégés, perfectly encapsulated the tensions between technological utopianism and Cosmism in the Soviet space fad of the 1920s. Like many in the Soviet avant-garde, the OST were taken with the wonders of technology and believed that art should mirror and interpret technological advancement in both mechanistic and abstract ways. Artists such as Vladimir Liushin, who produced *Station for Interplanetary Communications* (1922), seemed wholly beholden to the power of the machine to benefit society.[91] Yet Ivan Kudriashev, a Malevich protégé, eventually gravitated to a different view of the cosmos. Unlike other artists, Kudriashev had a direct connection to the space advocacy community: his father, a model builder, had been employed by Tsiolkovskii to build some of his conceptions. The younger Kudriashev accompanied his father on a visit to see the old man and translated Tsiolkovskii's technical terms for the model builder.[92] Kudriashev's philosophy, underlined in messianic essays about the expansion and settlement of humanity throughout the solar system, suggested a closer emotional affinity to Fedorov's mystic ideas than to earlier Suprematist works. Other Malevich followers, Lazar Lisitskii and Georgii Krutikov, explored a new type of architecture designed for "flying cities." These ideas stemmed not only from a fascination with space but also from the utilitarian view that because living space on the Earth was limited, one had to devise other spaces for habitation, a distinctively Fedorovian view of life.[93]

The most striking example of artistic fascination with space resulting from the meeting between the artistic avant-garde and the philosophy of Cosmism was in the work of the informal Soviet artists' group known as Amaravella. The self-contained contradictions characteristic of Russian Cosmist philosophy characterized their work: although they advocated a universal and cosmic consciousness to life and art, their art reflected deeply national influences (such as medieval Russian art), and their philosophy followed the tradition of a nationalist Russian approach to the cosmos, best

---

[90] Ibid., 57
[91] Vladimir Kostin, *OST (Obshchestvo stankovistov)* (Leningrad, 1976); John E. Bowlt, "The Society of Easel Artists (OST)," *Russian History/Histoire Russe* 9, nos. 2–3 (1982): 203–26.
[92] Kostin, *OST* (cit. n. 91), 24–6; Zhadova, *Malevich* (cit. n. 86), 129n19.
[93] S. O. Khan-Magomedov, "Proekt 'letaiushchego goroda,'" *Dekorativnoe iskusstvo*, 1973, no. 1:30–6; Kazus, "Idea of Cosmic Architecture and the Russian Avant-Garde" (cit. n. 88), 196–7.

underscored by many of Fedorov's followers. Superficially, the group aspired to combine the most modern aspects of both science and art, the progenitors of a long tradition during Soviet times, but on a deeper level, theirs was the lexicon of both "rational" and "irrational" science, of both modern and archaic art.[94]

Petr Fateev, a thirty-two-year-old painter, formed and led the original Amaravella around 1922. It reached a stable membership of a few energetic and inspired artists such as Viktor Chernovolenko, Aleksandr Sardan, Sergei Shigolev, and Boris Smirnov-Rusetskii by 1927–28, when the name Amaravella was coined, apparently derived by Sardan from a Sanskrit word meaning "bearing light" or "creative energy." The group, which operated as a commune, explored a remarkably wide range of ideas and approaches to art based on the members' nebulous philosophical ideas about cosmic harmony. Sardan, who was also a professional musician, produced compositions that were combinations of sound, painting, and architecture. His works such as *Sound in Space* (1920), *Lunar Sonata*, and *Cosmic Symphony* (both 1925) tried to represent the "sound" of architecture through vivid colorful hues that aspired toward a cosmic (aural) harmony. Other works such as *Earth, Ocean, Space* (1922) and *Cosmic Motive* (late 1920s) addressed his philosophical views, some of them borrowed from eastern philosophies, while *From the Moon to Space Way* (1930) and *Earthly Beacon and Signals from Space* (1926) elucidated technical ideas. The group exhibited their works several times, including once in New York in 1927, when six of Sardan's paintings were displayed at an exhibition organized by the Russian avant-garde artist Nikolai Rerikh. Rerikh, in turn, served as a link to the "other" space advocate community, centered on Tsiolkovskii: he befriended Aleksandr Gorskii, an influential Cosmist and occultist who himself moved to Kaluga, Tsiolkovskii's adopted hometown, in the 1930s.[95]

LINKING COMMUNITIES: BIOCOSMISTS

At the very extreme of the continuum from technological utopianism to Cosmism were those who were fully engaged in a spiritual and sometimes occultlike interest in space exploration. In the early 1920s, the most explicit mark of Cosmism's imprint emerged through scientific, cultural, and artistic icons such as Vladimir Vernadskii (the geochemist), Vladimir Zabolotskii (the poet), and Maksim Gor'kii (the writer) but also via short-lived groups such as the Anarchist-Biocosmists. The group (also known as the Biocosmist-Immortalists) coalesced in 1921 after the state's crackdown on anarchists following the funeral of famous Russian anarchist Petr Kropotkin. When the authorities arrested an anarchist group named the Universalists, a new collective, the Anarchist-Biocosmists, replaced them; adherents pledged their support to the Bolsheviks but also announced their goal of initiating a social revolution "in interplanetary space."[96] The group, which had factions in both Moscow and Petrograd, briefly published a journal, *Bessmertie* (Immortality), under the banner "Immortalism

---

[94] For survey of the vast literature on the union of science and art in the Soviet Union, see the special issue of *Leonardo* 27, no. 5 (1994), under the banner "Prometheus: Art, Science, and Technology in the Former Soviet Union."

[95] Iurii Linnik, *Amaravella: Put' k pleiadam; Russkie khodozniki-kosmisty* (Petrozavodsk, 1995), 82–145; Linnik, "Amaravella," *Sever*, 1981, no. 11:108–14.

[96] For the original Biocosmist manifesto, see A. Sviator, "Biokosmicheskaia poetika," in *Literaturnye manifesty ot simvolizma do nashikh dnei*, ed. S. B. Dzhimbinov (Moscow, 2000), 305–14, on 305.

and Interplanetarianism." In their manifesto, issued in 1921, they announced several goals, including victory over space ("not air navigation . . . but cosmic navigation"). They declared the two basic human rights to be the right to exist forever and the right to unimpeded movement in interplanetary space. Inspired by Fedorov's ideas, they wanted to abolish death, colonize the universe, and then resurrect those who had already died.[97] Just after Lenin's death, the Anarchist-Biocosmists published an official statement in *Izvestiia* arguing that all was not lost as the "[workers] and the oppressed all over the world could never be reconciled with the fact of Lenin's death."[98]

Devotees of Cosmism and Fedorov's philosophy were connected to the technological utopian spaceflight community via a network that highlighted the fine line between science and mysticism. Tsiolkovskii, someone who was equally at ease writing about propellant masses as about victory over death, was naturally the most obvious and important link between the two sides.[99] There were other, more famous links. During the 100th anniversary of Fedorov's birthday, Maksim Gor'kii, a devotee of Fedorov's, famously declared in an interview in *Izvestiia* that "freedom without power over nature—that's the same as freeing peasants without land."[100] It is less well known that Gor'kii, who also believed in the search for immortality, considered Tsiolkovskii to be an important scientific and philosophical thinker. During his exile, the writer had heard of Tsiolkovskii via the latter's 1925 work *Prichina kosmosa* (Reason for space), a meditation on humanity's spiritual calling to go into space. Although Gor'kii intended to visit Tsiolkovskii in Kaluga upon his return to the Soviet Union in 1928, the two never met. Tsiolkovskii, however, sent Gor'kii many of his brochures on Cosmist philosophy, and they evidently resonated deeply with the writer; Gor'kii sent a well-publicized congratulatory letter to the "interplanetary old man" (as he liked to call Tsiolkovskii) on his seventy-fifth birthday in 1932.[101]

Even at the extreme of mysticism, people remained connected with the technological utopians. One well-known Biocosmist member, Leonid Vasil'ev, who was also a respected researcher of telepathy, maintained a friendship with Aleksandr Chizhevskii, the young intellectual and well-known Cosmist who wrote extensively on the relationship between cosmic factors (such as sunspots) and social activity on Earth. Chizhevskii lived in Kaluga briefly and later wrote a massive memoir on his relationship with Tsiolkovskii.[102] Chizhevskii also holds a special place in the history of Soviet space exploration: he wrote the famous German-language introduction for the 1924 Tsiolkovskii monograph that effectively set off the Soviet space fad of the

---

[97] "Deklarativnaia rezoliutsiia," *Izvestiia VTsIK*, 4 Jan. 1922. The Biocosmists unsuccessfully tried to recruit such prominent scientists as Eugen Steinach and Albert Einstein. Michael Hagemeister, "Die 'Biokosmisten'—Anarchismus und Maximalismus in der frühen Sowjetzeit," in *Studia slavica in honorem viri doctissimi Olexa Horbatsch,* ed. Gerd Freidhof, Peter Kosta, and M. Schutrumpf, vol. 1, pt. 1 (Munich, 1983), 61–76; Hagemeister, "Russian Cosmism in the 1920s and Today" (cit. n. 16), 195–6.

[98] A. Sviatogor, N. Lebedev, and V. Zikosi, "Golos anarkhistov," *Izvestiia VTsIK*, 27 Jan. 1924.

[99] Tsiolkovskii also communicated with an international association, devotees of a philosophy similar to Russian Cosmism, known as the Association Internationale de Biocosmique, based in Lyon, France. Ass. Int. Biocosmique to Tsiolkovskii, [illegible but probably 16 April 1934], Lyon, f. 555, op. 3, d. 200, ll. 12–3, ARAN.

[100] A. Gornostaev, "N. F. Fedorov," *Izvestiia*, 29 Dec. 1928.

[101] Gor'kii to Tsiolkovskii, n.d., 1932, n.p., f. 555, op. 4, d. 183, l. 1, ARAN. For Gor'kii and Tsiolkovskii in general, see G. Chernenko, "Sorrento—Kaluga—Moskva," *Nauka i zhizn'*, 1972, no. 6:46–8.

[102] A. L. Chizhevskii, *Na beregu vselennoi: Gody druzhby s Tsiolkovskim; Vospominaniia* (Moscow, 1995).

1920s, enrapturing the technological utopians who wanted to build rockets to bring the Soviet Union into the modern world.[103]

UTOPIA ABANDONED?

The political, social, and cultural climate dramatically changed in the Soviet Union between the early 1920s, when the fad began, and the early 1930s, when the fad ended. The combined repercussions of the Cultural Revolution, the First Five-Year Plan, and nationwide collectivization completely transformed much of Soviet society. For those involved in scientific or technical work, the Shakhty trial and the Industrial Party affair redefined, with tragic consequences, the boundaries of "proper" behavior and expression. Party ideologues purged out of influential positions a huge number of old specialists, especially those with roots in prerevolutionary times.[104] They also removed "old influences" from the editorial boards of several popular science journals. The government absorbed P. P. Soikin's semiprivate publishing company, perhaps the most important promoter of space-related themes, and changed the profiles of several of its former journals. Although science popularization still remained a very important project for Bolsheviks, the tenor of outreach changed. The journal *Priroda i liudi*, for example, changed its name to *Revoliutsiia i priroda* (Revolution and nature) to reflect the explicitly utilitarian, socialist, and applied nature of its message. Its stated goal was now to popularize "technology for the masses." Similarly, the elite Academy of Sciences, although disconnected from the populist space fad, underwent a process of "Bolshevization" that significantly limited its independent voice in matters of science so that it could refocus attention to applied, rather than fundamental, science.[105]

The rise of the state (both government and party) as a ubiquitous and inescapable force in society at the turn of the 1930s profoundly affected the indigenously maintained space fad. In particular, the Bolshevik Party's effort to realign scientific and technical work in the country for socialist reconstruction proved decisive. After an explosion of media attention at the turn of decade, by 1933, the space fad was nearly over. The metamorphosis was striking. In 1931, the press published nearly two dozen articles on spaceflight; in 1932, less than a dozen; the following year—when there were no private popular science journals left—no more than a handful. The same journals that had popularized utopian discussions about space travel now devoted more attention to technical knowledge applicable to workers on the shop floor. Linking science to industrial productivity marginalized many seemingly outlandish ideas such as space exploration. Societies, exhibitions, media, and art on the topic either disappeared or mutated into new forms.

A few spaceflight supporters from the 1920s were casualties of the Great Terror, although it is important to underscore that none suffered *because* of their advocacy of space travel. Cosmist philosopher N. A. Setnitskii lost his life in the late 1930s,

---

[103] Alexander Tshijewsky, "Anstatt eines Vorworts," preface to K. E. Tsiolkovskii, *Raketa v kosmicheskoe prostranstvo* (Kaluga, 1924), unnumbered preface page.

[104] Kendall E. Bailes, *Technology and Society under Stalin: Origins of the Soviet Technical Intelligentsia, 1917–1941* (Princeton, N.J., 1978).

[105] Michael David-Fox and György Péteri, eds., *Academia in Upheaval: Origins, Transfers, and Transformations of the Communist Academic Regime in Russia and East Central Europe* (Westport, Conn., 2000); Andrews, *Science for the Masses* (cit. n. 45), 130–4.

while Tsiolkovskii's friend Aleksandr Chizhevskii was arrested in 1940 and eventually spent sixteen years in domestic exile. In 1939, the People's Commissariat of Internal Affairs (NKVD) shot Morris Leiteizen, former secretary of the Society for the Study of Interplanetary Communications and the son of an old Bolshevik who had been a friend of Lenin's. Mikhail Lapirov-Skoblo, one of the earliest advocates for spaceflight in the 1920s, also fell to the purges. After a very distinguished career as a vocal spokesperson for the Soviet scientific and technical intelligentsia, he was arrested in 1937, sentenced in 1941, and died in confinement in 1947 while working at a battery factory.[106]

Artists and writers also fell during the upheavals of the Cultural Revolution and the Great Terror. During the former, the Suprematists came under attack from the Association of Russian Revolutionary Painters (AKhRR) as part of a general move to discredit the artistic avant-garde.[107] Similarly, the Proletarian Writers' Association launched a campaign that discredited the genre of science fiction, calling the style a distraction to the problems at hand. By 1936, the government included *Aelita* on its list of banned movies; the NKVD arrested some science fiction writers in the late 1930s while the government removed even Jules Verne from children's literature. Soviet science fiction did not recover from the resultant consequences until the Khrushchev era.[108]

Most space advocates, however, survived. They successfully embraced the discursive shift from indefinite utopia to definite industrialization by changing their strategies. Popularizers and enthusiasts altered their lexicon rather than changing their vision. Many, for example, refocused their attention from rockets flying in space to the purer engineering problem of "reactive motion." Through the 1920s, interplanetary travel had always been connected to the development of reactive motion, that is, with rocket and jet engines. In the early 1930s, however, activists and enthusiasts disconnected reactive motion from interplanetary travel and connected it with more realistic goals that were part of the prevailing state culture of aviation. Although most space advocates never stopped aiming for outer space, they redefined the problem into smaller chunks, the first step being "conquering the stratosphere" using the principle of reactive motion. Stratospheric flight literally and metaphorically lowered the ceiling of ambition while locating the original idea of space exploration within prevailing aviation culture. Reactive motion implied a real engineering problem with real solutions; it also held immediate utility as such a principle could be used to propel airplanes. Many enthusiasts in Europe had already demonstrated the possibility. The limits of possibility moved downward from the cosmos to the clouds.

CONCLUSIONS

From the perspective of the Soviet state, the space fad was of no importance. During its existence, no major party or government official was involved in the activities of

---

[106] Semenova, "Russkii kosmizm" (cit. n. 15), 96–7; Roy Medvedev, *Let History Judge: The Origins and Consequences of Stalinism*, rev. ed. (New York, 1989), 444; E. N. Shoshkov, "Lapirov-Skoblo Mikhail Iakovlevich," in *Repressirovannoe ostekhbiuro* (St. Petersburg, 1995), 137.

[107] Fauchereau, *Malevich* (cit. n. 87), 31–3.

[108] McGuire, *Red Stars* (cit. n. 72), 13–5; Peter Kenez, *Cinema and Soviet Society, 1917–1953* (Cambridge, UK, 1992), 144; Britikov, *Russkii Sovetskii nauchno-fantasticheskii roman* (cit. n. 72), 137.

either the technological utopians or the mystically minded space advocates. The relatively loose controls over social, cultural, and economic activity during NEP allowed the ideas of space activists to flourish without notice or support from the party and the government. Trotskii's single public comment on the space fad was derisive and cautionary. In a section on proletarian culture and art in *Literature and Revolution*, he argued:

> Cosmism seems, or may seem, extremely bold, vigorous, revolutionary and proletarian. But in reality, Cosmism contains the suggestion of very nearly deserting the complex and difficult problems . . . on earth so as to escape into the interstellar spheres. In this way Cosmism turns out quite suddenly to be akin to mysticism . . . [and may] lead some . . . to the most subtle of matters, namely to the Holy Ghost.[109]

Interest in space, he argued, would lead enthusiasts from the useful to the useless and from science to religion—what Lenin had scorned as the opiate of the masses. Trotskii's comment (disingenuously?) avoided underscoring the connection between science and religion, represented in the space fad by the technological utopians and the mystics, respectively. Both rationales contributed in wholly different ways to the defining of the contours and flavor of the space fad in the 1920s but both also shared many deep-rooted rationales.

The most important contribution of the technological utopians—such as the societies and the popular media—was to link the cause of spaceflight with science and technology. Prior to the 1920s, in the public imagination, space exploration was part of the discourse of fantasy, speculation, and often mysticism. In the 1920s, by linking spaceflight with the sciences and suggesting that space travel was entirely plausible by means familiar to most people, the spaceflight advocacy community brought such ideas into the realm of possibility and the "rational." The link with science, which the Bolsheviks believed provided the way to modernization, also equated spaceflight with "being modern." After the late 1920s, spaceflight became, like aviation, one manifestation of the self-reflexive notion of twentieth-century modernization.

The approach of the technological utopians differed in important ways from that of their fellow Cosmists. Where technology-inspired space advocates looked to a future of many unknown possibilities for humanity, Cosmists looked to the past (the dead) as way station to a singular goal: the reanimation of humanity into a single universal organism. If the former tied their dreams of space exploration (however implausibly) with the modernizing exigencies of the day, the latter were not interested in modernization but the evolution of the species. It is tempting to argue that the tension between these seemingly contradictory ideas provided the charge for the creative outpouring on space exploration in the 1920s; or that both the "old" and the "new" appeal were necessary for mass interest in such an arcane idea as spaceflight. Such assertions would, however, be impossible to test since they raise counterfactual, rather than factual, questions.

A more analytically valuable perspective would be to view the two sensibilities as not altogether incompatible, especially as the boundaries between the two were not always clear. The nearly invisible web of connections via friendship or acquaintance that linked disparate believers in the cause of space travel muddled distinctions

[109] Trotsky, *Literature and Revolution* (cit. n. 11), 211.

between the differing rationales for space travel. Sometimes cold science and ill-defined mysticism existed in the same breath. The artists who emerged from the Suprematist umbrella embodied this duality without contradiction; they worked within the most avant-garde of artistic traditions—materialistic, forward thinking, urban—yet infused their work with Fedorovian views from the late nineteenth century rooted in a pastoral and antimaterialistic aesthetic.

Technological utopianism and Cosmism shared a number of basic elements: both were utopian, both relied on the notion that humanity needed complete control over nature, and both afforded technology a prominent role in the realization of their ultimate goal of transforming society. In their language and iconography, technological utopians spoke with the same evangelical tones as their spiritual compatriots. Like the Cosmists, utopians were obsessed with the future imperatives of humanity and paid fealty to technology, travel, and Tsiolkovskii. In advocating the science of space exploration in the 1920s, "believers" not only used the language of mysticism—the most obvious meeting point between science and religion—but also shared many of the same rationales, goals, and ideologies.

The case of spaceflight culture in the experimental climate of the NEP years provides a striking case in which the demarcations between science and mysticism were at best nebulous. Writing about Bolsheviks' fascination with technology, Anthony J. Vanchu noted that "[w]hile science and technology had the power to demystify religion and magic, they themselves came to be perceived as the locus of magical or occult powers that could transform the material world."[110] In effect, science and technology became a new cosmology in the Marxist-Bolshevik-Leninist context of the interwar years; they were both alternatives to religion and religions themselves. Spaceflight was one vibrant example of this conflation.

Through the decades after the 1930s, Soviet space advocates altered their strategies to fit the needs of practical science and industrialization. Still utopian, they abandoned the mystical for the technological. By the time that cosmonaut Titov declared that he had not found God nor angels in outer space, the religion of space travel could be distilled down to modernity, secularism, and progress. But statements such as Titov's obscured an alternate history of the Soviet space program that harked back to the 1920s, discarded and lost through much of the Soviet era. Titov's willful disengagement of Christ from the cosmos underscored the irony that his achievement had been made possible largely because of people such as Tsiolkovskii who had set out to do the exact opposite, that is, to integrate the mystical and the technological; the modern rocket with its new Communist cosmonaut was conceived as much in a leap of faith as in a reach for reason.

[110] Anthony J. Vanchu, "Technology as Esoteric Cosmology in Early Soviet Literature," in Rosenthal, *Occult in Russian and Soviet Culture* (cit. n. 12), 205–6.

Notes on Contributors:

Andy Byford is a Junior Research Fellow at Wolfson College, Oxford. He is the author of *Literary Scholarship in Late Imperial Russia: Rituals of Academic Institutionalisation* (Oxford, 2007). He has co-edited *Making Education Soviet, 1917–1953* (special issue of *History of Education* 35, nos. 4–5) and has published widely on Russian professions, academia and education. He is currently working on the "child study" movement in late imperial and early Soviet Russia (1881–1936).

Slava Gerovitch is Lecturer in Science, Technology and Society Program at MIT. He is the author of *From Newspeak to Cyberspeak: A History of Soviet Cybernetics* (Cambridge, Mass., 2002) and numerous articles on the history of Soviet astronautics, cybernetics, and computing. Currently he is working on a book on the technopolitics of automation in the Soviet space program.

Michael D. Gordin is Associate Professor of History at Princeton University. He has published two books: *A Well-Ordered Thing: D. I. Mendeleev and the Shadow of the Periodic Table* (New York, 2004), and *Five Days in August: How World War II Became a Nuclear War* (Princeton, N.J., 2007). He is currently writing a book about Heidelberg University and the origin of German and Russian nationalism in chemistry.

Karl Hall is Assistant Professor of History at Central European University in Budapest. He is completing a book on Soviet theoretical physics in the interwar period.

Alexei Kojevnikov teaches as Associate Professor in the Department of History, University of British Columbia, Vancover, Canada. He is the author of *Stalin's Great Science: The Times and Adventures of Soviet Physicists* (London, 2004) and the editor of *Science in Russian Contexts*, a special issue of *Science in Context* (2002).

Nils Roll-Hansen is Professor of History and Philosophy of Science, University of Oslo. He has published on Pasteur and spontaneous generation, early twentieth-century genetics, Scandinavian eugenics, environmental science, and reductionism in biology. He co-edited *Eugenics and the Welfare State: Sterilization Policy in Denmark, Sweden, Norway and Finland*, with Gunnar Broberg (second edition published by Michigan State University Press in 2005), and published *The Lysenko Effect: The Politics of Science* (Amherst, N.Y., 2005).

Kirill Rossiianov is a Senior Research Associate at the Institute of the History of Natural Sciences and Technology, Russian Academy of Sciences. His interests include the social history of Soviet and Russian science and the history of genetics and biomedical research in the twentieth century.

Sonja D. Schmid is a Social Science Research Associate at Stanford University's Center for International Security and Cooperation, and she teaches in the Program on Science, Technology, and Society. She received her PhD in Science and Technology Studies from Cornell University in 2005 and is currently working on a book about reactor-design choices for the Soviet civilian nuclear industry.

Asif A. Siddiqi is Assistant Professor of History at Fordham University in New York. His 2000 book, *Challenge to Apollo: The Soviet Union and the Space Race, 1945–1974*, was published by NASA. He is currently finishing a book, *The Rockets' Red Glare: Spaceflight and the Russian Imagination, 1857–1957*, on the social and cultural roots of space exploration in early twentieth-century Russia.

Olga Valkova is a Senior Researcher in the Institute for the History of Science and Techniques Russian Academy of Sciences, Moscow. Her 2006 book, *Olga Alexandrovna Fedchenko, 1845–1921*, was published by the Academy of Sciences publishing house Nauka in Russia. Today she is writing a book about women's participation in scientific explorations in Russia from the nineteenth century through the first three decades of the twentieth century.

Index

Alexander II, Tsar, liberalizing reforms of, 52
Alexandrov, Daniel, 26, 244
American Association of Scientific Workers, 124
American Committee for Democracy and Intellectual Freedom, 124
Anuchin, Dmitrii, 140, 148

Beketov, Andrei Nikolaevich, 151
Berdiaev, N. A., 6
Beria, Lavrentii: 87–88, 196; as atomic bomb committee head, 86; nuclear power program and, 106, 108
Bernal, J.D.: *12n,* 123, 128, 171; *Science in History,* 185; Society for Freedom in Science, 167
Blackett, P. M. S., 10
Bobrov, Evegenii, 61
Bogdanov, Aleksandr, 125–126, 128
Bogdanova, Elena Vasil'evna, 143–144
Bohr, Niels: correspondence principle, 245; discussions with Landau, 252–253; Gamow and, 249–250; impact on Russian, 254–258; as model for Landau, 234; paper with Léon Rosenfeld, 251–253; successes of quantum mechanics and, 248–249, 251
Bolshevik Revolution: 13; changes for women scientists and, 154–164; Higher Women's Schools decrees and, 154–155; honoring Tsiolkovskii, 261; overhaul of education system, 79–80, 129; science as motor of social progress, 167, 236, 248–249, 263–264, 266, 268, 285; state involvement in research and, 134
Bordet, Jules, 225
Borodin, Aleksandr Porfir'evich: 33–42, *34n;* funeral oration for Zinin, 48; Russian Chemical Society and, 44
Borscheid, Peter, 29
Bulgakov, S. N., 6
Bunsen, Robert Wilhelm, 29–32, 44
Butlerov, Aleksandr M., 39, 42, 48, 150, 240

Casimir, Hendrik, on Landau, 252–253
Charle, Christophe, 8
Chelpanov, Georgii, 70–73, 76
Cherenkov, Pavel, 129–131, 232
Chernobyl: 82–85, 98; assigning blame, 83, 103–105, 108–109; Communist Party investigation, 105, 108
China, education reformed on Soviet model, 122–123, 132
Chizhevskii, Aleksandr, 284, 286
civil service, professionalization of, 121, 124, 141
civil war of 1918-1919, 157–158
Cold War, 11
Communist Party: Association of Scientific Workers, 123–124; intelligentsia and, 86, 94, 103, 119–120; investigation of Chernobyl, 105, 108
Conant, James Bryant, 10
Congresses of Russian Naturalists, 149–151
Cosmism: 265–268; artistic connections with, 280–283; belief in power of technology, 18, 262; Biocosmist-Immortalists, 283–285; Tsiolkovskii's works and, 275
cosmonauts: 207; Cosmism and, 288; as emblems of new Russia, 261; spacecraft design and, 202–205
Council of Chief Designers, 197–199, *198,* 202–203

Darwin, Charles: 179, 218–219; Darwinism and war, 116; death as adaptation, 227; *Origin of Species* translated, 14; Russian authors on, 17; theories used to support racial inferiority, 218–220
Delianov, Ivan, 50
diphtheria vaccine, 225
Dollezhal, Nikolai, 93–94, 98
Dreigestirn, the, 31

Ehrlich, Paul, 213, 224–225
Elie (Il'ya Il'ich) Metchnikov: 17; cell state theories, 221–225; on civilization of races, 219–220; contributions of, 227–229; "developmental arrests," 217–218; harmonist ideas, 215–216; on Kalmucks, 218; on marriage, 144; phagocytic immunity, 226; work in immunology, 213–215; Engel'gardt, Aleksandr N., 42–43, 46, –47
Erlenmeyer, Emil, 32, 36, 39–40, 48–49
European studies critical to scientific training, 31–32, 237
experimental pedagogy, 73–79

Falkova, Esfir, 156
Fedchenko, Alexei Pavlovich, 145–146
Fedchenko, Olga (Armfeld), 145–146, 149, 151, 158
Fedorov, Aleksandr, 274–275
Fedorov, Nikolai Fedorovich, 265–267, 268
Fersman, A.E., 231
First Chief Administration, 87–88
Flerov, Georgii, 86
France: Caisse Nationale de la Recherche Scientifique (CNRS), 124; scientific research profession in, 124
Frank, S.L., 6

Gagarin, Yurii, 194, 206
Gamow, George: Bohr and, 249–250; on European study, 237; Jazz Band member, 247; visits

Gamow, George (*cont.*)
  Germany, 250; genetics, abolition of, 11; criticism of, 177; and eugenics, 127
Glagoleva-Arkadieva, Vera Aleksandrovna, 158
Glushko, Valentin, 190
GOELRO (state electrification plan), 92, 95, 119
Golitsyna, Eudoxia Ivanovna, 137–138
Goncharov, I. A., 39
Graham, Loren, 4, 169
Great Britain: brain drain to U.S., 133; social relations of science movement, 123–124
Great Reforms, 5, 24, 45–46, 116, 138
Gregory, Richard A., 171

Haeckel, Ernst, 222
Haldane, J.B.S., 185
Hall, Granville Stanley, 74
harmonists, 216
"He Lived among Us" (Pushkin), 1–2
Heidelberg: as appropriate for Russian students, 31–32; Badischer Hof, 36; chemical kruzhok, 33–37, 44; Dreigestirn, the, 31; importance to Russian Chemical Society, 47–49; Polish in, 40; women's commune of 1869, 147–148
Heidelberg University, 29–31
Helmholtz, Hermann von, 31–32
Herzen, Aleksander, 240
Hessen, Boris, 128
Higher Women's Courses, 152–155
History of the Warfare of Science with Theology in Christendom, 3
Hofmann, Karl, 36
Hungary, education modeled after Soviet, 131
Huxley, Julian, 168

Ianovskaia, Sofia Aleksandrovna, 162
Ianovskii, Kirill, 60
Il'in, N. P., 43–44
immunology: development of primitivity theory, 215–218; established as specialty, 213–214; immunity and primitivity link, 214–215; intellectual possibilities and limitations of, 221–224; Metchnikov on phagocyte/whole theory, 218–221; phagocytic and humoral immunity, 224–227; Imperial Amateurs' Society for Nature, Anthropology and Ethnography, 140, 143–145, 149
Institute of Plant Industry, 174, 177, 181
intellectual freedom, 183–186
intelligentsia: Lenin on, 8; Solzhenitsyn on, 9; splits into two groups, 6–7; as term, 4, 9–10; vs technical studies, 15–19; Ipatieff, Vladimir, 119
Ivanenko, Dmitrii, 247
Ivanovich Sreznevskii, Izmail, 142–143

Joffe, A. F., 231–232, 243–244
Joravsky, David, 168–169

Kaiser-Wilhelm-Gesellschaft, 120
Kashevarova- Rudneva, Olga, 149
Kekulé, August, 32, 39
Khrushchev, Nikita: instability and haphazard reforms under, 202; space program under, 190–193
Kirchhoff, Gustav, 30–32
Kojevnikov, Alexei, 170
Kol'tsov, Nikolai, 179, 181
Konradi, Evgeniia Ivanovna, 151
Konstantin, Konstantin, 261–262
Korolev, Sergei: 189–190; on death of Stalin, 193; management style, 200–201; seeks funding for space program, 196–197; and space program politics, 197–198; Korwin-Krukovskaia, Sofia, 146–147
Kots, Aleksandr Fedorovich, 163–164
Kovalevskaia, Sofia, 152
Kramarov, Grigorii, 270, 280
Kropotkin, Prince Peter, 3–4
Krutkov, Yuri, 244
kruzhki (kruzhok): 5; Borodin on, 37; defining, 26–27; importance in scientists' identity formation, 239–240; "mighty little heap" *[moguchaia kuchka]* and, 45; in physics field, 237–242; Russian national identification in the sciences, 25–26, 33, 37, 241; St. Petersburg, 243-244; Krzhizhanovsky, Gleb, 119
Kurchatov, Igor, 86, 89

Ladigina-Kots, Nadezhda Nikolaevna, 163–164
Landau, Lev: background, 230–235; and Bohr, on uncertainty principle, 242–254; and Peierls, on quantum mechanics, 251; summary of achievements, 258–259; visits Germany, 250, 251–252; work in Denmark, 252–253; Landau school and growth of theoretical physics: establishment of, 232–234; as Jazz Band member, 247
Lapirov-Skoblo, Mikhail, 286
Lavrenenko, Konstantin, 91
Lavrov, Petr, 5
Lazurskii, Aleksandr, 67, 71
Lebedev, P.N., growth of laboratory of, 241
Lebedev Physics Institute (FIAN), 231, 241–242
Leibniz, Gottfried, 23, 275
Leiteizen, Morris, 286
Lenin, Vladimir: on electrification and Communism, 119; and "proletarian science," 126; technological progress and, 90; Lenin Academy of Agricultural Science, 168
Leningrad Physico-Technical Institute (LFTI), 231–232
Lermontova, Iuliia, 150
Levins, Richard, 169
Lewontin, Richard, 169
Lysenko, Trofim D.: 11; 1948 genetics conference, 182–183; family background of, 171; Lysenko Affair, 13; plant development theory, 172–173
Lysenkoism: in France, 185–186; historical interpretations of, 168; Joravsky account of, 168–169; Lewontin and Levins on, 169; Marxist theory and, 169; Medvedev account of, 168; persistence of, 186–187; victory and consolidation of, 180–182; M

Makarov, General Apollon, 63
Maksimov, Nikolai, 172
Malevich, Kazimir, 280–281
Malia, Martin, 7
Malyshev, Viacheslav, 88
Mandelstam, L. I., 231
Medvedev, Zhores, 10, 168
Mendeleev, Dmitrii Ivanovich, 3, 33, 38–39, 240
Merton, Robert K., 124
Metchnikov, Elie (Il'ya Il'ich), 213: on cellular organisms, 221–224; on death as adaptation, 227; on human body as a harmonic whole, 216–218; on phagocytes, 214; on phagocytic immunity, 225–226; and racial inferiority, 218–220; Metchnikova, Olga Nikolaevna, 144
Mickiewicz, Adam, 2
Mikhailovskii, Nikolai, 215–216
Ministry of Energy and Electrification, 84
Ministry of Medium Machine Building (Sredmash), 83, 88–89, 95, 96–98, 100–103, 107–109
Mohr, Otto, 176–178
Moscow Archaeological Society, 140, 148
Moscow Power Engineering Institute, 91
Mozzhorin, Yurii, 196

Nazi Germany: armament production, 131; eugenics and, 127
Nechaev, Aleksandr, 63–64, 65, 71–72
Neporozhnii, Petr, 91–92
networking of scientific community, 192
Neustroeva-Knorring, Olga Evertovna, 156–157
Nicholas I, Tsar, 5
Nikolaevna, Maria, 140
Nikolaevna Vodovozova, Elena, 139
nuclear power program: First Chief Administration, 87–88; Ministry of Energy and Electrification (Minenergo), 90; NKVD and, 90; nuclear engineering as discipline, 94–95; Obninsk, 94; plant construction, 88–89, 94; promoted by ministers, 92–93; nuclear specialists and power engineers, 83, 93–94
nuclear weapons, 2, 11

Obrucheva, Maria Aleksandrovna (Bokova-Sechenova), 146
October Revolution: 90; as new era marker, 236; scientists view of, 236; women's rights and, 165; Olevinsky, Ladislaus, 39–40
Origin of Species, 14
Ostrogorskii, Aleksei, 59

Passek, T. P., 38
Pasteur Institute, 225
Pauli, Wolfgang, 250
Pavlov, I. P., 231
Pavlova, Maria Vasil'evna, 150, 156
Pedagogical Academy, 63, 65–66, 68
pedagogical circles, 61
Pedagogical Faculty of Vladimir Bekhterev's Psycho-Neurological Institute, 63, 66–67, 68
Pedagogical Museum, 64
"pedagogical psychology" conference, 65

pedagogy: Conferences in Experimental Pedagogy, 76; district pedagogical institutes as solution, 60; experimental, 73–79; growth as a science, 51–53, 64–65, 68, 78–79; independent teacher-training initiatives, 62–63; need for universities to control, 61; pedagogical circles, 61; pedology and, 75; proposals for reform, 60–61; Vipper on, 59; Peierls, Rudolf, 250
Pereiaslavtseva, Sophia, 149, 150
Perrin, Jean, 124
Pervukhin, Mikhail, 88
Peter the Great, Tsar, 23
physics: domestic institutionalization as a field, 233; Jazz Band group as social dynamic, 247–248; politicization of, 245–247; Physics Institute of the Academy of Sciences (FIAN), 86
Pirogov, Nikolai I., 32, 52
plutonium bomb, 87
Ponomareva, Valentina, 205
Popper, Karl, 128, 184
Potanin, Grigorii Nikolaevich, 145
Potanina, Alexandra Viktorovna, 145
Prianishnikov, D. N., 179–180
Prince Vladimir F. Odoevskii, 5
Pushkin, Aleksandr S., 2

Raevskaia, Anna Mikhailovna, 139–140
"renegadism" *(otshchepenstvo)*, 6
Rishavi, Liudvig Al'bertovich, 150
rocket engineers' personalities, 206–208
Röntgen, Wilhelm, 243
Roscoe, Henry, 30
Rosenfeld, Léon, 10, 251–253
Rossiter, Margaret W., 160
Rubinshtein, Moisei, 61
Russian academic intelligentsia, 120–123
Russian Chemical Society: 15, 149; Heidelberg genesis of, 42, 43–44, 44–48; kruzhok and, 25–27
Russian Geographical Society, 149

Sakharov, Andrei, 1–2, 169
Scherrer, Jutta, 8
science: challenges to professionalization of, 134–135; political control over, 179; professionalization affected by WWII events, 127–128
Science and Common Sense, 10
scientific freedom and liberal democracy, 167–168
Sdvizhkov, Denis, 8
Sechenov, I. M., 240–241
secondary teachers image, 54–55
Serebrovskii, A. S., 177
Seventh International Congress of Genetics, 176–178
Shtern, Lina Solomonovna, 163
Shulga-Nesterenko, Ivanovna, 158
Sigismundovich, Pavel (Paul Ehrenfest), 242–243
Signposts, 6, 7
Slavskii, Efim, 88–89, 92
social constructivism, 125–126, 128–129
Society for Freedom in Science, 167

Society for the Study of Interplanetary Communications, 268–271
Sokolov, Nikolai N., 42–43
Sokolov and Engel'gardt's Chemical Journal, 43
Solzhenitsyn, Aleksandr, 9
Soshkina, Elizaveta Dmitirievna, 162–163
Soviet Academy of Sciences, 230
Soyfer, Valery, 169–170
space industry: cosmic societies, 265–271; Cosmism, 265–268; cosmonaut role compared to U.S. astronauts, 203–204; Council of Chief Designers, 197–199; differing from intelligentsia, 262; Experimental Design Bureau No. 1, 194; lack of recognition for engineers, 206–207; in NEP era, 260–288; organizational problems of, 193–196, 262; production problems in, 194; space travel, public's fascination with, 270–271; technological utopianism, 262–265; Vostok spacecraft, 199–200
space program: 1927 exhibition, 275–277; banned literature and art, 285–286; cultural support for, 261–262; effect of rise of state on, 285–286; female cosmonauts, 205; importance to Russian psyche, 260–261; under Kruschev, 190–191; media treatment of, 271–274; Sector for Propaganda and Popularization of Astronautics, 274–275; under Stalin, 189–190; Special Committee for Reactive Technology, 193
Special Committee on the Atomic Bomb, 86–87
Speranskii, Mikhail, 28
Sputnik: 11, 132–134, 190; as culmination of progress, 261; effects of, 206
Sreznevskaia, Olga Izmailovna, 142–143
St. Petersburg Academy of Sciences, 150
Stalin, Joseph: 11; authorizes execution of engineers, 189; establishes Committee for Reactive Technology, 193; ideas of and school reform, 80; nuclear research and, 86–87; oppressive policies towards rocket engineers, 189–190; space program and, 193–196; Stalinist view of pedagogy, 80–81
State Astronomical Institute, 163
Steklov, V. A., 241
Stoletov, Aleksandr Grigorievich, 239
Struve, P. B., 6
Suprematism, 280–281
Suslova, Nadezhda Prokofievna, 146, 148
Szent-Györgyi, Albert, 131

Tamm, Igor, 231, 235
teaching profession: mentorship, 56; teacher as psychologist, 68–73; training, 57. *See also* pedagogy.
Tereshkova, Valentina, 205
testing, mental and personality, 76–77
Theory of Relativity (Einstein), 126–127
Timiriazev Agricultural Academy, 169
Tolstoi, Dmitrii, 53

Trotsky, Leon (Lev Trotskii): and "proletarian science," 126; on space program, 287
Tsiolkovskii, Konstantin, 274
Turchaninova, Anna Alexandrovna, 137

United States: effect of Sputnik on, 133–13; political awakening of scientists in, 124–125
universities: growth of, 141; separate, for women, 149, 151
Ushinskii, Konstantin, 52, 64, 75
Uvarov, Count Alexei Sergeevich, 148
Uvarova, Countess Praskovia Sergeevna, 148

Vagner, Vladimir, 58
Vannikov, Boris, 88
Vavilov, N. I.: 159; arrest rumor, 178; arrested, 181; family background of, 171
Vavilov, S. I., 232
Vavilov, Sergei, 86, 128, 131
Vera Figner, 147–148
Vernadskii, Vladimir I., 5, 86, 236
vernalization theory, 171–173
Vigel, Philip, 137
Vipper, Robert, 59
Voeikov, Dmitrii, 42
von Behring, Emil, 225
von Helmholtz, Hermann, 31
von Liebig, Justus, 29
Vostok spacecraft: 199–200; design of, 202–203; manual control in, 203–205
Vygotskii, Lev, 80

Weismann, August, 227
wheat, growing, 175
White, Andrew Dickson, 3–4
women in science, 1860-1940: admitted to universities, 140–142; background, 136–148, 164–165; civil war's effects on, 157–158; clerking for male family member, 142–144; European studies of, 147; expeditionary researches by, 157t; first Russian educational courses, 152–154; growth of, 158–162, 161t; growth of geology, 161–162; Heidelberg women's commune of 1869, 147–48; Higher Women's Courses, 152–155; marriage and, 144–148, 163–164; as members of scientific societies, 149; paying jobs for, 149–151; as political activists, 162; as revolutionaries, 147–148; self-education of, 142–143; on separate universities for women, 151; social changes, 138–140; World War II, effects of on scientific profession, 127–128

Zaitsev, Varfolomei, 218
Zapol'skaia, Liubov' Nikolaevna, 153–154
Zenchenko, Sergei, 58–59
Zhitinskii, Nikolai, 39
Zinin, Nikolai N., 35, 47
Zirkle, Conway, 10

SUGGESTIONS FOR CONTRIBUTORS TO OSIRIS

OSIRIS is devoted to thematic issues, conceived and compiled by guest editors who submit volume proposals for review by the OSIRIS Editorial Board in advance of the annual meeting of the History of Science Society in November. For information on proposal submission, please write to the Editor at Osiris@georgetown.edu.

1. Manuscripts should be submitted electronically in Rich Text Format using Times New Roman font, 12 point, and double-spaced throughout, including quotations and notes. Notes should be in the form of footnotes, also in 12 point and double-spaced. The manuscript style should follow *The Chicago Manual of Style*, 15th ed.

2. Bibliographic information should be given in the footnotes (not parenthetically in the text), numbered using Arabic numerals. The footnote number should appear as superscript. "Pp." and "p." are not used for page references.

   a. References to books should include the author's full name; complete title of book in *italics*; place of publication; date of publication, including the original date when a reprint is being cited; and, if required, number of the particular page cited (if a direct quote is used, the word "on" should precede the page number). *Example*:

   [1] Mary Lindemann, *Medicine and Society in Early Modern Europe* (Cambridge, 1999), 119.

   b. References to articles in periodicals or edited volumes should include the author's name; title of article in quotes; title of periodical or volume in *italics*; volume number in Arabic numerals; year in parentheses; page numbers of article; and, if required, number of the particular page cited. Journal titles are spelled out in full on the first citation and abbreviated subsequently according to the journal abbreviations listed in *Isis Current Bibliography*. *Example*:

   [2] Lynn K. Nyhart, "Civic and Economic Zoology in Nineteenth-Century Germany: The 'Living Communities' of Karl Möbius," *Isis* 89 (1999): 605–30, on 611.

   c. Journal articles are given in full in the first reference. For succeeding citations, use an abbreviated version of the title with the author's last name. *Example*:

   [3] Nyhart, "Civic and Economic Zoology" (cit. n. 2), 612.

3. Special characters and mathematical and scientific symbols should be entered electronically.

4. A small number of illustrations, including graphs and tables, may be used in each volume. Hard copies should accompany electronic images. Images must meet the specifications of The University of Chicago Press "Artwork General Guidelines" available from the Editor.

5. Manuscripts are submitted to OSIRIS with the understanding that upon publication copyright will be transferred to the History of Science Society. That understanding precludes consideration of material that has been previously published or submitted or accepted for publication elsewhere, in whole or in part. OSIRIS is a journal of first publication.

OSIRIS (SSN 0369-7827) is published once a year.

Single copies are $33.00.

Address subscriptions, single issue orders, claims for missing issues, and advertising inquiries to *Osiris*, The University of Chicago Press, Journals Division, PO Box 37005, Chicago, IL 60637.

Postmaster: Send address changes to *Osiris*, The University of Chicago Press, Journals Division, PO Box 37005, Chicago, IL 60637.

OSIRIS is indexed in major scientific and historical indexing services, including *Biological Abstracts*, *Current Contexts*, *Historical Abstracts*, and *America: History and Life*.

Copyright © 2008 by the History of Science Society, Inc. All rights reserved. The paper in this publication meets the requirements of ANSI standard Z39.48-1984 (Permanence of Paper).♾

Paperback edition, ISBN 978-0-226-30457-1

Osiris

A RESEARCH JOURNAL DEVOTED TO THE HISTORY OF SCIENCE AND ITS CULTURAL INFLUENCES

A PUBLICATION OF THE
HISTORY OF SCIENCE SOCIETY

EDITOR
KATHRYN OLESKO
Georgetown University

MANUSCRIPT EDITOR
JARELLE S. STEIN

PROOFREADER
JENNIFER PAXTON

OSIRIS EDITORIAL BOARD

SONJA BRENJTES
Aga Khan University

ANN JOHNSON
University of South Carolina

PAMELA O. LONG
Independent Scholar

MICHAEL GORDIN
Princeton University

MORRIS LOW
Johns Hopkins University

BERNARD LIGHTMAN
York University
EX OFFICIO

HSS COMMITTEE ON PUBLICATIONS

KAREN PARSHALL
University of Virginia

KEN ADLER
Northwestern University

PHIL PAULY
Rutgers University

PAULA FINDLEN
Stanford University

PAUL FARBER
Oregon State University

ELIZABETH G. MUSSELMAN
Southwestern University

EDITORIAL OFFICE
BMW CENTER FOR GERMAN & EUROPEAN STUDIES
SUITE 501 ICC
GEORGETOWN UNIVERSITY
WASHINGTON, D.C. 20057-1022 USA
osiris@georgetown.edu